# Plasticity and Ductile Fracture in Structural Engineering

**Scrivener Publishing**
100 Cummings Center, Suite 541J
Beverly, MA 01915-6106

*Publishers at Scrivener*
Martin Scrivener (martin@scrivenerpublishing.com)
Phillip Carmical (pcarmical@scrivenerpublishing.com)

# Plasticity and Ductile Fracture in Structural Engineering

**Yong Bai**

Scrivener Publishing

WILEY

This edition first published 2026 by John Wiley & Sons, Inc., 111 River Street, Hoboken, NJ 07030, USA and Scrivener Publishing LLC, 100 Cummings Center, Suite 541J, Beverly, MA 01915, USA

For more information about Scrivener publications please visit www.scrivenerpublishing.com.



**Wiley Global Headquarters**
111 River Street, Hoboken, NJ 07030, USA

For details of our global editorial offices, customer services, and more information about Wiley products visit us at www.wiley.com.

The manufacturer's authorized representative according to the EU General Product Safety Regulation is Wiley-VCH GmbH, Boschstr. 12, 69469 Weinheim, Germany, e-mail: Product_Safety@wiley.com.



**AI disclosure statement:** The author confirms that no artificial intelligence (AI)-generated content or AI-assisted writing tools were used in the preparation of this chapter (including summaries and the abstract). All text, analysis, and conclusions are the result of the author's original work.

***Library of Congress Cataloging-in-Publication Data***

ISBN 9781394434442

Cover image: Generated with AI using Adobe Firefly
Cover design by Russell Richardson

Set in size of 11pt and Minion Pro by Manila Typesetting Company, Makati, Philippines

Printed in the USA

10 9 8 7 6 5 4 3 2 1

# Contents

# Preface

Plasticity and ductile fracture are some of the most important subjects for mechanical engineers and civil engineers who work on the structural design.

During my Ph.D. period, I did some work on the plastic hinge method and plastic node method and I extended the elastic-perfect plastic theory to account for the effect of strain hardening and strain rate hardening.

In the early 90s, I worked on ductile fracture criteria in the analysis of the tubular member under tension.

Back to ten years ago when I visited MIT, I recognized the excellent works of professor Tomasz Wierzbicki and his students.

Since then, I found the importance of their works, so I decided to implement the model with my Ph.D. student Dr. Mohsen Saneian, and this becomes the foundation of this book.

Recently, we have been working on the finite element analysis of the ultimate strength of composite structures such as polymers, which is a very important subject in marine engineering especially. We would like to apply the existing model to composite pipes and test the applicability of that; this is presented as the last part of this book.

We would like to summarize the practice in the industry and academy into this new book of plasticity and ductile fracture.

It is useful because we modified our finite element model and derived ultimate strength equations based on plasticity and ductile fracture, and our work considers the effect of strain hardening and strain rate hardening; this is the scope of our book.

In summary, this book is the accumulation of the knowledge we have gained in the past years in China, Japan, Norway, and the USA.

This book is useful for the analysis of metal structures under different types of loadings, and some examples have been done to test the models in this way. We wish this book will be worldwide and be valuable for engineers and graduate students; we hope the advances of plasticity and ductile fracture have been promoted by writing this book.

**Prof. Yong Bai**
July 2025

# Acknowledgement

I would like to sincerely acknowledge Dr. Mohsen Saneian for his invaluable contributions throughout the development of this book. His deep technical expertise, rigorous research efforts, and exceptional commitment played a central role in shaping the structure, depth, and quality of the material presented. From the initial conceptual framework to the detailed analyses, modeling, and drafting of each chapter, he brought a high level of professionalism, clarity, and originality that significantly enhanced the technical rigor and coherence of this work.

Dr. Mohsen Saneian earned his Ph.D. in Structural Engineering from Zhejiang University, where he further pursued postdoctoral research focused on the fracture response of flexible pipes. His academic journey also includes advanced investigations into marine operations, expanding his breadth of expertise in offshore structural engineering.

I would also like to acknowledge my graduate students at Zhejiang University who were involved in the development process of this book.

Finally, sincere appreciation is extended to the flexible pipe manufacturing company Ningbo OPR Inc. for their invaluable support and contribution to the publication of this work.

# 1

# Stress and Strain

***Abstract***

Chapter 1 introduces the fundamental concepts of stress and strain in the context of plasticity in structural engineering. It highlights the limitations of the classical theory of elasticity when dealing with complex stress states and justifies the need for plasticity theory. The chapter discusses the history and development of plasticity theory, from early formulations by Tresca and von Mises to modern hardening rules and numerical methods. The mechanical behavior of metals, including ductility, strain hardening, and anisotropy, is examined. Stress and strain tensors are introduced, along with their decomposition into hydrostatic and deviatoric components, enabling the definition of yield criteria. The chapter covers essential concepts such as equivalent stress, principal stresses, Lode parameters, and strain invariants, and explains how these relate to material deformation and failure. By offering a conceptual framework and mathematical tools, this chapter lays the foundation for realistic structural analysis and design using plasticity theory.

***Keywords:*** Plasticity, stress-strain relationship, yield criterion, deviatoric stress, principal stress, strain hardening, Bauschinger effect

## 1.1 Plasticity in Structural Engineering

The engineering design of large structures often involves a two-stage process: first, the internal force field acting on the structural material must be defined, and second, the response of the material to that force field must be determined. The first stage involves an analysis of the stresses acting within the structural elements; the second involves a knowledge of the properties of the structural material. The linear relationship between stress and strain in an idealized material forms the basis of the *mathematical theory of elasticity*, which has in turn been applied widely in practice to actual materials to estimate stress or strain in the structural elements under a specified working load condition. These stresses are restricted to be less than the specified *working* or *allowable* stress, which is chosen as some fraction of the yield strength of the material. A safe design thus evolves, not

Yong Bai. Plasticity and Ductile Fracture in Structural Engineering, (1–34) © 2026 Scrivener Publishing LLC

due to the adequacy of the structural analysis and the understanding of the properties of the material, but by reliance upon the experience of decades or centuries.

An actual structure is a very complex body with an extremely complicated state of stress. Many secondary stresses arise owing to fabrication, eruption, and localization. The combination of unknown initial stress, secondary stresses, and stress concentration and redistribution due to discontinuities of the structure defies an idealized evaluation based on the theory of elasticity. The *theory of plasticity* represents a necessary extension of the theory of elasticity and is concerned with the analysis of stresses and strains in the structure in the plastic as well as the elastic ranges. It furnishes more realistic estimates of the load-carrying capacities of structures and provides a better understanding of the reaction of the structural elements to the forces induced in the material. An understanding of the role of the relevant mechanical variables that define the characteristic reaction of the material to the applied force is therefore essential to the engineer designing structures.

Both the theory of elasticity and the theory of plasticity are phenomenological in nature. They are the formalization of experimental observations of the macroscopic behavior of a deformable solid and do not inquire deeply into the physical and chemical basis of this behavior. A complete account of the theory and application of plasticity must deal with two equally important aspects: (1) the general technique used in the development of stress–strain relationships for inviscid elastic–plastic materials with work hardening as well as strain softening and (2) the general numerical solution procedure for solving an elastic–plastic structural problem under the action of loads or displacements, each of which varies in a specified manner.

The first task of plasticity theory is to set up relationships between stress and strain under a complex stress state that can adequately describe the observed plastic deformation. This is a difficult task. However, deformational rules for metals that, in general, agree well with experimental evidence have been firmly established and successfully used in engineering applications. Moreover, in recent years, the methods of plasticity have also been extended and applied to study the deformational behavior of geological materials, such as rocks, soils, and concretes. The extension of plasticity theory to nonmetallic materials is probably the most active research subject in the field of mechanics of materials at present, and various material models have been developed.

The second task of the theory is to develop numerical techniques for implementing these stress–strain relationships in the analysis of structures.

Because of the nonlinear nature of the plastic deformation rules, solutions of the basic equations of solid mechanics inevitably present considerable difficulties. However, in recent years, the rapid development of high-speed computers and modem techniques of finite-element analysis has provided the engineer with a powerful tool for the solution of virtually any nonlinear structural problem. It has also provoked newer developments and wider applications of the classical plasticity theory. Research activity in this field has increased tremendously during the last decade or so.

### 1.1.1 Introduction

The theory of plasticity is the branch of mechanics that deals with the calculation of stresses and strains in a body made of ductile material, permanently deformed by a set of applied forces. The theory is based on certain experimental observations on the macroscopic behavior of metals in uniform states of combined stresses. The observed results are then idealized into a mathematical formulation to describe the behavior of metals under complex stresses. Unlike elastic solids, in which the state of strain depends only on the final state of stress, the deformation that occurs in a plastic solid is determined by the complete history of the loading. The plasticity problem is, therefore, essentially incremental in nature, the final distortion of the solid being obtained as the sum total of the incremental distortions following the strain path.

Metal may be regarded as macroscopically homogeneous and isotropic when the small crystal grains forming the aggregate are distributed with random orientations. As a result of plastic deformation, the crystallographic directions gradually rotate toward a common axis, producing a *preferred orientation*. An initially isotropic material thereby becomes anisotropic, and its mechanical properties vary with direction. The development of anisotropy with progressive cold work and the resulting strain-hardening is too complex to be successfully incorporated into the theoretical framework. In the mathematical theory of plasticity, it is generally assumed that the material remains isotropic throughout the deformation, irrespective of the degree of cold work. Since the strain-hardening characteristic of metal in a complex state of stress can be related to that in uniaxial tension or compression, it is necessary to examine the uniaxial stress–strain behavior in some detail before considering the general theory of plasticity.

The plastic deformation in a single crystal is generally produced by slip, which is the sliding of adjacent blocks of the crystal along definite crystallographic planes, called slip planes. The boundary line separating the slipped region of a crystal from the neighboring unclipped region is called a *dislocation*. The movement of the dislocation, which is responsible for the slip, is initiated by a line defect causing a local concentration of stress. Slip usually occurs on those planes which are most densely packed with atoms. The magnitude and direction of the relative movement in a slip are specified by a vector known as the *Burgers vector*. A dislocation is said to be of unit strength when the magnitude of the Burgers vector is equal to one atomic spacing. The terms *edge dislocation* and *screw dislocation* are used to describe the situations where the Burgers vector is normal and parallel respectively to the dislocation line. In general, a dislocation is partly edge and partly screw-in character, and the dislocation line forms a curve or a closed loop.

In a polycrystalline metal, the crystallographic orientation changes from one grain to the next through a narrow transition zone, or grain boundary, which acts as an effective barrier to slip. Dislocations pile up along the active slip planes at the grain boundaries, the effect of which is to oppose the generation of new dislocations. When the applied stress is increased to a critical value, the shear stress developed at the head of the dislocation pile-up becomes large enough to cause dislocation movement across the boundary. The dislocation pile-up is mainly responsible for the strain-hardening of the metal in the early stages of plastic deformation. The rate of hardening of the polycrystalline metal is always higher than that of the single crystal, where the increase in yield stress is caused by dislocations interacting with one another and with foreign atoms serving as barriers. The dislocation interactions control the yield strength of a polycrystalline metal only in the later stages of the deformation.

If the temperature of the strain-hardened metal is progressively increased, the cold-worked state becomes more and more unstable, and the material eventually reverts to the unstrained state. The overall process of heat treatment that restores the ductility to the cold-worked metal is known as *annealing*. The temperature at which there is a marked decrease in the hardness of the metal is known as the *recrystallisation temperature*. The dislocation density decreases considerably on recrystallization, and the cold-worked structure is replaced by a set of new strain-free grains. The greater the degree of cold work, the lower the temperature necessary for recrystallization, and the smaller the resulting grain size.

In ductile metals, under favorable conditions, plastic deformation can continue to a very large extent without failure by fracture. Large plastic strains do occur in many metal-working processes, which constitute an important area of application of the theory of plasticity. While elastic strains may be neglected in such problems, the continued change in geometry of the workpiece must be allowed in the theoretical treatment. Severe plastic strains are produced locally in certain mechanical tests, such as the hardness test and the notch tensile test. The significance of these tests cannot be fully appreciated without a knowledge of the extent of the plastic zone and the associated state of stress. Situations in which elastic and plastic strains are comparable in magnitude arise in some important structural problems when the loading is continued beyond the elastic limit. Structural designs based on the estimation of collapse loads are more economical than elastic designs since the plastic method takes full advantage of the available ductility of the material.

### 1.1.2 History

It could be mentioned that Tresca proposed the first *yield condition*, which states that a metal yields plastically when the maximum shear stress attains a critical value. This theory was presented by a series of his papers from 1864 to 1872, and that was the origin of plasticity, as a branch of the mechanics of continua on the extrusion of metals. St. Venant introduced the basic *constitutive relations* for rigid (especially plastic materials' in-plane stress in 1870), which is the actual formulation of plasticity theory. This formulation was outstanding because of the suggestion of a *flow rule*, which stated that the principal axes of the stress occur simultaneously with the principal axes of strain rate. Subsequently, the general equations in three dimensions were obtained by Levy in 1870. Then, von Mises in a landmark paper in 1913 accompanied the results of Levy by his J2-theory (pressure-insensitive yield criterion). The St. Venant-Levy-von Mises equations, for the plane continuum problem to include the elastic component of strain, were extended in 1924 by Prandtl. Later, this extension was performed to three dimensions by Reuss in 1930.

In order to include a general yield function and discuss the relation between the direction of the plastic strain rate and the *regular or smooth* yield surface, von Mises generalized his previous work for a rigid (entirely plastic solid) in 1928. Hence, the concept of using the yield function as a *plastic potential* in the incremental stress–strain relations of flow theory was presented. The von Mises yield function is very prominent today, and it may be regarded as a plastic potential for the Saint-Venant-Levy-von Mises-Prandtl-Reuss stress–strain relations. In 1932 and 1933, Reuss introduced the proper flow rule related to the Tresca yield condition, which includes singular rules. Heretofore 1940, most of the attention was on the problems with flow or entire plasticity, so the development of incremental constitutive relationships for hardening materials progressed slowly.

Therefore, the basic concepts and fundamental ingredients in the *classical theory of metal plasticity* were drastically developed after 1940.

A noteworthy paper was published by Prager in 1949, which presented a general framework of the plastic constitutive relations for *hardening materials* with regular yield surfaces. The *loading function* and the *loading–unloading* conditions were accurately formulated. These conditions, such as the *continuity* condition (near neutral loading), the *consistency* condition (for loading from plastic states), the *uniqueness* condition, and the condition of the *irreversibility* of plastic deformation, were formulated and studied. Besides, the interrelationship between the *convexity* of the yield surface, the *normality* to the yield surface, and the uniqueness of the associated boundary-value problem was known. Prager extended this general framework in 1958 to contain thermal effects by allowing the yield surface to change its shape with temperature. In 1951, Drucker presented the *material stability postulate*, which is an important notion of work hardening. Afterwards, Drucker improved his postulate in 1959 to include time-dependent phenomena like linear viscoelasticity and creep.

### 1.1.3 Plasticity of Metals

The widespread use of metals in structural engineering is largely due to their combination of properties of strength, weldability, and ductility. *Strength* is desirable in making structures that must withstand severe loading conditions. *Weldability* is an obvious attraction from the construction point of view because it is a means to the effective joining together of components into "continuous" structures. The word *ductility* describes, in general, the ability of a bar to be "drawn" into a longer, thinner bar, usually with the aid of a die. The adjective "plastic" simply describes the idea of molding a shape, as a potter might a jug from a lump of clay. This analogy is not at all inappropriate, because dimensional changes of tenfold can easily be obtained with a ductile metal, by compression or shearing, for example.

Many metals also possess the very useful property of *elasticity* or resilience if they are only deformed a small amount. An important element of the analogy between the plastic deformation of metals and the behavior of potter's clay is that the molding takes place *quickly* and *permanently*. It is desirable that a jug, once formed, should not "sag" before it is baked, and the potter, therefore, makes sure that the clay is mixed with the right amount of water so that it can be molded by sufficient pressure and yet will not deform appreciably under its own weight in the time which elapses before it hardens. The plastic deformation of metals at room temperature has both of these characteristics; for example, a piece of mild-steel rod or strip at room temperature may be bent into an angle practically instantaneously, and it will retain the same shape indefinitely. In technical language, we say that this sort of deformation is *time-independent.*

This use of the word "plastic" as an adjective should not be confused with its use as a noun to describe a large class of nonmetallic materials, "plastics"—so-called because they are moldable under certain chemical and environmental conditions. True, there are some obvious similarities in mechanical behavior between metals and plastics, but there is an important difference in that the deformation of "plastics" is generally *time-dependent.*

Thus, a nylon thread will "creep" for a while if it sustains a constant pull, and then gradually "recover" over some time if it is later unloaded. This is a marked contrast with the plastic behavior of metals at room temperature and, indeed, at higher temperatures. This kind of time-dependent behavior is although increasingly important in the design of structures made of plastics.

Ductility is an important property of metals as far as structural engineering is concerned for two reasons. First, it makes it possible to re-form the cast ingots of metal into a wide variety of shapes. Complex shapes such as crankshafts can be made by *forging*; prismatic members like I-beams and sheets can be made by successive continuous *rolling* operations; rods and wire can be *reduced* to a smaller diameter by *drawing* through a die; sections of various shapes can be formed by *extrusion* through a die; sheet can be *pressed* into complex shapes like car-body panels.

The list could be extended considerably. The technology of these processes is outside the scope of this book: we shall, however, study briefly the *mechanics* of some simple metal-forming processes. Because metals are softer at elevated temperatures (but below the melting point), many of the processes listed above are performed on *hot* billets. In fact, at these temperatures, metals *creep* under sustained load, i.e., exhibit some time-dependent behavior, but they can nevertheless be regarded satisfactorily as plastic in these processes because the imposed deformations are quick in the time-scale of creep.

Secondly, ductility is a useful property of metals in structural engineering from the point of view of structural failure. A structure made of ductile material will deform considerably if overloaded and will usually still sustain a considerable load in its "bent" condition. Such a structure thus gives a helpful warning of impending collapse. In contrast, a structure made of a brittle material would fail suddenly and catastrophically by "snapping" an overloaded component. Another way of looking at this is to note that with ductility goes the capacity for absorbing considerable amounts of energy.

Metals are thus suitable materials for the construction of vehicles, such as cars and ships, which may be involved in unexpected collisions.

Moreover, it is not only at failure that ductility plays an important part in structural integrity. In all structures, there are geometrical features that produce a local "concentration of stress." Such features include holes of all kinds, angular corners, notches (whether present by design or accident), and abrupt changes in the thickness of sections. Broadly, the action of a ductile material at a stress concentration is to "yield" a little, and in so doing to make the stress concentration more diffuse and thus less intense. Multiple-riveted joints provide a good example of this sort of behavior; the load transmitted through the joint is unlikely to be shared equally between the rivets, but the overloaded rivets will tend to yield a little and produce a more uniform load sharing.

This sort of behavior was well understood by those who pioneered riveting as a means of joining components: after all, the rivets underwent very considerable plastic deformation when they were being put in, so a little more in the course of their useful life would be unlikely to impair their integrity.

### 1.1.4 Plasticity Theory

The task of plastic theory is twofold: first, to set up relationships between stress and strain which describe adequately the observed plastic deformation of metals, and second, to develop techniques for using these relationships in the study of the mechanics of metal forming processes, and the analysis and design of structures. We shall see that in the accomplishment of both of these tasks, there is plenty of scope for the exercise of individual judgment and taste. The theory is a more powerful tool for engineering analysis and design in the hands of a worker who is well aware of the scope, within the theory, for individual initiative.

It seems worthwhile at this stage to digress a little and make some remarks about the problems which are encountered in the construction of *any* theory. The general aim of these remarks is to elucidate the nature of the relationship between objects and experimental observations on the one hand and ideas on the other.

Our problem is, essentially, how the human mind can come to grips with the physical world and manipulate it. Now, one of the most interesting achievements of the human mind has been the development of *mathematics*, which is fundamentally an abstract subject dealing in particular with relationships between symbols, including numbers. We might guess, therefore, that mathematics would be a useful tool in our task of obtaining an understanding of the physical world. But how can mathematics, which is so clear and precise, and in many ways simple, be applied to the physical world, which is apparently *consistent.*

For example, in discussing the position of a ship at sea, we regard the ship as a *point.* The bows and the stern actually occupy different positions in space and so have different latitude and longitude coordinates. However, it would clearly be a waste of everyone's time to think of quoting two positions for two different parts of the ship at sea, simply because the ship is extremely small compared to the size of the ocean. The situation is reversed, however, if we are thinking of bringing the ship into a dock: here, the pilot will be very much concerned with the positions of both bow and stern concerning the dock.

We can extend these considerations about ships almost as far as we please. The hydrodynamics expert whose job it is to decide the shape of the hull of a ship will regard the ship as a *rigid body* that is moving through the water. On the other hand, the structural engineer will certainly regard the ship's plating as *flexible* and indeed will probably be concerned with the distortion of the ship as a whole in heavy seas, if for no other reason than to make sure that the propeller shaft bearings remain in sufficiently good alignment. The mechanical engineer will be conscious of the *vibrations* of the ship's structure, which may be caused by the engines, so he too will consider the ship as being made of flexible material. Yet again, the interior designer of a large liner will picture the ship primarily as a floating *hotel*, and so on.

The point of this rather obvious analysis is that *any* mental picture of the ship we may have depends on what questions we are asking about the ship. If we ask "where is the ship?" we think of it as a *point* if it is out at sea, and as an object of a particular *shape* and *size* if it is near a dock. To answer other obviously important questions in the design of a ship, we consider it successively as a *rigid body*, a *flexible body*, a *hotel* etc. There is no real contradiction between any of these different idealizations, not even between a point and a body of a particular shape and size, or between a rigid and a flexible body, although, at first sight, these are absolutely contradictory. Quite simply, in each case, the idealization is the best one for the particular question at hand.

The process of idealization of the physical world is one in which we set up conceptual *models* in which only the *main features* from a particular standpoint are reproduced. Now, we are all familiar with *physical* models. An architect's model of a proposed building, for example, indicates the main features of the building without including a lot of detail.

We are not misled by the fact that the model, perhaps, is made of balsa-wood and has unglazed windows into thinking that the finished building will also be made of wood and have simple holes for windows; we understand easily that the sole object of a model of this sort is to give an indication of the layout and proportions of the building in a three-dimensional representation. Precisely, the same sort of relationship between reality and model holds for *conceptual models*; the only difference is that conceptual models only exist in the human mind.

Science is rich with examples of conceptual models, and indeed, the progress of science is largely a result of the invention of appropriate conceptual models to fit particular experimental observations. A good example of a conceptual model is Maxwell's kinetic theory of gases. In this theory, Maxwell uses as a model of gas an empty space containing a number of small, fast-moving, perfectly elastic, heavy balls which bounce indefinitely between the walls of the container, obeying Newton's laws of motion. This simple and imaginative model explains satisfactorily a number, but by no means all, of the phenomena of gases and mixtures of gases. In discussing this theory, we are not committing ourselves to a belief that atoms or molecules *are* small spherical particles:

We are simply saying that *if* we regard them as such, then in certain circumstances, the predictions of the model agree well with experimental observations. In other circumstances, for example, at very high pressures, this model becomes inadequate, and we have to seek another.

Nearer to our own field, the useful subject of rigid-body mechanics provides many good examples of conceptual models; indeed, the subject owes its existence to the conceptual model of the ***rigid*** body. Now, no real object is actually rigid: sufficiently large forces will always produce deformation or fracture. However, in many situations, the forces tending to distort the body are relatively unimportant compared to those tending to accelerate it, and so, it is reasonable to set up and use the concept of the perfectly rigid body. We are all familiar with the language of this branch of mechanics at the schoolboy level: "A heavy particle rests on a smooth flat horizontal table. It is attached to a point on the table by a light, inextensible string." An idealized situation is described very clearly. The corresponding situation in the physical world might be described as follows: "A lump of metal lies on a table. It is attached by means of a piece of string to a nail that has been driven into the table. On investigation, it turns out that the coefficient of friction between the lump and the table is 0–1, that the mass of the string is only 1/20 of that of the lump of metal, and that when the mass is suspended by the string in the gravitational field, the string extends by about 1 percent." By removing from our model the extensibility and mass of the string and all friction, we arrive at a situation that is amenable to relatively simple analysis using Newton's laws, and we hope that the main features of the behavior of the model will be broadly similar to those of the real system. We can test the predictions of our model by performing an experiment and comparing the predicted and observed behavior. If this comparison reveals large discrepancies, we must, of course, reconsider the features of the model. For example, in this particular situation, the absence of friction from the model would lead to large differences in behavior between the model and the physical set-up in long-term motion, but probably not in short-term motion. If instead of discarding these "secondary" effects in setting up our model, we had retained them, all the mathematics would have been very much more complex, and all clarity would have disappeared.

Thus, in making idealizations and setting up conceptual models, we run, in general, the risk of removing some feature which is in fact rather important. We cannot, of course, make any *direct* test of our models against the real physical world: the only test we can make of the validity of our models is to see whether or not predictions based on their use agree with experimental observations made on real physical systems.

As we must *necessarily* make idealizations in setting up our simple models, we cannot, of course, ever expect *complete* agreement between "theory" and experiment, and we must be content with broad agreement on the main features of behavior. The question of whether the agreement is satisfactory or not is, thus, to some extent, a question of taste. In engineering, we often have to judge between an extremely crude model which leads to very simple calculations and answers which may agree with the experiment to within, say, ±20 percent, and a much more sophisticated model which leads to complex and lengthy calculations and produces answers which are to within, perhaps, ±5 percent of experimental observations.

It is implicit in the preceding discussion that one of the barriers to the general progress of science and engineering is the difficulty of choosing appropriate conceptual models. This is one of the points at which the intellect and the imagination are worked hardest, but once the key has been found, so to speak, new realms of understanding are opened up. The theory of plasticity is based, as we shall see, on an exceptionally simple idealization of material behavior, and it has enabled engineers to make enormous advances in their understanding of certain kinds of structural action and in the setting up of rational design procedures. The development of the theory over the past 30 years has marked a pronounced maturing process in the field of structural engineering, which has led to increased confidence and a growing awareness of the value of imagination and intuition in the processes of structural design.

This brings us to a fundamental and rather subtle point that is not at all well understood by some engineers. Because a lack of appreciation of this point apparently acts as a barrier to the acceptance of the plastic theory as relevant to structural design, it seems worthwhile to look into the question in some depth.

In the theory of elasticity, there is a unique relationship between stress and strain, which results in the behavior of elastic structures being expressed in the form of *equations*. As we shall see later, plastic theory, on the other hand, has an essential *discontinuity* in the stress–strain relationship, and the theory involves, in consequence, discontinuities and inequalities. This might appear at first sight to make the theory difficult, but it turns out that the situation may be exploited to produce some extremely simple results. Part of the price to be paid for this simplicity is that there is often an apparent degree of indeterminacy about solutions to problems in plasticity. In fact, the answers to the *main* questions which are asked are unambiguous, but there is often a degree of flexibility in solutions in the sense that the values of some variables may be chosen arbitrarily within limits without affecting the answers to the main questions. To those brought up on elasticity theory, this *minor indeterminacy* may well be puzzling because it is unfamiliar: nothing like it appears in elastic theory, where, once a problem has been posted, the values of all the variables may be determined uniquely, in principle, at least.

Now, the apparently complete determinacy of elasticity solutions is due in part to the fact that elastic structures are normally idealized as being *stress-free* initially. Actual structures are very rarely stress-free as built because, quite apart from having to carry their own weight, they sustain "locked-in" stresses due to slight misfit of the components with each other, due to thermal expansion on welding, and because as-rolled sections invariably have high residual stresses due to differential cooling in the manufacturing process. Some especially critical structures may be put through a "stress-relieving" procedure, which aims at removing initial stresses by a sort of annealing process, but in general, it can be stated that structures, as-built, contain random unknown residual stresses of significant magnitude. In general, therefore, the "initially stress-free" idealization of structures in elastic theory cannot be justified, and the determinism of the stress quantities in the theory can properly be regarded only as a consequence of an unjustifiable idealization.

The plastic theory deals with the initial stress problem in a quite different way. It concentrates on the state of the structure at *collapse* and, in fact, concludes that initial stresses do not affect the collapse strength of the structure. In other words, plastic theory, instead of *ignoring* initial stresses, shows positively that in some important respects, their magnitude is *irrelevant.*

The emphasis in plastic theory on the importance of some quantities and the relative insignificance of others makes the theory particularly valuable for structural design. Anyone who has attempted any engineering design or project work will be well aware that in the creative work of this sort, there are so many variables that to consider them all would be quite impossible. The practice must pick out what appear to be the important variables and assign reasonable values to the others by intuition if by no other means. In other words, there are usually so many facets to a design problem that there is no unique solution. There is, so to speak, a lot of room for free will in design, and experienced engineers will often express the profound idea that engineering design is, to a large extent, an *art.* In an important sense, plasticity theory harmonizes with this overall design situation, in that it places a premium on the use of imagination and intuition.

## 1.2 Tension and Compression

The simplest type of loading is represented by the uniaxial stress condition, e.g., the simple tension test, for which $\sigma_3 = \sigma_2 = 0$, $\sigma_1 > 0$, or the simple compression test, for which $\sigma_1 = \sigma_2 = 0$, $\sigma_3 < 0$. The well-known uniaxial stress–strain diagram, in which the axial principal stress $\sigma_1$ (or $\sigma_3$) is plotted against the axial strain $\varepsilon_1$ (or $\varepsilon_3$), affords a useful representation of the plastic as well as the elastic behavior.

### 1.2.1 Uniform Loading

A typical curve for a simple tension specimen of mild steel is demonstrated in Figure 1.1. The initial elastic region generally appears as a straight line *OA*, where *A* defines the *limit of proportionality.* On further straining, the relation between stress and strain is no longer linear, but the material is still elastic, and upon release of the load, the specimen reverts to its original length. The maximum stress point *B* at which the load can be applied without causing any permanent deformation defines the *elastic limit.* Point *B* is also called the *yield point*, for it marks the initiation of plastic or irreversible deformation. Usually, there is little difference between the proportional limit, *A*, and the elastic limit, *B*. Mild steel exhibits an upper yield point, *B*, and a lower yield point, C. Beyond point C, there is an extension at approximately constant load. The behavior in the flat region CD is

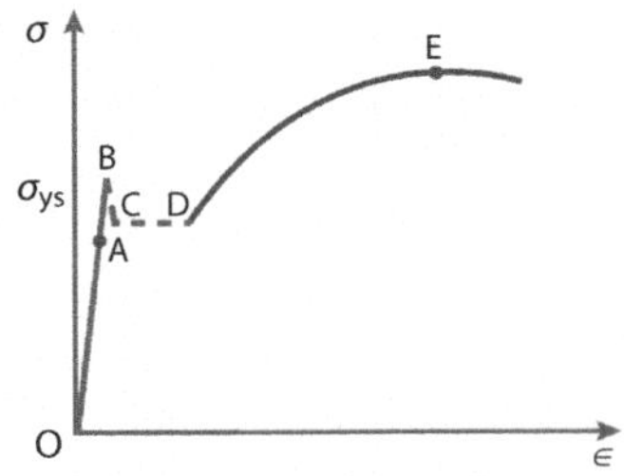


**Figure 1.1** Mild steel stress-strain curve.

generally referred to as *plastic flow*. For most metals, as illustrated in Figure 1.2, however, neither a sharp yield point nor plastic flow is discernible, and yield strength is generally defined by an *offset yield stress* ($\sigma_{ys}$) corresponding usually to a 0.1% definition of strain. This offset yield stress is defined as the *initial yield stress.*

Above the yield point, the response of the material is both elastic and plastic. The slope of the curve decreases steadily, monotonically, and eventually, failure of the specimen occurs at point E. A *ductile material* like mild steel, as displayed in Figure 1.1, is able to incur comparatively large strains without failure.

On the other hand, as shown in Figure 1.2, cast iron is a brittle material since it fails after very little straining. It is generally considered when discussing the failure of metals that there are two types of failure modes, cleavage types such as exhibited by cast iron, and shear-type failures such as exhibited by mild steel. The failure characteristics of geological materials are much more complicated. They also depend on loading type; for example, concrete exhibits brittle behavior under tensile loadings, but under

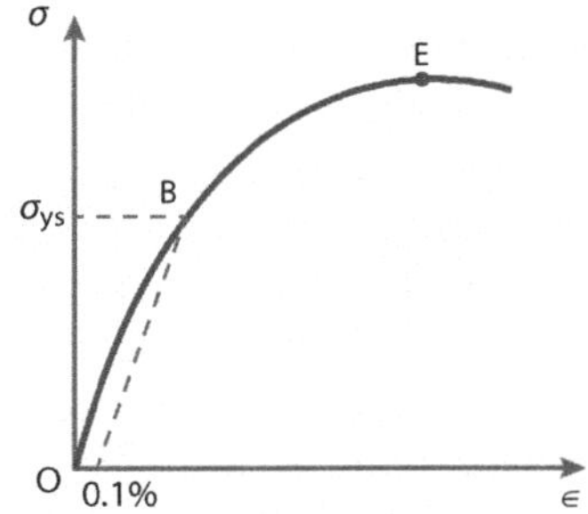


**Figure 1.2** Material steel stress-strain curve.

compressive loadings with confining pressure, it may exhibit a certain degree of ductility before failure.

### 1.2.2 Loading Function

Now, consider the test in which the specimen is first loaded monotonically to some value beyond the initial yield point and then completely unloaded. Figure 1.3 shows the loading, unloading, and preloading paths. So, as can be seen, when the stress is reduced, the strain decreases along an almost elastic unloading line *AB*, which is parallel to the initial linear region of the curve. When the load is again zero at the end of unloading, the strain is not zero; there remains a *residual strain OB*. The irrecoverable strain *OB* is referred to as *plastic strain*, while the recoverable strain *BC* is the *elastic strain*. Now, if this specimen is reloaded, the stress–strain curve follows the reloading path *BA*, which is identical to the unloading path *AB*. The material is therefore elastic until the previous maximum stress at point *A* is reached again. The stress $\sigma_A$ is regarded as the *subsequent yield stress*, beyond which further plastic deformation is indeed, and the stress–strain curve again follows the original one for uniform loading.

For most materials, after the initial yield point has been reached, the stress–strain curve continues to rise, although the slope becomes progressively less until the slope falls to zero as failure occurs. Thus, the subsequent yield stress increases with further straining. This effect of the material being able to withstand greater stress after plastic deformation is known as *strain hardening* or *work hardening*, in the sense that the material gets stronger the more it is strained or worked.

For some materials, such as concrete or rock, in a simple compression test, there is a region beyond the failure or peak point in which the slope of the curve is negative. Such behavior is called *strain softening*. This type

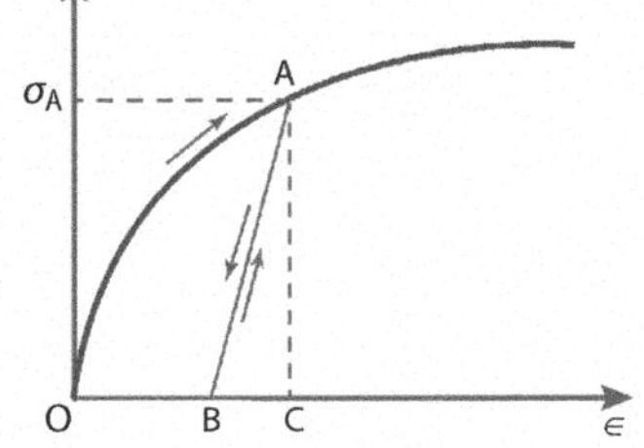


**Figure 1.3** Paths of loading, unloading, and reloading.

of material gets weaker with continuous straining beyond a certain limit or peak stress.

### 1.2.3 Opposite Loading

If we perform a simple compression test on metal, we will obtain an almost identical stress–strain curve as in a simple tension test. However, after a plastic pretraining of a specimen, the stress–strain curve of this specimen in compression differs considerably from the curve which would be obtained on reloading this specimen in tension or on loading an undisturbed specimen in compression. Figure 1.4 displays that for the specimen with a preloading $\sigma'_y$ in tension, its corresponding compressive yielding occurs at a stress level $\sigma''_y$, which is less than the initial yield stress $\sigma_y$ and much less than the subsequent yield point $\sigma'_y$. This phenomenon is known as the *Bauschinger effect* and is usually present whenever there is a stress reversal.

It is evident from the previous discussion that there is no one-to-one correspondence between stress and strain in a plastically deformed solid. In other words, the strain is not a function of stress alone but depends on the previous *loading history*. Thus, the material is *load path-dependent*. This can be illustrated by the simple case of zero stress when residual strains of different magnitudes can be established by varying the loading history, with the stress starting and finishing at zero.

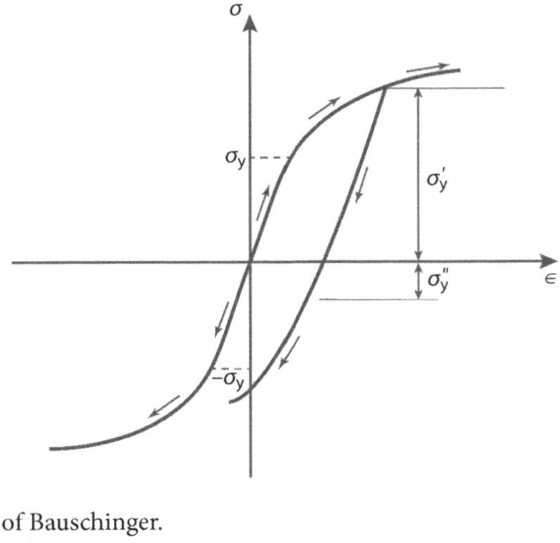


**Figure 1.4** Effect of Bauschinger.

In this discussion so far, we have assumed that there is a single stress–strain curve for tension or compression, independent of the rate of straining. This assumption is referred to as *time independence*. It is reasonably true for structural metals at room temperature under a static loading condition. Rate effects are very important for materials under dynamic loading conditions.

## 1.3 Stress Tensor and Deviatoric Stress Tensor

When a body is subjected to a set of external forces, internal forces are produced in different parts of the body so that each element of the body is in a state of static equilibrium. As shown in Figure 1.5, through any point $O$ within the body, consider a small surface element $\delta S$ whose orientation is specified by the unit vector $\mathbf{l}$ along the normal drawn on one side of the element. The material on this side of $\delta S$ may be regarded as exerting a force $\delta\mathbf{P}$ across the surface element upon the material on the other side. The limit of the ratio $\delta\mathbf{P}/\delta S$ as $\delta S$ tends to zero is the stress vector $\mathbf{T}$ at $O$ associated with the direction $\mathbf{I}$. Forgiven external loading, the stress acting across any plane passing through a given point $O$ depends on the orientation of the plane. The resolved component of the stress vector along the unit normal $\mathbf{l}$ is called the direct or normal stress, denoted by $\sigma$, while the component tangential to the plane is known as the shear stress, denoted by $\tau$.

As displayed in Figure 1.6, consider that a set of rectangular axes $Ox$, $Oy$, and $Oz$ emanating from a typical point $O$ and imagine a small rectangular parallelepiped at $O$ having its edges parallel to the axes of reference. The normal stresses across the faces of the block are denoted by $\sigma_x$, $\sigma_y$, and $\sigma_z$, where the subscripts denote the directions of the normal to the faces.

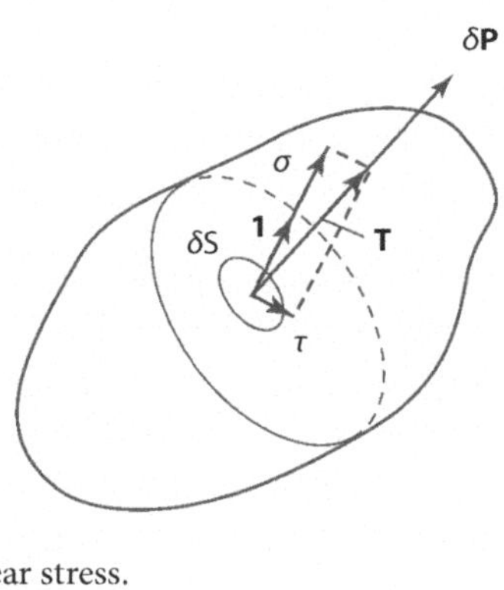


**Figure 1.5** Normal and shear stress.

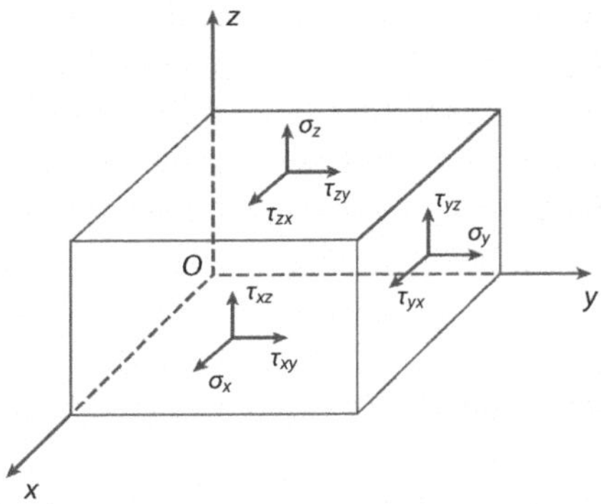


**Figure 1.6** Components of the stress tensor.

The shear stress acting on the faces normal to the $x$-axis is resolved into the components $\tau_{xy}$ and $\tau_{xz}$ parallel to the $y$ and $z$ axes, respectively. The first suffix denotes the direction of the normal to the face, and the second suffix is the direction of the component. Similarly, the shear stresses on the faces normal to the $y$-axis are denoted by $\tau_{yx}$ and $\tau_{yz}$ and those on the faces normal to the $z$-axis by $\tau_{zx}$ and $\tau_{zy}$. The stresses are taken as positive if they are directed when the outward normal to the faces is in the positive directions of the coordinate axes.

The positive directions are all reversed on the remaining faces of the block where the outward normals are in the negative directions of the axes of reference. The nine components of the stress at any point form a second-order tensor $\sigma_{ij}$, known as the stress tensor, where $i$ and $j$ take integral values 1, 2, and 3. The stress components may be displayed as elements of the square matrix:

$$\sigma_{ij} = \begin{bmatrix} \sigma_x & \tau_{xy} & \tau_{xz} \\ \tau_{yx} & \sigma_y & \tau_{yz} \\ \tau_{zx} & \tau_{zy} & \sigma_z \end{bmatrix} = \begin{bmatrix} \sigma_{11} & \sigma_{12} & \sigma_{13} \\ \sigma_{21} & \sigma_{22} & \sigma_{23} \\ \sigma_{31} & \sigma_{32} & \sigma_{33} \end{bmatrix} \qquad (1.1)$$

The forces acting on the faces of the parallelepiped are clearly in equilibrium. To examine the couple equilibrium, let $\delta x$, $\delta y$, and $\delta z$ denote the lengths of these faces along the respective coordinate axes. Then the resultant couple about the $z$-axis is $(\tau_{xy} - \tau_{yx})\ \delta x\ \delta y\ \delta z$, which must vanish for equilibrium. This gives $\tau_{xy} = \tau_{yx}$. Similarly, the conditions for couple equilibrium about the other two axes give $\tau_{yz} = \tau_{zy}$ and $\tau_{zx} = \tau_{xz}$. These identities may be expressed as $\sigma_{ij} = \sigma_{ji}$, implying that the stress tensor is symmetric concerning its subscripts. Thus, there are six independent stress components; three normal components $\sigma_x$, $\sigma_y$, and $\sigma_z$; and three shear components $\tau_{xy}$, $\tau_{yz}$, and $\tau_{zx}$, which completely specify the state of stress at each point of the body. The matrix representing the stress tensor is symmetrical.

The mean of the three normal stresses, equal to $(\sigma_x + \sigma_y + \sigma_z)/3$, is known as the *hydrostatic stress*, denoted by $\sigma_0$. A *deviatoric* or *reduced stress* tensor $s_{ij}$ is defined as that which is obtained from $\sigma_{ij}$ by reducing the normal stress components by $\sigma_0$. This gives the deviatoric stresses as:

$$s_{ij} = \begin{bmatrix} s_x & s_{xy} & s_{xz} \\ s_{yx} & s_y & s_{yz} \\ s_{zx} & s_{zy} & s_z \end{bmatrix} = \begin{bmatrix} (\sigma_x - \sigma_0) & \tau_{xy} & \tau_{xz} \\ \tau_{yx} & (\sigma_y - \sigma_0) & \tau_{yz} \\ \tau_{zx} & \tau_{zy} & (\sigma_z - \sigma_0) \end{bmatrix} \tag{1.2}$$

The deviatoric normal stresses are therefore given by:

$$\begin{cases} 3s_x = 2\sigma_x - \sigma_y - \sigma_z \\ 3s_y = 2\sigma_y - \sigma_z - \sigma_x \\ 3s_z = 2\sigma_z - \sigma_x - \sigma_y \end{cases} \tag{1.3}$$

The deviatoric shear stresses are the same as the actual shear stresses. Since $s_x + s_y + s_z = 0$, the deviatoric normal stresses cannot all have the same sign. The difference between any two normal components of the deviatoric stress is the same as that between the corresponding components of the actual stress. Expressed in suffix notation, the relationship between $s_{ij}$ and $\sigma_{ij}$ is:

$$s_{ij} = \sigma_{ij} - \sigma_0\delta_{ij} = \sigma_{ij} - \frac{1}{3}\sigma_{kk}\delta_{ij} \tag{1.4}$$

where $\delta_{ij}$ is the *Kronecker delta*, whose value is unity when $i = j$ and zero when $i = j$. Evidently, $\delta_{ij} = \delta_{ji}$. Any repeated or dummy suffix indicates a summation of all terms obtainable by assigning the values 1, 2, and 3 to this suffix in succession. Thus $\sigma_{kk} = \sigma_x + \sigma_y + \sigma_z$. It follows from the definition of the delta symbol that $\sigma_{ij}\delta_{jk} = \sigma_{ik}$, where $j$ is a dummy suffix and $i$, $k$ are free suffixes. Each term of a tensor equation must have the same free suffixes, but a dummy suffix can be replaced by any other letter different from the free suffixes.

### 1.3.1 Principal Stress, Principal Shear Stress, and Invariants of Stress Tensor

In the state of spatial stress, if the coordinate axis is properly selected, the shear stress at one point will be zero, and only the normal stress will be left. The direction of the three mutually perpendicular coordinate axes is called the principal direction of the stress tensor (or the principal axis). The plane perpendicular to the principal direction is called the principal plane, and the normal stress on the plane is called the principal stress. The three principal stresses are $\sigma_1, \sigma_2, \sigma_3$. The value of the principal stress $\sigma_N$ can be determined by the determinant as follows:

$$\begin{vmatrix} \sigma_x - \sigma_N & \tau_{xy} & \tau_{zx} \\ \tau_{xy} & \sigma_y - \sigma_N & \tau_{yz} \\ \tau_{zx} & \tau_{yz} & \sigma_z - \sigma_N \end{vmatrix} = 0 \tag{1.5}$$

Then,

$$\sigma_N^3 - I_1 \sigma_N^2 - I_2 \sigma_N - I_3 = 0 \tag{1.6}$$

Where,

$$\begin{aligned} I_1 &= \sigma_x + \sigma_y + \sigma_z \\ I_2 &= -\left(\sigma_x \sigma_y + \sigma_y \sigma_z + \sigma_z \sigma_x\right) + \left(\tau_{xy}^2 + \tau_{yz}^2 + \tau_{zx}^2\right) \\ I_3 &= \sigma_x \sigma_y \sigma_z + 2\tau_{xy}\tau_{yz}\tau_{zx} - \sigma_x \tau_{yz}^2 - \sigma_y \tau_{zx}^2 - \sigma_z \tau_{xy}^2 \end{aligned} \tag{1.7}$$

In the coordinate transformation, the stress component will change, but the value of the principal stress remains unchanged. Therefore, the values $I_1, I_2, I_3$ are independent of the orientation of the coordinate axis, and they are respectively called the first, second, and third invariants of the stress tensor. If the principal axis is taken as the coordinate axis, the stress invariant can be expressed by the principal stress:

$$\begin{aligned} I_1 &= \sigma_1 + \sigma_2 + \sigma_3 \\ I_2 &= -(\sigma_1\sigma_2 + \sigma_2\sigma_3 + \sigma_3\sigma_1) \\ I_3 &= \sigma_1\sigma_2\sigma_3 \end{aligned} \tag{1.8}$$

In addition, the shear stress on a plane with a 45° angle from one principal direction and the other two principal directions is an important stress value. The shear stress on these planes is called the principal shear stress, and obviously, they have three pairs, denoted by $\tau_1, \tau_2, \tau_3$. The relationship between the principal shear stress and the principal stress is:

$$\begin{aligned} \tau_1 &= \frac{1}{2}(\sigma_2 - \sigma_3) \\ \tau_2 &= \frac{1}{2}(\sigma_3 - \sigma_1) \\ \tau_3 &= \frac{1}{2}(\sigma_1 - \sigma_2) \end{aligned} \tag{1.9}$$

Choose the value with the largest absolute value $\tau_1, \tau_2, \tau_3$, which is the value of the maximum shear stress. If the three principal stresses are arranged in order $\sigma_1 \geq \sigma_2 \geq \sigma_3$, then:

$$\tau_{\max} = \frac{1}{2}(\sigma_1 - \sigma_3) \tag{1.10}$$

In this chapter, only the relevant conclusions are introduced. For detailed derivation and analysis, please refer to the relevant textbook of elasticity.

### 1.3.2 Decomposition of Stress Tensor and Deviatoric Stress Tensor

In order to establish a yield criterion under a complex stress state, it is necessary to analyze the stress state in depth. According to Bridgman's experiments, the hydrostatic pressure does not affect the yield, and the deformation corresponding to the hydrostatic stress state is the change of elastic volume, but not the change of shape. It can be assumed that, if the stress tensor describing a point stress state is decomposed into two parts, one part is the average normal stress, namely, the hydrostatic pressure, which has nothing to do with the plastic deformation. The other part is the residual part after deducting the average stress, which is directly related to the shape change and plastic deformation. Mean normal stress $\sigma_m$ can be defined as:

$$\sigma_m = \frac{1}{3}\left(\sigma_x + \sigma_y + \sigma_z\right) \tag{1.11}$$

It can be known from the first invariant of the stress tensor $I_1 = \sigma_1 + \sigma_2 + \sigma_3$ that the average normal stress is also invariant, which does not change with the choice of the coordinate axis. Now, the stress tensor is decomposed into two parts, namely:

$$\begin{aligned}\begin{bmatrix} \sigma_x & \tau_{xy} & \tau_{zx} \\ \tau_{xy} & \sigma_y & \tau_{yz} \\ \tau_{zx} & \tau_{yz} & \sigma_z \end{bmatrix} &= \begin{bmatrix} \sigma_m & 0 & 0 \\ 0 & \sigma_m & 0 \\ 0 & 0 & \sigma_m \end{bmatrix} + \begin{bmatrix} \sigma_x - \sigma_m & \tau_{xy} & \tau_{zx} \\ \tau_{xy} & \sigma_y - \sigma_m & \tau_{yz} \\ \tau_{zx} & \tau_{yz} & \sigma_z - \sigma_m \end{bmatrix} \\ &= \begin{bmatrix} \sigma_m & 0 & 0 \\ 0 & \sigma_m & 0 \\ 0 & 0 & \sigma_m \end{bmatrix} + \begin{bmatrix} S_x & S_{xy} & S_{zx} \\ S_{xy} & S_y & S_{yz} \\ S_{zx} & S_{yz} & S_z \end{bmatrix}\end{aligned} \tag{1.12}$$

The second tensor on the right side of the equation is called the deviatoric stress tensor, where $S_x = \sigma_x - \sigma_m$, $S_y = \sigma_y - \sigma_m$, $S_z = \sigma_z - \sigma_m$, $S_{xy} = \tau_{xy}$, $S_{zx} = \tau_{zx}$, $S_{yz} = \tau_{yz}$. The stress sphere tensor corresponds to the uniform stress state, which causes only the change of elastic volume but no change of shape. The stress bias tensor reflects the degree to which an actual stress state deviates from a uniform stress state, and the stress state it represents will only cause the shape change of the material, but not the volume change. It is precisely because the stress deviator shows the part of plastic deformation related to shape change that the concept of stress deviator is of great significance in plastic mechanics.

The deviatoric stress tensor is also a kind of stress state; it represents a special stress state, so it has a principal direction and an invariant. We can see that $S_{ij}$ has the same principal direction as $\sigma_{ij}$, and its invariant can be expressed as:

$$
\begin{aligned}
J_1 &= S_x + S_y + S_z \\
J_2 &= -\left(S_x S_y + S_y S_z + S_z S_x\right) + \left(S_{xy}^2 + S_{yz}^2 + S_{zx}^2\right) \\
&= \frac{1}{2}\left[S_x^2 + S_y^2 + S_z^2 + 2\left(S_{xy}^2 + S_{yz}^2 + S_{zx}^2\right)\right] \\
&= \frac{1}{2} S_{ij} S_{ij} \\
J_3 &= S_x S_y S_z + 2 S_{xy} S_{yz} S_{zx} - S_x S_{yz}^2 - S_y S_{zx}^2 - S_z S_{xy}^2 \\
&= \frac{1}{3}\left\{S_x^3 + S_y^3 + S_z^3 + 6 S_{xy} S_{yz} S_{zx} + 3\left[S_{xy}^2\left(S_x + S_y\right) + S_{yz}^2\left(S_y + S_z\right) + S_{zx}^2\left(S_z + S_x\right)\right]\right\} \\
&= \frac{1}{3} S_{ij} S_{jk} S_{ki}
\end{aligned}
\tag{1.13}
$$

The invariant can also be expressed by the principal deviatoric stress tensor $S_1, S_2, S_3$:

$$
\begin{aligned}
J_1 &= S_1 + S_2 + S_3 = 0 \\
J_2 &= -\left(S_1 S_2 + S_2 S_3 + S_3 S_1\right) = \frac{1}{2}\left(S_1^2 S_2^2 + S_3^2\right) \\
J_3 &= S_1 S_2 S_3
\end{aligned}
\tag{1.14}
$$

In the theory of plasticity, the second invariant $J_2$ is a significant value. $J_2$ can be expressed by $\sigma_{ij}$ as:

$$
\begin{aligned}
J_2 &= \frac{1}{6}\left[\left(\sigma_x - \sigma_y\right)^2 + \left(\sigma_y - \sigma_z\right)^2 + \left(\sigma_z - \sigma_x\right)^2 + 6\left(\tau_{xy}^2 + \tau_{yz}^2 + \tau_{zx}^2\right)\right] \\
&= \frac{1}{6}\left[\left(\sigma_1 - \sigma_2\right)^2 + \left(\sigma_2 - \sigma_3\right)^2 + \left(\sigma_3 - \sigma_1\right)^2\right]
\end{aligned}
\tag{1.15}
$$

### 1.3.3 Several Variables Related to $J_2$

(1) Normal stress $\sigma_s$ and shear stress $\tau_s$ on the octahedral plane

Now let's talk about stresses on a particular plane. The three stress directions of the coordinate axes z, y, and z and the investigated point are selected to coincide. Consider an inclined plane with normal line **N**, where **N** is the unit vector and the direction cosines are $1/\sqrt{3}$.

This plane is called an isoclinic plane (see Figure 1.7). Obviously, there are eight such planes (one in each quadrant), and they form an octahedron, so an isoclinic plane is also called an octahedral plane. The octahedral normal stress OS (or denoted as OC) on an octahedral plane is shown as:

$$\sigma_{oct} = \frac{1}{3}(\sigma_1 + \sigma_2 + \sigma_3) = \sigma_m \tag{1.16}$$

The total stress on the octahedral plane can be expressed as:

$$F_{oct} = \sqrt{\frac{1}{3}\left(\sigma_1^2 + \sigma_2^2 + \sigma_3^2\right)} \tag{1.17}$$

Then, the octahedral shear stress is:

$$\begin{aligned}\tau_{oct} &= \sqrt{F_{oct}^2 - \sigma_{oct}^2} = \sqrt{\frac{2}{9}\left(\sigma_1^2 + \sigma_2^2 + \sigma_3^2 - \sigma_1\sigma_2 - \sigma_2\sigma_3 - \sigma_3\sigma_1\right)} \\ &= \frac{1}{3}\sqrt{(\sigma_1 - \sigma_2)^2 + (\sigma_2 - \sigma_3)^2 + (\sigma_3 - \sigma_1)^2} \\ &= \sqrt{\frac{2}{3}J_2}\end{aligned} \tag{1.18}$$

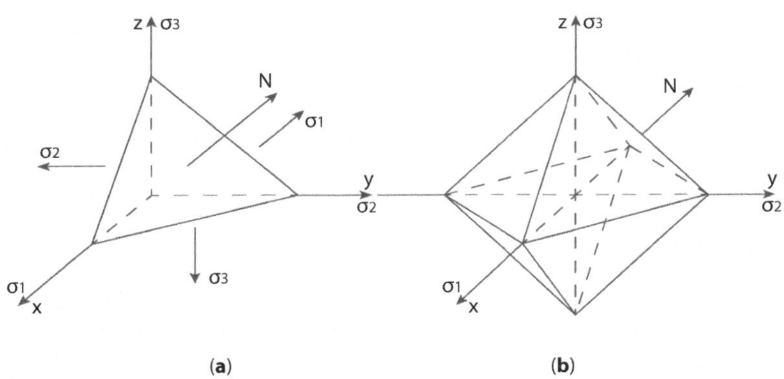


**Figure 1.7** Stresses on the octahedral plane.

It can be seen that the stress on the octahedral plane can be decomposed into two parts, in which the octahedral shear stress is very useful in plastic mechanics.

(2) Equivalent stress $\bar{\sigma}$

The concept of equivalent stress is introduced to compare the strength effects under different stress states. The unidirectional stress state can reflect its strength only by the magnitude of stress. The strength of complex stress states should be characterized by the combined action of stress components. A comprehensive quantity is introduced to make it comparable to unidirectional stress, that is, equivalent stress. The equivalent stress is defined as

$$\begin{aligned}\bar{\sigma} = \frac{3}{\sqrt{2}}\tau_{oct} &= \frac{\sqrt{2}}{2}\sqrt{(\sigma_x-\sigma_y)^2+(\sigma_y-\sigma_z)^2+(\sigma_z-\sigma_x)^2+6\left(\tau_{xy}^2+\tau_{yz}^2+\tau_{zx}^2\right)} \\ &= \frac{\sqrt{2}}{2}\sqrt{(\sigma_1-\sigma_2)^2+(\sigma_2-\sigma_3)^2+(\sigma_3-\sigma_1)^2}\end{aligned} \tag{1.19}$$

It is easy to prove that $\bar{\sigma}$ can also be expressed by deviatoric stress as:

$$\bar{\sigma} = \sqrt{\frac{3}{2}}\sqrt{S_x^2+S_y^2+S_z^2+2\left(S_{xy}^2+S_{yz}^2+S_{zx}^2\right)} = \sqrt{\frac{3}{2}}\sqrt{S_{ij}S_{ij}} = \sqrt{3J_2} \tag{1.20}$$

Unidirectional stress state can be regarded as a special case of the complex stress state. For unidirectional stretching, $\sigma_1 \neq 0, \sigma_2 = \sigma_3 = 0$. Then $\bar{\sigma} = \sigma_1$.

Obviously, $\bar{\sigma}$ is also an invariant, independent of the choice of coordinates. From the function of the form, you can see that the normal stress component to increase or decrease hydrostatic stress, equivalent stress values remain unchanged, meaning that it has nothing to do with the hydrostatic stress tensor and only with the deviatoric stress tensor.

## 1.4 Stress Space, π Plane, and Lode Parameters

### 1.4.1 Principal Stress Space and Plane

The stress state of a certain point can be expressed by six independent stress components, three principal stresses, and their corresponding three

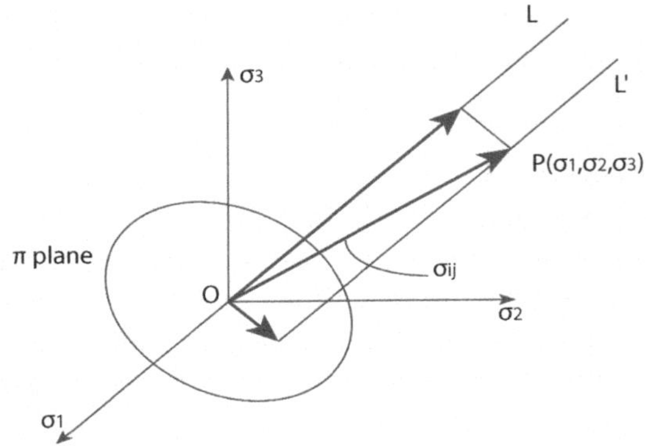


**Figure 1.8** Decomposition of stress vectors on the isoclinal plane.

principal directions, three invariants of the stress tensor, and the principal direction of the stress. Just as the three coordinate values of x, y, and z in a three-dimensional space can determine the position of a point in space, the six independent stress components that determine a point's stress state can also be expressed in the imaginary six-dimensional stress space. The so-called six-dimensional stress space is an imaginary space with six stress components as six coordinate axes. Any point in the stress space represents a stress state. Considering that the material is initially isotropic, it is sufficient to know the magnitude of the three principal stresses to establish yield conditions. Therefore, if the three principal stresses $\sigma_1, \sigma_2, \sigma_3$ are taken as three perpendicular rectangular axes to form a spatial rectangular coordinate system (see Figure 1.8), then point $P$ in the space corresponds to the stress state of a point in the object $(\sigma_1, \sigma_2, \sigma_3)$, or the stress state of the point is expressed by the vector $\overrightarrow{OP}$, which is called the stress state vector. This space is called the principal stress space and was proposed by Haigh-Westergaard. In the process of load change, each point in the object is changing all the time, the stress state in the stress space corresponding stress point is in constant change of its position, then the stress state vector of the vector end trace that tracks the stress path of the corresponding points is called a stress history, namely, a curve in the stress space has a little bit of the changing process of the stress state. Let ***i***, ***j***, and ***k*** represent the unit vectors of the three axes in the principal stress space. Let's make a line $L$ in the stress space, which passes through the origin 0 and is equal to the angle between the three coordinate axes. It is called the isoclinal line $L$. Since any point $L$ has the same coordinates on all three axes, the equation of line $L$ is:

$$\sigma_1 = \sigma_2 = \sigma_3 \tag{1.21}$$

In other words, points on the $L$ line represent the state of uniform stress, and their deviatoric stress is zero. The equation of the plane perpendicular to $L$ in which the point is observed in the stress space is:

$$\sigma_1 + \sigma_2 + \sigma_3 = 0 \tag{1.22}$$

Since the average stress at all points on this plane is zero and only the deviatoric stress is presented, it is called the $\pi$ plane. The point on the $\pi$ plane corresponds to the deviatoric stress.

Now, the vector $\overrightarrow{OP}$ is decomposed as follows:

$$\begin{aligned}\overrightarrow{OP} &= \sigma_1 \boldsymbol{i} + \sigma_2 \boldsymbol{j} + \sigma_3 \boldsymbol{k} \\ &= (\sigma_1 - \sigma_m)\boldsymbol{i} + (\sigma_2 - \sigma_m)\boldsymbol{j} + (\sigma_3 - \sigma_m)\boldsymbol{k} + \sigma_m \boldsymbol{i} + \sigma_m \boldsymbol{j} + \sigma_m \boldsymbol{k} \\ &= (S_1 \boldsymbol{i} + S_2 \boldsymbol{j} + S_3 \boldsymbol{k}) + (\sigma_m \boldsymbol{i} + \sigma_m \boldsymbol{j} + \sigma_m \boldsymbol{k}) \\ &= \overrightarrow{OQ} + \overrightarrow{ON}\end{aligned} \tag{1.23}$$

where $\overrightarrow{OQ}$ is the stress deviator, $\overrightarrow{ON}$ which stands for hydrostatic stress. Since the projection $\overrightarrow{ON}$ on the three axes is equal, it must have coincided with the isoclinal lines $L$. And since $S_1 + S_2 + S_3 = 0$ we know that $\overrightarrow{OQ}$ must be on the $\pi$ plane.

Note that if you take a line $L'$ that crosses any point in the plane and is perpendicular to the plane, the deviatoric stress at each point on the line is the same. Since the hydrostatic stress has no effect on the plastic properties of the material, when considering the relationship between the plastic deformation and stress states, we can focus on the plane and only consider the projection (such as $\overrightarrow{OQ}$) on the plane, because the deviant part of the stress state represented by any point in the stress space must fall on the $\pi$ plane.

### 1.4.2 Two-Dimensional Expression of Deviatoric Stress

We are only interested in deviatoric stress, so in the stress space, we only need to study the stress state represented by each point on the plane. The position of any point on the plane can be represented by two parameters as follows:

$$\left(\sigma_1,\sigma_2,\sigma_3\right) \Leftrightarrow \left(x,y\right) or \left(r,\theta\right) \tag{1.24}$$

Project the unit vector $(\boldsymbol{i}, \boldsymbol{j}, \boldsymbol{k})$ onto the $\pi$ plane, and write it as $(\boldsymbol{i}', \boldsymbol{j}', \boldsymbol{k}')$. Take the rectangular coordinate *Oxy* on $\pi$ the plane and make the *y*-axis coincide with $\boldsymbol{j}'$, as shown in Figure 1.9.

In order to establish the relationship between a point $(\sigma_1,\sigma_2,\sigma_3)$ of the principal stress space and a point $(x, y)$ of the rectangular coordinate space, we first need to figure out the projection length of $\boldsymbol{i}$, $\boldsymbol{j}$, and $\boldsymbol{k}$ on a plane. Let the endpoint of vector $\boldsymbol{i}$ be $a$ on the $\sigma_1$ axis, the projection onto the $\pi$ plane be $a'$, and $\boldsymbol{j}$ be $b$ on the $\sigma_2$ axis; the projection is $b'$ on the $\pi$ plane, so the length of the vector $\overrightarrow{ab}$ is:

$$\left|\overrightarrow{ab}\right| = \sqrt{\left|\boldsymbol{i}\right|^2 + \left|\boldsymbol{j}\right|^2} = \sqrt{2} \tag{1.25}$$

Since $\overrightarrow{ab}$ is parallel to the $\pi$ plane, then:

$$\left|\overrightarrow{a'b'}\right| = \sqrt{2} \tag{1.26}$$

Therefore,

$$\left|\overrightarrow{Oa'}\right| = \frac{\left|\overrightarrow{a'b'}\right|}{2} \frac{1}{\cos 30°} = \sqrt{\frac{2}{3}} \tag{1.27}$$

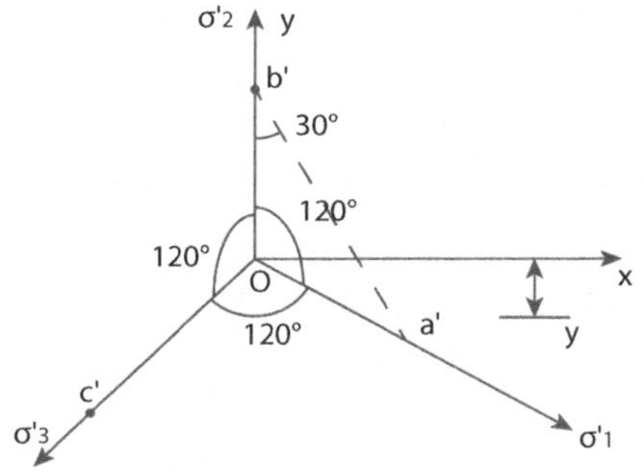


**Figure 1.9** Deviatoric stress on the $\pi$ plane.

Therefore, $a'$ in the $Oxy$ coordinate system is $(x=\sqrt{2}/2, y=-1/\sqrt{6})$, and the coordinate of the projection on the π plane of point $(\sigma_1,0,0)$ of the principal stress space is $(\sqrt{2}/2\sigma_1,-1/\sqrt{6}\sigma_1)$. Similarly, the coordinate of $b'$ in the $Oxy$ coordinate system is $(x=0, y=\sqrt{2/3})$, so the coordinate of $(0,\sigma_2,0)$ in the π plane is $(0,\sqrt{2/3}\sigma_2)$. The coordinates of $c'$ in the $Oxy$ system are $(-\sqrt{2}/2,-1/\sqrt{6})$, and the coordinate of $(0,0,\sigma_3)$ in the π plane is $(-\sqrt{2}/2\sigma_3,-1/\sqrt{6}\sigma_3)$.

Thus, the coordinates of $(\sigma_1,0,0)+(0,\sigma_2,0)+(0,0,\sigma_3)=(\sigma_1,\sigma_2,\sigma_3)$ under a rectangular $Oxy$ coordinate system in the $\pi$ plane are:

$$
\begin{aligned}
x &= \frac{\sqrt{2}}{2}(\sigma_1-\sigma_3)=\frac{\sqrt{2}}{2}(S_1-S_3) \\
y &= \frac{2\sigma_2-\sigma_1-\sigma_3}{\sqrt{6}}=\frac{2S_2-S_1-S_3}{\sqrt{6}}
\end{aligned} \tag{1.28}
$$

Set the module of the vector $\overrightarrow{OQ}$ as $r_\sigma$ and the angle between it and the axis $\theta_\sigma$. Then, the coordinates in the plane polar coordinate system can be obtained.

$$
\begin{aligned}
r_\sigma &= \sqrt{x^2+y^2}=\sqrt{2J_2}=\sqrt{\frac{2}{3}}\bar{\sigma} \\
\tan\theta_\sigma &= \frac{y}{x}=\frac{1}{\sqrt{3}}\frac{2\sigma_2-\sigma_1-\sigma_3}{\sigma_1-\sigma_3}=\frac{1}{\sqrt{3}}\mu_\sigma
\end{aligned} \tag{1.29}
$$

where $\theta_\sigma$ is called Lode angle, which reflects the proportional characteristics of each stress deviator. $r_\sigma$ then reflects the quantitative characteristics between them. The projection onto the $\pi$ plane of a stress state is equal to its equivalent stress times $\sqrt{2/3}$.

$$
\mu_\sigma = \frac{2\sigma_2-\sigma_1-\sigma_3}{\sigma_1-\sigma_3}=2\frac{\sigma_2-\sigma_3}{\sigma_1-\sigma_3}-1 \tag{1.30}
$$

The $\mu_\sigma$ is called the Lode parameter, which reflects the ratio of the intermediate principal stress to the other two principal stresses and is independent of the average stress $\sigma_m$, so it is a quantity reflecting the characteristics of deviatoric stress. If there are $\sigma_1 \geq \sigma_2 \geq \sigma_3$, then:

$$\begin{aligned} -1 &\le \mu_\sigma \le 1 \\ -30^\circ &\le \theta_\sigma \le 30^\circ \end{aligned} \tag{1.31}$$

Here are the $\mu_\sigma$ values for a few special cases.

1) Uniaxial tensile state: $\sigma_2 = \sigma_3 = 0, \sigma_1 > 0$, then $\mu_\sigma = -1$
2) Pure shear state: $\sigma_1 = -\sigma_3, \sigma_2 = 0$, then $\mu_\sigma = 0$
3) Uniaxial compression state: $\sigma_1 = \sigma_2 = 0, \sigma_3 < 0$, then $\mu_\sigma = 1$

The above discussion shows that $(\sigma_1,\sigma_2,\sigma_3),(S_1,S_2,S_3)$ can be determined; furthermore, $(x, y)$ and $(r_\sigma,\theta_\sigma)$ can be determined. On the contrary, given $(x, y)$ or $(r_\sigma,\theta_\sigma)$, we can get $(S_1,S_2,S_3)$. Using $S_1 + S_2 + S_3 = 0$, it is easy to get:

$$\begin{aligned} S_1 &= \frac{1}{\sqrt{2}}x - \frac{1}{\sqrt{6}}y = \sqrt{\frac{2}{3}}r_\sigma \sin\left(\theta_\sigma + \frac{2}{3}\pi\right) \\ S_2 &= \sqrt{\frac{2}{3}}y = \sqrt{\frac{2}{3}}r_\sigma \sin(\theta_\sigma) \\ S_3 &= -\frac{1}{\sqrt{2}}x - \frac{1}{\sqrt{6}}y = \sqrt{\frac{2}{3}}r_\sigma \sin\left(\theta_\sigma - \frac{2}{3}\pi\right) \end{aligned} \tag{1.32}$$

Therefore $(\sigma_1,\sigma_2,\sigma_3) \to (S_1,S_2,S_3) \Leftrightarrow (x,y) \Leftrightarrow (r_\sigma,\theta_\sigma)$.

## 1.5 Deviatoric Strain and Equivalent Strain

### 1.5.1 The Relationship between Displacement and Strain

In the case of small deformation, the displacement and strain are determined by the Cauchy formula

$$\begin{aligned} &\varepsilon_x = \frac{\partial u}{\partial x}, \quad \varepsilon_y = \frac{\partial v}{\partial y}, \quad \varepsilon_z = \frac{\partial w}{\partial z}, \\ &\varepsilon_{xy} = \frac{2}{3}\left(\frac{\partial u}{\partial y} + \frac{\partial v}{\partial x}\right), \quad \varepsilon_{yz} = \frac{1}{2}\left(\frac{\partial v}{\partial z} + \frac{\partial w}{\partial y}\right), \\ &\varepsilon_{zx} = \frac{1}{2}\left(\frac{\partial w}{\partial x} + \frac{\partial u}{\partial z}\right) \end{aligned} \tag{1.33}$$

The above formula is called the geometric equation, whose tensor form is:

$$\varepsilon_{ij} = \frac{1}{2}\left(u_{i,j} + u_{j,i}\right) \tag{1.34}$$

where $\varepsilon_{ij}$ is the Cauchy linear strain tensor. The comma between i and j indicates the derivative, for example $u_{x,y} = \partial u_x / \partial y$. In practice, engineering strain is often applied, and its expression is:

$$\begin{aligned}
\varepsilon_x &= \frac{\partial u}{\partial x}, \quad \varepsilon_y = \frac{\partial v}{\partial y}, \quad \varepsilon_z = \frac{\partial w}{\partial z}, \\
\gamma_{xy} &= \frac{\partial u}{\partial y} + \frac{\partial v}{\partial x}, \quad \gamma_{yz} = \frac{\partial v}{\partial z} + \frac{\partial w}{\partial y}, \\
\gamma_{zx} &= \frac{\partial w}{\partial x} + \frac{\partial u}{\partial z}
\end{aligned} \tag{1.35}$$

That is, the engineering shear strain $\gamma_{xy}, \gamma_{yz}, \gamma_{zx}$ is twice the strain tensor components $\varepsilon_{xy}, \varepsilon_{yz}, \varepsilon_{zx}$. Where the positive strain $\varepsilon_{xy}, \varepsilon_{yz}, \varepsilon_{zx}$ represents the relative expansion of the unit length of the differential line segment, and the shear strain $\gamma_{xy}, \gamma_{yz}, \gamma_{zx}$ represents the change in the right angle between the differential lines (expressed in radians). They're both dimensionless quantities. The deformation state of a certain point can be determined from these six strain components.

Although the elastic-plastic deformation is larger than the pure elastic deformation, for most applications, the small deformation hypothesis is still applicable before the ultimate load.

For large deformation or finite deformation, there are two ways of describing it: (1) the Lagrange method described in the initial coordinates of the object before the deformation and (2) the Euler method described in the instantaneous coordinates after deformation. The strain tensor caused by the two methods is not the same. If the former is adopted and the displacement vector is expressed by $U$, the Green strain tensor $E_{ij}$ and the displacement-strain relation can be written as:

$$E_{ij} = \frac{1}{2}\left(U_{i,j} + U_{j,i} + U_{k,i}U_{k,j}\right) \tag{1.36}$$

If the latter is adopted, the Euler strain tensor $\varepsilon_{ij}^{*}$ is obtained; it can be written as:

$$\varepsilon_{ij}^{*} = \frac{1}{2}\left(u_{i,j} + u_{j,i} + u_{k,i}u_{k,j}\right) \tag{1.37}$$

It can be seen from the above two equations that when small deformation occurs, they are completely equivalent without considering small high-order quantities.

### 1.5.2 Decomposition of the Strain Tensor and Deviatoric Strain

Similar to the stress tensor, the strain tensor can be decomposed into the volume strain tensor and the deviatoric strain tensor, i.e.,

$$\begin{bmatrix} \varepsilon_x & \varepsilon_{xy} & \varepsilon_{zx} \\ \varepsilon_{xy} & \varepsilon_y & \varepsilon_{yz} \\ \varepsilon_{zx} & \varepsilon_{yz} & \varepsilon_z \end{bmatrix} = \begin{bmatrix} \varepsilon_m & 0 & 0 \\ 0 & \varepsilon_m & 0 \\ 0 & 0 & \varepsilon_m \end{bmatrix} + \begin{bmatrix} \varepsilon_x - \varepsilon_m & \varepsilon_{xy} & \varepsilon_{zx} \\ \varepsilon_{xy} & \varepsilon_y - \varepsilon_m & \varepsilon_{yz} \\ \varepsilon_{zx} & \varepsilon_{yz} & \varepsilon_z - \varepsilon_m \end{bmatrix} \tag{1.38}$$

The first tensor at the right end of the above equation is called the volume strain tensor, which is represented by $\varepsilon_m\delta_{ij}$, where the average strain is:

$$\varepsilon_m = \frac{1}{3}\left(\varepsilon_x + \varepsilon_y + \varepsilon_z\right) = \frac{1}{3}\varepsilon_{kk} \tag{1.39}$$

The second tensor is called the deviatoric strain tensor, denoted by *e*.

$$e_{ij} = \begin{bmatrix} e_x & e_{xy} & e_{zx} \\ e_{xy} & e_y & e_{yz} \\ e_{zx} & e_{yz} & e_z \end{bmatrix} = \begin{bmatrix} \varepsilon_x - \varepsilon_m & \varepsilon_{xy} & \varepsilon_{zx} \\ \varepsilon_{xy} & \varepsilon_y - \varepsilon_m & \varepsilon_{yz} \\ \varepsilon_{zx} & \varepsilon_{yz} & \varepsilon_z - \varepsilon_m \end{bmatrix} \tag{1.40}$$

Thus, the strain tensor can be abbreviated as:

$$\varepsilon_{ij} = \varepsilon_m \delta_{ij} + e_{ij} \tag{1.41}$$

Volume strain tensor is the expansion strain of the same in all directions; it only causes the change of the volume (such as the volume of expanding or contracting) and does not produce a change in shape, volume strain $\theta = \varepsilon_x + \varepsilon_y + \varepsilon_z = 3\varepsilon_m$. The sum of the three positive deviatoric strain tensors is zero, indicating that it has no volume deformation and only reflects the part of the shape change in the deformation.

Similar to the stress tensor and the deviatoric stress tensor, the strain tensor and the strain tensor also have invariants. The three invariants of the strain tensor are $I_1'$, $I_2'$, and $I_3'$, and the three invariants of the deviatoric strain tensor are $J_1'$, $J_2'$, and $J_3'$. The expression of the latter one is:

$$\begin{aligned}
J_1' &= e_x + e_y + e_z = e_1 + e_2 + e_3 = 0 \\
J_2' &= \frac{1}{2} e_{ij} e_{ij} = \frac{1}{2}\left(e_1^2 + e_2^2 + e_3^2\right) \\
&= \frac{1}{6}\left[\left(\varepsilon_x - \varepsilon_y\right)^2 + \left(\varepsilon_y - \varepsilon_z\right)^2 + \left(\varepsilon_z - \varepsilon_x\right)^2 + \frac{3}{2}\left(\gamma_{xy}^2 + \gamma_{yz}^2 + \gamma_{zx}^2\right)\right] \\
&= \frac{1}{6}\left[\left(\varepsilon_1 - \varepsilon_2\right)^2 + \left(\varepsilon_2 - \varepsilon_3\right)^2 + \left(\varepsilon_3 - \varepsilon_1\right)^2\right] \\
J_3' &= \left|e_{ij}\right| = e_1 e_2 e_3 \\
&= e_x e_y e_z + 2 e_{xy} e_{yz} e_{zx} - e_x e_{yz}^2 - e_y e_{zx}^2 - e_z e_{xy}^2
\end{aligned} \tag{1.42}$$

## Bibliography


1. Bai, Y., Yu, B., Cheng, P., Offshore Installation of Reinforced Thermoplastic Pipe (RTP). *Proceedings of the Twenty-second International Offshore and Polar Engineering Conference*, Rhodes, Greece, 17–22, June 2012.
2. Bonnemaire, B., Response of an Armoured Riser for Arctic Offshore Loading. *Proceedings of the Fourteenth International Offshore and Polar Engineering Conference*, Toulon, France, 23–28, May 2004.
3. Calladine, C.R., *Plasticity for Engineers: Theory and Applications*, Woodhead, Cambridge, 2011.

4. Chen, W.F., *Plasticity in Reinforced Concrete*, McGraw-Hill, New York, 1982.
5. Chen, W.F. and Baladi, G.Y., *Soil Plasticity: Theory and Implementation*, Elsevier, Amsterdam, 1985.
6. Chen, W.F. and Han, D.J., *Plasticity for Structural Engineers*, Springer-Verlag, New York, 1988.
7. Chen, W.F. and Saleeb, A.F., *Constitutive Equations for Engineering Materials, Volume 1: Elasticity and Modeling*, Wiley-Intercedence, New York, 1982.
8. Sun, L. and Qi, B., Global analysis of a flexible riser. *J. Mar. Sci. Appl.*, 10, 4, 478–4845, 2011.

# 2
# Yielding Conditions

***Abstract***
Chapter 2 presents the foundational concepts of yielding in materials subjected to complex stress states. It begins by introducing the definition of stress and yield conditions, followed by the development of the yield surface and π-plane representation in principal stress space. The chapter thoroughly examines classical yield criteria such as Tresca and von Mises, discussing their physical basis, mathematical formulation, and graphical interpretation. Yield loci and their symmetries are explained using the concept of the C-curve, emphasizing the importance of isotropy and the absence of the Bauschinger effect. Advanced topics include anisotropic yield conditions (Hill and Hosford criteria), and hardening models—both isotropic and kinematic—along with their combined forms. The discussion extends to the concepts of loading, unloading, and neutral processes in hardening materials. It concludes with a detailed explanation of plastic deformation mechanisms, the derivation of flow rules, and the fundamental principle of maximum plastic work. These concepts collectively provide a comprehensive framework for predicting material behavior beyond the elastic limit in engineering design and analysis.

***Keywords*:** Yield surface, deviatoric stress, Tresca criterion, Mises criterion, hardening rule, plastic flow, normality rule

## 2.1 Introduction

Stress is defined as the intensity of internal forces acting between particles of a body across imaginary internal surfaces. Figure 2.1 displays a surface area $\Delta A$ passing through a point $P_O$ with a unit vector n normal to the area $\Delta A$. Let $F_n$ be the resultant force due to the action across the area $\Delta A$ of the material from one side onto the other side of the cut plane **n**. Then, the stress vector at point $P_O$ associated with the cut plane **n** is defined by:

Yong Bai. Plasticity and Ductile Fracture in Structural Engineering, (35–64) © 2026 Scrivener Publishing LLC

$$\lim_{\Delta A \to 0} \frac{F_n}{\Delta A} = T = \overset{n}{T} \quad (2.1)$$

The state of stress at a point is defined as the totality of *all* stress vectors $\overset{n}{T}$ at that point.

Under the action of external force, the body initially shows an elastic response. When the load reaches a certain degree, the body begins to enter a plastic state where the internal stress is higher or reaches the yield point. The question is, under what conditions does the stress state satisfy that the material will yield? The relation to the stress at which a point within an object begins to yield is called the initial yield condition, sometimes referred to simply as the yield criterion, also known as the plastic condition, and the surface to which it corresponds in the stress space is called the yield surface.

For simple stress states, yield conditions are easy to determine. For example, for uniaxial stretching, when tensile stress is applied. When the material yield limit reaches 0, the yield begins. Therefore, the yield condition can be written as:

$$\sigma = \sigma_s \text{ or } \sigma - \sigma_s = 0 \quad (2.2)$$

For the pure shear state, when the shear stress reaches the shear yield limit of the material, the yield begins, and the yield condition is:

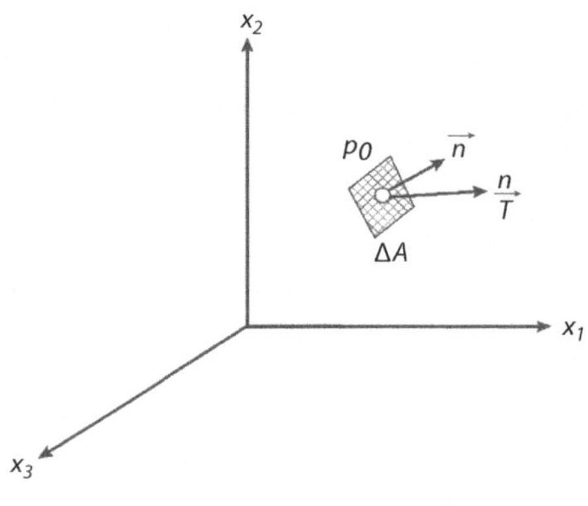


**Figure 2.1** Stress vector $\overset{n}{T}$ at point $P_0$ associated with cut $n$.

$$\tau = \tau_s \text{ or } \tau - \tau_s = 0 \tag{2.3}$$

In general, the state of stress is determined by 6 independent stress components, which obviously cannot be simply taken as a criterion for judging whether to start yielding or not. Moreover, these 6 components are also related to the choice of coordinate axes. However, it can be confirmed that in general, the yield condition should be related to the six stress components and the properties of the material; that is, the yield condition under complex stress state can be written as the following function:

$$f\left(\sigma_x, \sigma_y, \sigma_z, \tau_{xy}, \tau_{yz}, \tau_{zx}\right) = 0 \text{ or } f\left(\sigma_{ij}\right) = 0 \tag{2.4}$$

Now, the goal is to figure out what the function $f$ is. For single tensile or pure shear states, $f$ can be determined by experiments, but for complex stress states, a research method combining theoretical analysis and experimental testing should be adopted.

## 2.2 Yield Locus

Since the yield locus is the boundary of the elastic domain, it is *path-independent* in the sense that a point on it may be approached by many different loading paths within the elastic region, as illustrated schematically in Figure 2.2.

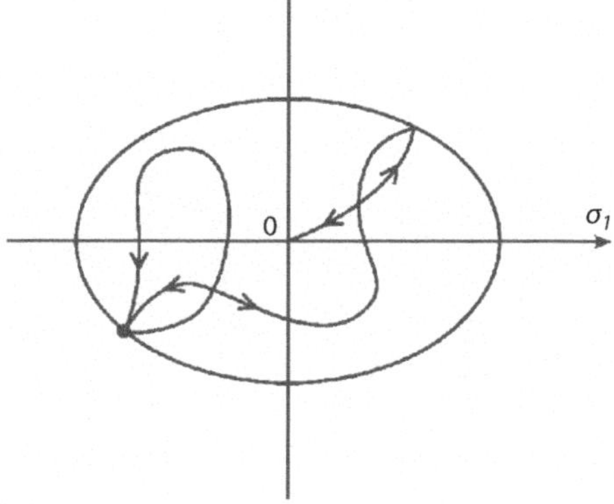


**Figure 2.2** Yield locus in a two-dimensional stress space.

Experimental work on many metals has amply justified the concept of the yield locus. Insofar as it is necessary for an experiment to decide, somewhat arbitrarily, precisely what level of plastic straining marks the end of the elastic range, yield surfaces obtained for the same material by different experimenters may differ by small amounts. However, such differences do not detract from the validity of the concept of the yield locus.

The yield locus in stress-space may be an *initial* yield locus for a previously unloaded specimen or a *subsequent* yield locus for a specimen with a previous history of plastic straining. In simple tension or compression of a strain-hardening material, the yield stress is altered by any plastic deformation, and so in general, we expect the yield locus to be *altered* if any plastic deformation takes place. In particular, if the stress point goes *outside* a current yield locus, it will lie *on* a new yield locus, so that locally at least, the yield locus will be carried along with the stress point.

## 2.3 Yield Surface

In general, the state of stress at a point in a body is specified by the values of the six independent stress components $\sigma_x$, $\sigma_y$, $\sigma_z$, $\tau_{yz}$, $\tau_{zx}$, and $\tau_{xy}$ referred to as an arbitrary set of orthogonal *x, y, z* axes, as shown in Figure 2.3. It might appear, therefore, that we would need a six-dimensional stress space in which to represent a general yield locus. However, it is *always* possible to choose directions of orthogonal axes 1, 2, 3 for which the shear stresses $\tau_{23}$, $\tau_{31}$, $\tau_{12}$ all vanish, as indicated in Figure 2.3. These are called *principal axes,* and the corresponding stresses ($\sigma_1$, $\sigma_2$, $\sigma_3$) are the *principal stresses.* It is therefore convenient to choose the principal axes as the reference

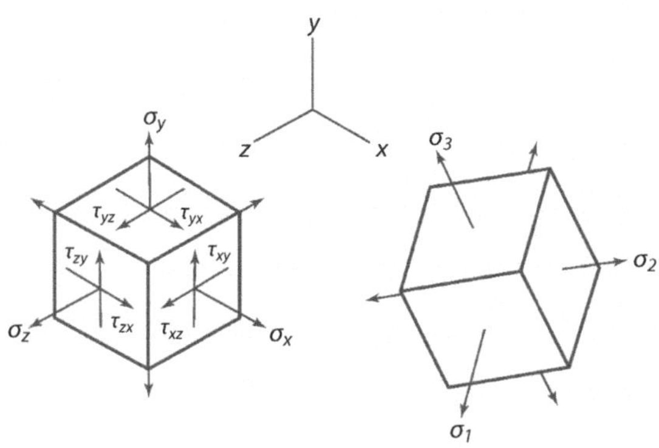


**Figure 2.3** Three-dimensional state of stress.

axes, and this incurs no loss of generality *if the material is isotropic,* i.e., has the same properties in all directions. As all of our visual experience is in the three-dimensional space, it will obviously be desirable in working out the geometry of yield surfaces to work in terms of a *three-dimensional principal stress space.* However, we can justify the necessary idealization of our material as isotropic. At first sight, this might appear unlikely, because the arrangement of atoms in a crystal lattice, being orderly and having well-defined directions, is clearly *geometrically* anisotropic, and thus probably also mechanically isotropic. However, metallographic examination indicates that unless special precautions are taken in preparation, a block of undeformed metal contains in general many grains or crystals with the respective lattice directions higgledy-piggledy. We would thus expect that the anisotropy of individual grains would be lost and that the material in bulk would be isotropic. However, we might expect that even if a material is *initially* isotropic, subsequent plastic deformation may add a measure of order to the random orientation of grains and hence produce a geometrical anisotropy in the bulk material. Experiments show that the degree of mechanical anisotropy developed during moderately large plastic deformation is not great. It is, therefore, reasonable to build mechanical isotropy into our idealized simple plastic material.

We should note, however, that the degree of mechanical isotropy found, for example, in metal plates that have been reduced substantially in thickness by cold rolling in one direction may, in some circumstances, demand the introduction of a degree of anisotropy in the theory. We shall not explore this aspect of the theory.

Returning then to the three-dimensional principal stress space, the nature of the yield surface for the idealized material needs several different pieces of information.

First, we introduce the experimental observation that the addition of an equal-all-around (hydrostatic) pressure to a specimen of polycrystalline metal does not affect its yield behavior, at least for pressures of the same order as the tensile yield stress; this has been demonstrated by many direct and indirect experiments. By this statement, we mean that if the point $(\sigma_1, \sigma_2, \sigma_3) = (\alpha, b, c)$ lies on the yield surface, then so also does the point $(a + h, b + h, c + h)$ for all $h$, at least within fairly wide limits.

This observation is not surprising if we think about the atomic-level picture. The response of metal to hydrostatic pressure by itself is purely elastic because there is no tendency for dislocations to move if there is no question of change of *shape* of the specimen. This insensitivity of the yield behavior to hydrostatic pressure is thus broadly associated with the plastic incompressibility of the material.

We ought to note that metals may become brittle under hydrostatic tension because under these conditions, the mode of deformation might change from shear to cleavage, and that some normally brittle materials, like rocks, may flow plastically under shear stress in the presence of high hydrostatic compression. For the present purposes, we shall assume that the material is ductile under all relevant stress conditions.

It is also worth noting, in passing, that material like saturated clay, which is *irreversibly* compressible, exhibits plastic yield behavior under shear stresses which are a function of the hydrostatic pressure. The explanation for this difference between the plastic behavior of metals and clays undoubtedly lies in their different microstructure. Clays are composed of flake-shaped particles of the order of $10^{-6}$ m long, and the cohesive action is electrochemical in the presence of water.

Returning to our idealized metal, it follows that any *point (a, b, c)* on the yield surface generates a *line* passing through the point and parallel to the line $\sigma_1 = \sigma_2 = \sigma_3$, which is in the first octant of principal stress space and equally inclined to all three principal stress axes. Hence, the yield surface is a *prism* which may be thought of as being generated by sliding a curve along with the "space diagonal" $\sigma_1 = \sigma_2 = \sigma_3$.

This being so, all the additional information necessary for us to specify the yield surface completely is the shape and size of the cross-section of the prism. Now, by elementary solid geometry, all planes perpendicular to the line $\sigma_1 = \sigma_2 = \sigma_3$ have the equation $\sigma_1 + \sigma_2 + \sigma_3$ = constant. As it is immaterial which cross-section of the prism we consider, we can set the constant to any arbitrary value. The most convenient value is zero, which corresponds to the plane perpendicular to the axis of the prism and passing through the origin of principal stress space. We call this special plane the π-plane and the intersection of the yield prism with it the C-curve.

It is a matter of elementary algebra to "reduce" any point on the yield surface to the corresponding point on the C-curve. Let *(a, b, c)* lie on the yield surface. The corresponding point on the *C* curve has coordinates *(a + h, b + h, c + h)*, and in addition, since it lies on the π-plane, the sum of the three principal stresses is zero; so we gave the following:

$$h = \frac{1}{3}(a + b + c) \tag{2.5}$$

We can thus write, for any point on the yield surface, the identity:

$$(\sigma_1,\sigma_2,\sigma_3) = (\sigma_1',\sigma_2',\sigma_3') + (\sigma_0,\sigma_0,\sigma_0) \tag{2.6}$$

where

$$\sigma_0 = \frac{\sigma_1 + \sigma_2 + \sigma_3}{3} \tag{2.7}$$

and

$$\begin{cases} \sigma_1' = \sigma_1 - \sigma_0 \\ \sigma_2' = \sigma_2 - \sigma_0 \\ \sigma_3' = \sigma_3 - \sigma_0 \end{cases} \tag{2.8}$$

Note that:

$$\sigma_1' + \sigma_2' + \sigma_3' = 0 \tag{2.9}$$

The stress $(\sigma'_1,\sigma'_2,\sigma'_3)$ is known as the *deviatoric stress* or *stress deviation,* and $(\sigma_0,\sigma_0,\sigma_0)$ is known as the *hydrostatic* stress. In geometrical terms, we have broken down the stress vector in three-dimensional stress space into two components, one (the deviatoric stress) lying in the π-plane, the other (hydrostatic stress) being perpendicular to the π-plane.

As we have now focused our interest in the nature of the three-dimensional yield surface onto the C-curve in the π-plane, it is now advantageous to draw a proper plane diagram, shown in Figure 2.4. The drawing shows a view (without perspective) along with the space diagonal of the three axes $\sigma_1$, $\sigma_2$, $\sigma_3$ (positive and negative senses) together with what appears as a triangular mesh inscribed on the π-plane but which may equally well be thought of as a view of a cubic lattice in the first octant of the stress space.

In Figure 2.4, the stress-point (6, 0, 0) is represented by point $A_1$. The same point also represents the corresponding deviatoric stress (4, -2, -2), which is as it should be by virtue of the properties of the π-plane. We are now in a position to investigate the nature of the C-curve and particularly some aspects of its symmetry. First, suppose that point $A_1$ (6,0,0) lies on the yield surface. Then, if the material is *isotropic,* there is no reason why we should not relabel the axes in all possible ways; we thus conclude

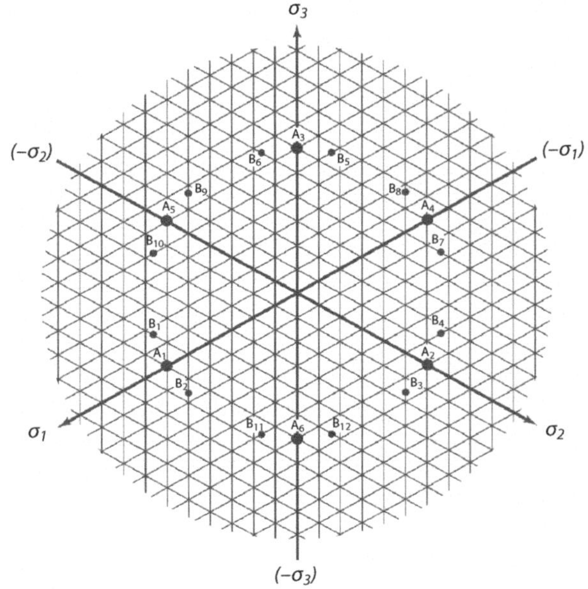


**Figure 2.4** π-plane.

that points (0, 6, 0) and (0,0, 6) also lie on the yield surface, as shown in Figure 2.4; $A_2$, $A_3$. Furthermore, if the material has no Bauschinger effect, the point (−6, 0, 0) lies on the surface, and so, similarly, do points (0, −6,0) and (0,0, −6); $A_4$-$A_6$. Thus, by appealing to isotropy and no Bauschinger effect, one point on the surface has become six. Restrictions of this sort are clearly likely to lead to some sort of symmetry in general. In fact, the point considered was special in the sense that $\sigma_2 = \sigma_3$. If $\sigma_1 = \sigma_2 = \sigma_3$, a single point, by virtue of isotropy and no Bauschinger effect, is multiplied into *twelve* points. For example, if point $B_1$ (7,0.5, 2) is on the yield locus, so are $B_2$ (7, 2, 0.5), $B_3$ (2, 7, 0.5), $B_4$ (0.5, 7, 2), $B_5$ (0.5, 2, 7), and $B_6$ (2, 0.5, 7). The other six points are found by multiplying throughout by (−1). Now, considering $B_1$ and $B_2$, we see that these are mirror images of each other about the projection of the $\sigma_1$ axis onto the π-plane. Also, $B_2$ and $B_{11}$ are mirror images about a line bisecting the projections of the $\sigma_1$ and $-\sigma_3$ axes on the π-plane. The problem of specifying the C-curve thus becomes (subject, of course, to the idealizations of isotropy and no Bauschinger effect) the problem of specifying the C-curve in one 30-degree sector of the π-plane. The curve within this sector can then be used as a "template" for all the other sectors, as indicated in Figure 2.5.

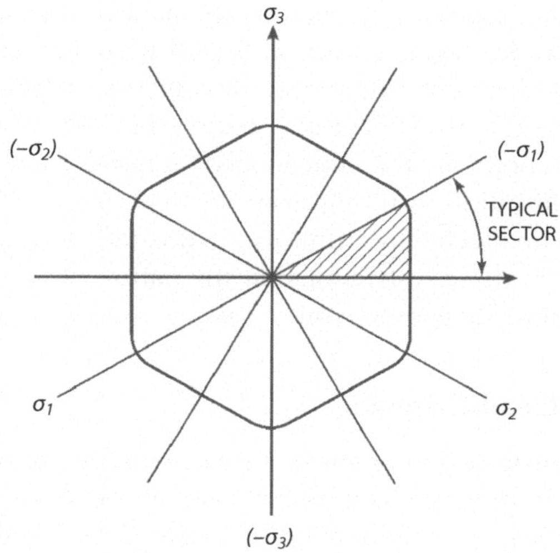


**Figure 2.5** C-curve in the π-plane.

In the π-plane, three of the six axes of symmetry for the C-curve consist of the projections of the uniaxial tension/compression axes. The other three correspondences are states of pure shear for which, in general, the principal stresses are of the form $(b, -b, 0)$ or, in general, projections of states of pure shear plus arbitrary hydrostatic pressure. It may be shown that the angular range of one sector may be covered by combined tension/torsion tests on thin-walled tubes. This is the basis for the classical experiments on tubes of aluminum, copper, and mild steel.

## 2.4 Yield Criterion

Suppose that an element of material is subjected to a system of stresses of gradually increasing magnitude. The initial deformation of the element is entirely elastic, and the original shape of the element is recovered on complete unloading. For certain critical combinations of the applied stresses, plastic deformation first appears in the element. A law defining the limit of elastic behavior under any possible combination of stresses is called the *yield criterion*. The law applies not only to loading directly from the annealed state but also to reloading of an element unloaded from a previous plastic state. In developing a mathematical theory, it is necessary to take into account a number of idealizations at the outset. Firstly, it is assumed that the conditions of loading are such that all strain rate and thermal effects can be neglected. Secondly, the Bauschinger effect and the hysteresis loop, which arise from nonuniformity on the microscope scale, are disregarded. Finally, the material is assumed to be isotropic, so that its properties at each point are the same in all directions. There is a useful and immediate simplification resulting from the experimental fact that yielding is practically unaffected by a uniform hydrostatic tension or compression. The effects of these restrictions on the nature of the yield criterion will first be examined in geometrical terms.

### 2.4.1 Geometrical Criterion

Figure 2.6 demonstrates a system of three mutually perpendicular axes with the principal stresses taken as rectangular coordinates. The state of stress at any point in a body may be represented by a vector emanating from the origin *O*. Imagine a line *OH* equally inclined to the three axes, so that its direction cosines are $(1/\sqrt{3}, 1/\sqrt{3}, 1/\sqrt{3})$. The stress vector **OQ**, whose components are $(\sigma_1, \sigma_2, \sigma_3)$, may be resolved into a vector **OG** along with *OH* and a vector **OP** perpendicular to *OH*. The vector **OG** is of magnitude $\sqrt{3}\sigma_0$ and represents the hydrostatic stress with components $(\sigma_0, \sigma_0, \sigma_0)$.

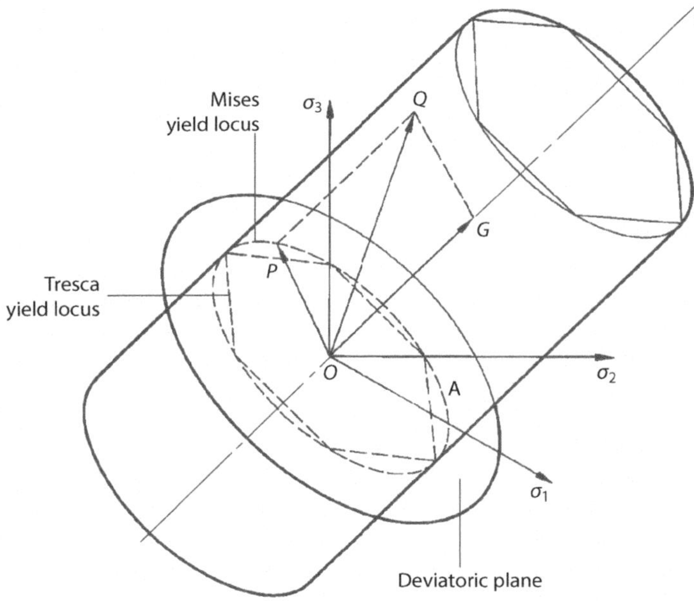


**Figure 2.6** Geometrical representation of the yield criterion in the principal stress space.

The vector **OP** represents the deviatoric stress with components ($s_1$, $s_2$, $s_3$), and its magnitude is $\sqrt{2J_2}$. For any given state of stress, the deviatoric stress vector will lie in the plane passing through $O$ and perpendicular to $OH$. This plane is known as the *deviatoric plane,* and its equation is $\sigma_1 + \sigma_2 + \sigma_3 = 0$ in the principal stress space. Since uniform hydrostatic stress does not affect yielding, it follows that yielding can depend only on the magnitude and direction of the deviatoric stress vector **OP**. The yield surface is therefore a right cylinder whose generators are perpendicular to the deviatoric plane. Any stress state in which the stress point lies on the surface of the cylinder corresponds to a state of yielding. Any point inside the cylinder represents an elastic state of stress. The curve $A$, in which the yield surface is intersected by the deviatoric plane, is called the yield locus. The equation of a possible yield locus, which is assumed to be *convex* (i.e., concave to the origin), is a possible yield criterion.

Figure 2.7 illustrates the yield locus together with the orthogonal projections of the stress axes on the deviatoric plane, which is taken in the plane of the paper. The yielding of isotropic materials can depend only on the values of principal stresses and not on their directions. Thus, if the stresses ($p$, $q$, $r$) cause yielding, so will the stresses ($p$, $r$, $q$), implying that the yield locus is symmetrical about the projected $\sigma_1$ axis. It similarly follows that the yield locus is symmetrical concerning the projections of the $\sigma_2$ and $\sigma_3$ axes. This amounts to the fact that the yield criterion is a function of the invariants of the deviatoric stress tensor. A further restriction on the form

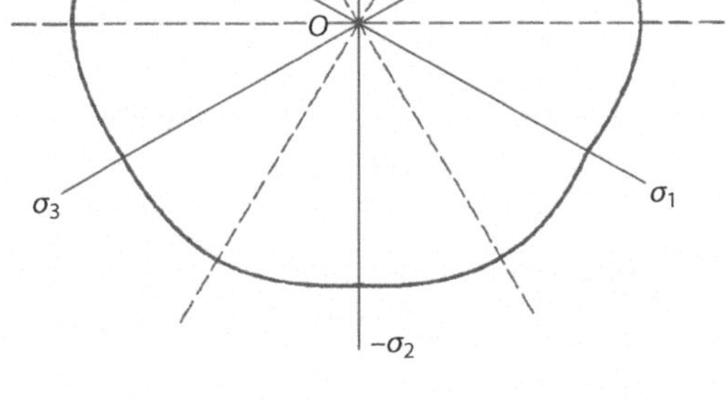


**Figure 2.7** The general appearance of the deviatoric yield locus has six axes of symmetry.

of the yield locus is imposed by the assumption that the Bauschinger effect is absent. Thus, if $(p, q, r)$ is a plastic state, $(-p, -q, -r)$ is also a plastic state. In other words, a radial line drawn from any point on the yield locus must meet the locus again at the same distance from the origin. Hence, the yield locus must also be symmetrical about the lines orthogonal to the projected axes. The shape of the yield locus is therefore repeated over the twelve 30° segments formed by the six diameters as shown in Figure 2.7. Stated mathematically, the yield criterion is expressible in the following form:

$$f(J_2, J_3) = const \tag{2.10}$$

where $f$ is an even function of $J_3$, which changes sign with the stresses. The yield criterion is called *regular* if the locus has a continuously turning tangent everywhere and *singular* when the locus has sharp corners. The yield surface is considered strictly convex when a straight line joining any two points on the yield locus lies completely inside the locus. A flat on the yield surface can be regarded as a limiting state of convexity.

### 2.4.2 General Criterion

Figure 2.8 shows that the deviatoric stress vector **OP** may be regarded as the sum of the projections on the deviatoric plane of the component vectors of **OQ** along the axes of reference. Since each stress axis is inclined to the deviatoric plane at an angle of $\sin^{-1}(1/\sqrt{3})$ in the original stress space, each projected length along the axes is $\sqrt{2/3}$ times the actual length. Hence, the lengths of the component vectors in the deviatoric plane are $OL = \sqrt{2/3}\sigma_1$, $LM = \sqrt{2/3}\sigma_2$, $MP = \sqrt{2/3}\sigma_3$. From geometry, the rectangular components of **OP** with respect to the horizontal and vertical through $O$ are as follows:

$$ON = \frac{\sigma_1 - \sigma_3}{\sqrt{2}} = r\cos\theta \tag{2.11}$$

$$PN = \frac{2\sigma_2 - \sigma_3 - \sigma_1}{\sqrt{6}} = r\sin\theta \tag{2.12}$$

where $(r, \theta)$ are the polar coordinates of $P$. For the experimental determination of the yield criterion, it is convenient to introduce the Lode parameter $\mu$ defined as

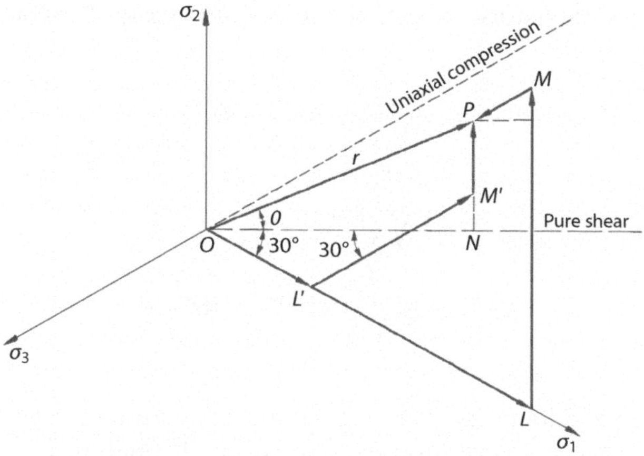


**Figure 2.8** Deviatoric stress vector and its components along the projected axes.

$$\mu = \frac{2\sigma_2 - \sigma_3 - \sigma_1}{\sigma_3 - \sigma_1} = -\sqrt{3}\tan\theta, \sigma_1 > \sigma_2 > \sigma_3 \tag{2.13}$$

To obtain the yield locus, it is only necessary to apply stresses for which $\theta$ covers the range from 0 to $\pm\pi/6$, $\mu$ varying from 0 to $\mp1$. When $\mu = 0$, $\sigma_2 = 0.5\ (\sigma_3 + \sigma_1)$ and we have a state of pure shear denoted by 12 ($\sigma_3 - \sigma_1$, $\sigma_1 - \sigma_3$, 0) together with hydrostatic stress $0.5(\sigma_3 + \sigma_1)$. When $\mu = -1$, $\sigma_1 = \sigma_2$ and the state of stress is equivalent to a uniaxial compression ($\sigma_3 - \sigma_1$, 0, 0) and a hydrostatic stress $\sigma_1$. Let the yield stresses in pure shear and uniaxial tension (or compression) be denoted by $k$ and $Y$, respectively. Then, it follows from Equations (2.11) and (2.12) that $r = \sqrt{2}k$ at $\theta = 0$ and $r = \sqrt{2/3}Y$ at $\theta = \pi/6$. For most metals, $k$ lies between $Y/2$ and $Y/\sqrt{3}$. Equations (2.11) and (2.12) may be written alternatively in the form

$$s_1 - s_3 = \sqrt{2}r\cos\theta \tag{2.14}$$

$$s_1 + s_3 = -s_2 = -\sqrt{\frac{2}{3}}r\sin\theta \tag{2.15}$$

from which the deviatoric principal stresses can be expressed as

$$\begin{cases} s_1 = \sqrt{\dfrac{2}{3}} r \cos\left(\dfrac{\pi}{6} + \theta\right) \\ s_2 = \sqrt{\dfrac{2}{3}} r \sin\theta \\ s_3 = -\sqrt{\dfrac{2}{3}} r \cos\left(\dfrac{\pi}{6} - \theta\right) \end{cases} \tag{2.16}$$

When the yield locus is given, $r$ is a known function of $\theta$, and Equation (2.16) defines the yield criterion parametrically through $\theta$.

When the vector **OP** is given, the deviatoric stresses can be obtained graphically by noting the fact that $s_2 = \sqrt{2/3}PN$. Hence, if a point $M'$ is located on $PN$ such that $PM'/NM' = 2$, then $PM' = \sqrt{2/3}\, s_2$. Let $L'M'$ be drawn parallel to $O\sigma_3$. Since $L'M' = OL' + PM'$ by geometry, it follows that $OL' = \sqrt{2/3}\, s_1$ and $L'M' = -\sqrt{2/3}\, s_3$. The points $L'$ and $M'$, therefore, define the deviatoric stresses. The hydrostatic stress is of magnitude $\sqrt{2/3}LL'$, but it is not defined by the vector **OP**. The ratios of the deviatoric stresses are as follows:

$$s_1 : s_2 : s_3 = (\sqrt{3} - \tan\theta) : 2\tan\theta : -(\sqrt{3} + \tan\theta) \tag{2.17}$$

By multiplying Equation (2.16) and remembering that $s1s2s3 = J3$ and $r = \sqrt{2J2}$, we obtain

$$J_3^2 = \frac{4}{27} J_2^3 \sin^2 3\theta \tag{2.18}$$

If the polar equation of the yield locus is given, $J2$ is a known function of $\theta$. Eliminating $\theta$ between Equation (2.18) and the yield locus, it is possible to obtain the yield criterion in Equation (2.10). Alternatively, if the yield criterion is given in terms of $J_2$ and $J_3$, the equation for the yield locus may be derived from it. It may be noted that as $\theta$ varies from 0 to $\pi/6$, the value of $J_2$ varies from $k^2$ to $Y^2/3$, and that of $J_3$ varies from 0 to $-2/27Y^3$. Plastic yielding is predominantly influenced by the magnitude of $J_2$.

## 2.5 Common Yielding Conditions

### 2.5.1 Tresca Yield Condition

In 1864, H. Tresca, a French engineer, conducted a series of experiments in which ductile metals were squeezed through molds of different shapes. He found that there were very fine lines on the surface of the deformed metals, which were close to the direction of the maximum shear stress. According to this phenomenon, he believed that the plastic deformation of the metal was caused by the slip of crystals in the metal, caused by shear stress. From these metal extrusion test results, Tresca proposed the following yield conditions. When the maximum shear stress reaches a certain fixed value inherent in the material, the material begins to enter the plastic state, that is, to yield. This condition is called the maximum shear stress condition, also known as the Tresca condition, and its mathematical expression is as follows:

$$\tau_{\max} = \frac{1}{2}k \tag{2.19}$$

Here, $k$ is a constant relative to the material. If we know that the order of principal stress is $\sigma_1 \geq \sigma_2 \geq \sigma_3$, the above equation can be written as

$$\tau_{\max} = \frac{\sigma_1 - \sigma_3}{2} = \frac{k}{2} \tag{2.20}$$

Then,

$$\sigma_1 - \sigma_3 = k \text{ or } (\sigma_1 - \sigma_3) - k = 0 \tag{2.21}$$

In general, it is often impossible to determine the order of the three principal stresses at each point in advance, so the condition is usually written as follows:

$$\begin{aligned} |\sigma_1 - \sigma_2| &= k \\ |\sigma_2 - \sigma_3| &= k \\ |\sigma_3 - \sigma_1| &= k \end{aligned} \tag{2.22}$$

When at least one of the above equations is true, the material begins to deform plastically. Because $k > 0$, these three things can't be taken at the same time. It can be expressed as a complete expression, and Equation (2.22) can be rewritten as

$$\left[(\sigma_1-\sigma_2)^2-k^2\right]\left[(\sigma_2-\sigma_3)^2-k^2\right]\left[(\sigma_3-\sigma_1)^2-k^2\right]=0 \tag{2.23}$$

According to $x=\frac{\sqrt{2}}{2}(\sigma_1-\sigma_3)=\frac{\sqrt{2}}{2}k=const$, the yield curve is a straight line in the range of $-30°\le\theta_\sigma\le 30°$ on the $\pi$ plane. According to the symmetry discussed above, it can be concluded that the yield condition of Tresca is a regular hexagon in the $\pi$ plane, which is called the Tresca hexagon. In the three-dimensional stress space, the yield surface is a regular six-prism surface with the $L$ axis. The radius of the external circle of the Tresca hexagon is $r=\frac{\sqrt{2}}{2}k$, and the radius of its inscribed circle is $r=\frac{k}{\sqrt{2}}$, as shown in Figure 2.9.

The characteristics of Tresca yield conditions are as follows:

1) The physical concept is clear, and there is a clear physical explanation.
2) When we know the order of the three principal stresses $\sigma_1\ge\sigma_2\ge\sigma_3$; it is a linear function of the principal stress, and the expression is simple.

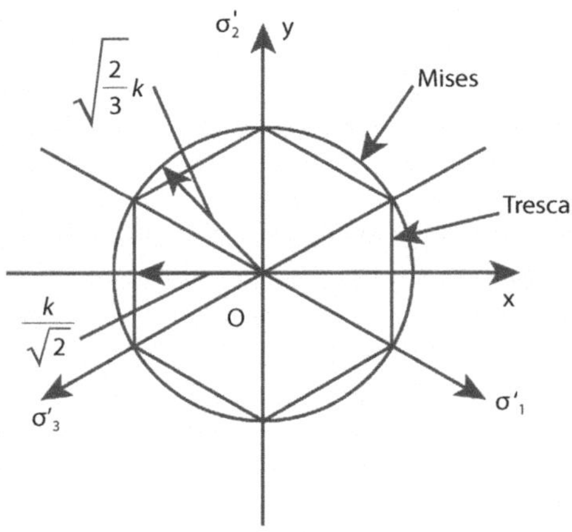


**Figure 2.9** Tresca and von Mises yield conditions.

3) No intermediate principal stress is considered.
4) When the order of principal stress is unknown, the form is too complex.
5) There are angular points on the yield curve, which is a non-smooth curve, which brings difficulties to the mathematical processing.

### 2.5.2 Mises Yield Condition

When the order of magnitude of the principal stresses is unknown, expressions for the Tresca condition are mathematically difficult to solve for spatial problems. German scientist R. von Mises proposed in 1913 that the external circle of the hexagonal shape of Tresca could be used as a yielding surface:

$$x^2 + y^2 = \left(\frac{\sqrt{2}}{3}k\right)^2 \tag{2.24}$$

Where $x = (\sigma_1 - \sigma_3)/\sqrt{2}$, $y = (2\sigma_2 - \sigma_1 - \sigma_3)/\sqrt{6}$, then:

$$(\sigma_1 - \sigma_2)^2 + (\sigma_2 - \sigma_3)^2 + (\sigma_3 - \sigma_1)^2 = 2k^2 \tag{2.25}$$

When Mises proposed this yield condition, he thought that the condition he proposed was only approximate, not an accurate yield condition. But the experimental results show that for ductile metal materials, it is closer to the experimental situation. This condition indicates that when the stress intensity reaches a certain value, the material will enter the yield state, which is called the invariant condition of stress intensity, also known as the Mises yield condition (also known as the Huber-von Mises yield condition). In the stress space, the yield surface represented by the Mises condition is a cylindrical surface perpendicular to the $\pi$ plane with $L$ as the axis.

The Mises yield condition takes into account the influence of the intermediate stress on the yield, overcomes the disadvantages of the Tresca condition in form, and can be expressed by a unified formula.

The Mises condition can also be written as

$$(\sigma_x - \sigma_y)^2 + (\sigma_y - \sigma_z)^2 + (\sigma_z - \sigma_x)^2 + 6\left(\tau_{xy}^2 + \tau_{yz}^2 + \tau_{zx}^2\right) = 2k^2 \tag{2.26}$$

### 2.5.3 Anisotropic Yield Condition

The anisotropy in yield can be divided into two types: initial anisotropy and deformation-induced anisotropy. The former occurs in anisotropic materials of various structures and exists before any plastic deformation. The latter occurs after strain strengthening occurs, even if the material is initially isotropic. Hill yield condition and Hosford yield condition are commonly used anisotropic yield conditions.

#### *2.5.3.1 Hill Yield Criterion*

In 1948, Hill extended the Mises yield conditions to introduce anisotropy into yield criteria. The form of the yield condition is given as follows:

$$\begin{aligned} 2f(\sigma_{ij}) = {} & F(\sigma_y - \sigma_z)^2 + G(\sigma_z - \sigma_x)^2 + H(\sigma_x - \sigma_y)^2 \\ & + 2L\tau_{yz}^2 + 2M\tau_{zx}^2 + 2N\tau_{xy}^2 = 1 \end{aligned} \tag{2.27}$$

where *F*, *G*, *H*, *L*, *M*, and *N* are 6 independent parameters related to the material, which can be confirmed according to the tensile (compression) yield stress and shear yield stress in the direction of the anisotropic principal axis. Such as

$$\begin{aligned} F &= \frac{1}{2}\left[\frac{1}{(\sigma_y^s)^2} + \frac{1}{(\sigma_z^s)^2} - \frac{1}{(\sigma_x^s)^2}\right] \\ G &= \frac{1}{2}\left[\frac{1}{(\sigma_z^s)^2} + \frac{1}{(\sigma_x^s)^2} - \frac{1}{(\sigma_y^s)^2}\right] \\ H &= \frac{1}{2}\left[\frac{1}{(\sigma_x^s)^2} + \frac{1}{(\sigma_y^s)^2} - \frac{1}{(\sigma_z^s)^2}\right] \\ L &= \frac{1}{2(\tau_{yz}^s)^2}, M = \frac{1}{2(\tau_{zx}^s)^2}, L = \frac{1}{2(\tau_{xy}^s)^2} \end{aligned} \tag{2.28}$$

where $\sigma_x^s, \sigma_y^s$, and $\sigma_z^s$ are respectively the tensile (compression) yield stress relative to each anisotropic spindle; $\tau_{xy}^s, \tau_{yz}^s$, and $\tau_{zx}^s$ are shear yield relative to each anisotropic principal axis.

The above criterion is still widely used and is called the Hill'48 yield criterion. In 1979, Hill extended the above equation, namely the so-called Hill'79 yield criterion.

$$\begin{aligned} &F\left|\sigma_2-\sigma_3\right|^m+G\left|\sigma_3-\sigma_1\right|^m+H\left|\sigma_1-\sigma_2\right|^m+L\left|2\sigma_1-\sigma_2-\sigma_3\right|^m \\ &+M\left|2\sigma_2-\sigma_1-\sigma_3\right|^m+N\left|2\sigma_3-\sigma_1-\sigma_2\right|^m=\sigma_s^m \end{aligned} \tag{2.29}$$

where $\sigma_1, \sigma_2$, and $\sigma_3$ are 3 principal stresses; $\sigma_s$ is the yield stress; *F*, *G*, *H*, *L*, *M*, and *N* are constants; and $m$ is related to the degree of anisotropy of the material and must be greater than 1 to ensure the convexity of the yield surface.

### 2.5.3.2 *Hosford Yield Criterion*

In 1972, Hosford proposed isotropic yield conditions without shear stress in the form as follows:

$$\frac{1}{2}\left|\sigma_2-\sigma_3\right|^n+\frac{1}{2}\left|\sigma_3-\sigma_1\right|^n+\frac{1}{2}\left|\sigma_1-\sigma_2\right|^n=\sigma_s^n \tag{2.30}$$

In the case of anisotropy, the Hosford yield condition can be written as:

$$F\left|\sigma_2-\sigma_3\right|^n+G\left|\sigma_3-\sigma_1\right|^n+H\left|\sigma_1-\sigma_2\right|^n=\sigma_s^n \tag{2.31}$$

where *F*, *G*, and *H* are constants; $\sigma_1, \sigma_2$, and $\sigma_3$ are three principal stresses. Index $n$ is relative to material lattice type (fcc, bcc, hcp, etc.), usually $n > 2$. For bcc materials, $n = 6$. For fcc materials, $n = 8$.

Hosford series yield conditions are mainly proposed for the deformation behavior of aluminum alloy and other cubic structural metals, and the description of such metal plates is relatively accurate.

## 2.6 Hardening Rule

Due to the material's hardening effect, the boundary of the successive elastic range of the material after plastic deformation is the loading surface, which is related not only to the instantaneous stress but also to the loading history of the material. This problem is very complicated, and it is still one of the basic problems that should be studied. At present, there are two kinds of hardening models widely used in practice, namely, the isotropic hardening model and the kinematic hardening model.

### 2.6.1 Isotropic Hardening Model

The isotropic hardening model assumes that the loading surface of the material is the uniform expansion (not reduction) of the initial yield surface in the stress space, but the central position remains unchanged. The scale of the expansion of the loading surface can be represented by a parameter $\kappa$, which represents the degree of hardening load function can be defined as:

$$\varphi\left(\sigma_{ij},\kappa\right)=f\left(\sigma_{ij}\right)-H\left(\kappa\right)=0 \tag{2.32}$$

The initial yield function is

$$\varphi\left(\sigma_{ij}\right)=f\left(\sigma_{ij}\right)-C=0 \tag{2.33}$$

where $C$ is a constant and $H\left(\kappa\right)$ is a monotonically increasing positive function of the strengthening parameter $\kappa$. It can be represented by different mechanical quantities. Since $f\left(\sigma_{ij}\right)$ is a homogeneous function of the stress component, the loading surface is the expansion of the initial yield surface. Figure 2.10 shows the loading curves of the Tresca condition and Mises condition on the $\pi$ plane.

### 2.6.2 Kinematic Hardening Model

According to kinematic hardening theory, the initial yield surface maintains the same size and shape during the loading process but shifts in the stress space. The instantaneous position of the initial yield surface in the stress space is the loading curve of the material. Set $\sigma_{ij}'$ as the displacement (tensor) of the center of the yield surface in the stress space, which is a

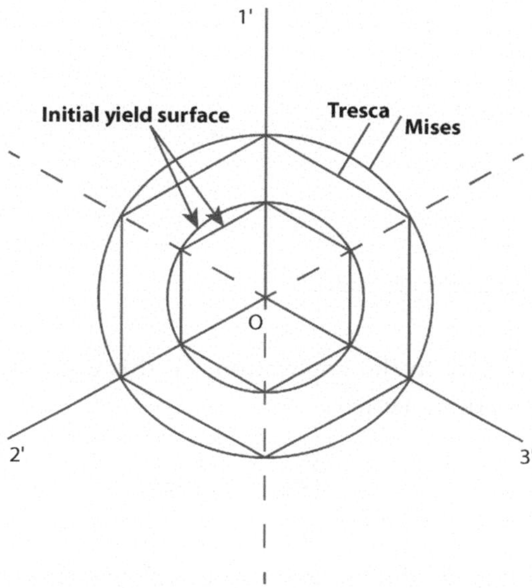


**Figure 2.10** Isotropic hardening model.

parameter of the stress history, then the loading function of the follow-up reinforcement can be written as follows:

$$\varphi\left(\sigma_{ij},\sigma_{ij}^{'}\right)=f\left(\sigma_{ij},\sigma_{ij}^{'}\right)-C=0 \tag{2.34}$$

where $C$ is a constant and is usually assumed:

$$d\sigma_{ij}^{'}=cd\varepsilon_{ij}^{p} \tag{2.35}$$

If $c$ is constant, the material is linearly reinforced. If the stress point is located at the "corner point" of the yield surface, then:

$$d\sigma_{ij}^{'}=cd\varepsilon_{ij}^{p}=d\sigma_{ij} \tag{2.36}$$

Figure 2.11 is a kinematic hardening model of the yield conditions of Tresca on the $\pi$ plane.

When the hardening model in the stress space degenerates to the lower-dimensional stress subspace, the loading surface will deform, so it will

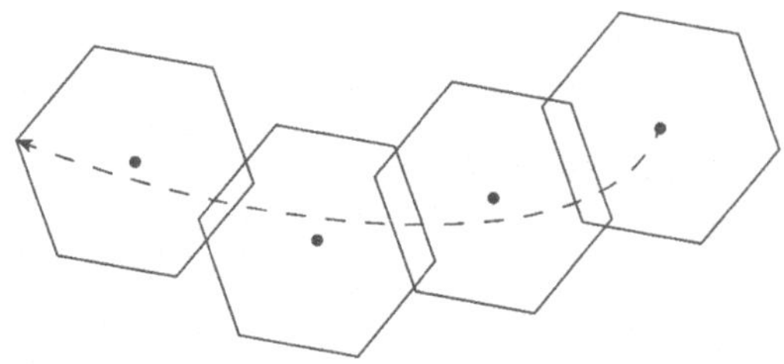

**Figure 2.11** Kinematic hardening model.

bring some complexity to the application. To this end, Ziegler proposed the following correction, namely,

$$d\sigma'_{ij} = \left(\sigma_{ij} - \sigma'_{ij}\right) d\mu \tag{2.37}$$

$d\mu > 0$ indicates that the direction of motion of the yield surface is parallel to the line between the stress point and the center of the surface. On the plane, Ziegler's modified model is shown in Figure 2.12(a) (Tresca condition). For comparison, Figure 2.12(b) shows the Prager model.

Obviously, the isotropic hardening model not only does not take into account the Bauschinger effect, but also thinks that the material has a negative Bauschinger effect; it tends to be absolute, so some people call this theory "ideal anisotropic hardening." In order to make the theory consistent with the practice, the two models can be combined to obtain the so-called combination model, whose loading function can be written as follows:

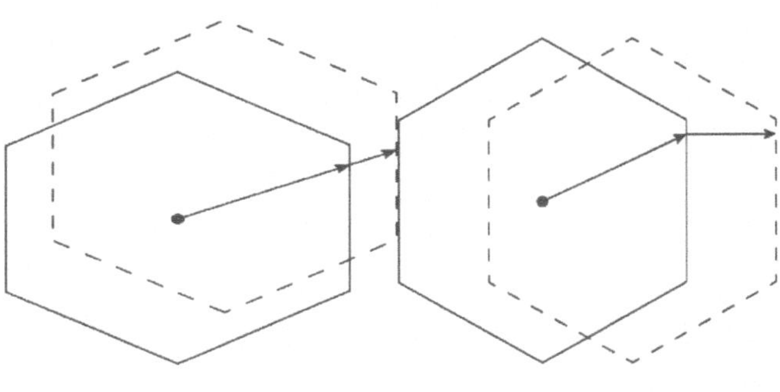

(a) (b)

**Figure 2.12** Ziegler model and Prager model.

$$\varphi\left(\sigma_{ij},\varepsilon_{ij}^{p},\kappa\right)\left(\sigma_{ij},\sigma_{ij}^{'}\right)-H\left(\kappa\right)=0 \tag{2.38}$$
$$d\sigma_{ij}^{'}=cd\varepsilon_{ij}^{p}$$

When $\sigma_{ij}^{'}=0$, the above formula becomes an isotropic hardening model; when $H\left(\kappa\right)=const$, the above formula becomes a kinematic hardening model.

## 2.7 Loading and Unloading of Hardening Materials

The loading function can be generally written as:

$$\varphi\left(\sigma_{ij}^{'},H_{a}\right)=0 \tag{2.39}$$

where $H_a$ is a parameter representing the stress history, such as $cd\varepsilon_{ij}^{p}$. $\kappa$ is a fixed hypersurface in $\left(\sigma_{ij},H_{a}\right)$ space. Although stress points can vary, they can only be above or within the surface, not outside it. Otherwise, equilibrium cannot be maintained.

The loaded mathematical can also be regarded as a family of surfaces $H_a$ as the parameter in the stress space. For the hardening material, $H_a$ changes, and the surface moves outward. The surface doesn't change, $H_a$ doesn't change. Therefore, when the stress point changes, there may be three cases (all discussed in the stress space):

1) The stress point moves inward from the instantaneous loading, the material enters the elastic state from the plastic state, and the plastic strain remains unchanged.
2) Although the stress point moves, it remains at the original instantaneous loading surface, i.e., $H_a$, with no change, and the plastic strain remains unchanged.
3) The stress point moves outward from one instantaneous loading face to another adjacent loading surface, and the material enters another affinity from a plastic state. During this process, both the plastic strain and $H_a$ change.

The above three kinds of changes in stress and strain are called the unloading process, neutral load changing process, and loading process, respectively. The conditions can be expressed as follows:

$$\begin{aligned} &\frac{\partial \varphi}{\partial \sigma_{ij}} d\sigma_{ij} < 0, \textit{unloading} \\ &\frac{\partial \varphi}{\partial \sigma_{ij}} d\sigma_{ij} = 0, \textit{neutral change} \\ &\frac{\partial \varphi}{\partial \sigma_{ij}} d\sigma_{ij} > 0, \textit{loading} \end{aligned} \tag{2.40}$$

In the process of unloading and neutral change, the plastic deformation remains unchanged, and $d\varepsilon_{ij}^{p} = 0$, so $H_a = 0$. In the process of loading, $d\varepsilon_{ij}^{p} \neq 0$, $H_a \neq 0$ are obtained.

If the material is ideal plastic, then the yield surface and the loading surface coincide, known as the limit surface. At this time, the stress point can only be on the limit surface and cannot be outside the limit surface. And the limit function is independent of the stress:

$$\varphi\left(\sigma_{ij}^{'}\right) = 0 \tag{2.41}$$

Then, we have

$$\begin{aligned} &\frac{\partial \varphi}{\partial \sigma_{ij}} d\sigma_{ij} < 0, \textit{unloading}, d\varepsilon_{ij}^{p} = 0 \\ &\frac{\partial \varphi}{\partial \sigma_{ij}} d\sigma_{ij} = 0, \textit{neutral change}, d\varepsilon_{ij}^{p} \neq 0 \end{aligned} \tag{2.42}$$

## 2.8 Plastic Deformation

It is immediately clear that there will be a *change* in the plastic strain, and clearly, we cannot say anything about the *total* plastic strain because this comprises all the plastic strain which has taken place in the previous loading history of the specimen. We must always, therefore, think in terms of the *strain increments* which occur during plastic flow. For a strain-hardening material, the *magnitude* of the plastic strain increment at yield will depend on the magnitude of the stress increment and the slope of the strain-hardening curve.

However, it turns out that for a material that remains isotropic as it hardens (which is our implicit assumption), the *direction* of the plastic strain increment vector in a three-dimensional strain increment space is *independent* of the degree of strain hardening and depends only on the *shape* of the yield surface.

In discussing strain increments, we need to know the directions of the axes of principal strain increments. For an isotropic material, we expect these to coincide with the axes of principal stress. In other words, a rectangular parallelepiped of isotropic material, loaded on its faces by principal stresses, would be expected during any plastic deformation to deform in such a way that its faces remained mutually perpendicular. For example, it would be surprising, given the loading shown in Figure 2.13, for there to be any shear deformation like that shown if the material were *isotropic.*

Next, suppose that during the plastic deformation, according to Tresca's condition, the principal stresses happen to be in the order $\sigma_1 > \sigma_2 > \sigma_3$. The planes on which the shear stress reaches the critical value $k$ are at 45° to the faces of the block as shown in Figure 2.14(a), and in particular, the directions of the maximum shearing stresses are perpendicular to axis 2. It seems reasonable to assume, therefore, that the deformation increment consists of sliding on the inclined planes shown, and on parallel planes, thus involving, in particular, *no change of dimension* in the 2-direction. From the plane diagram of Figure 2.14(b), we can see that, for small plastic strain increments, the decrease in length of the block in the 3-direction equals the increase in length in the 1-direction.

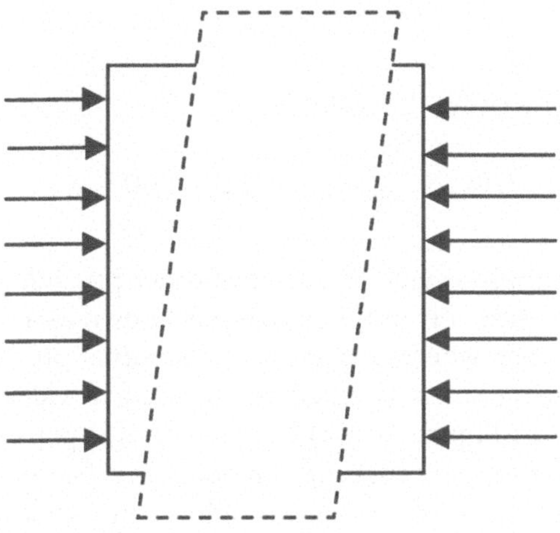

**Figure 2.13** Principal stress and strain-increment axes coincide.

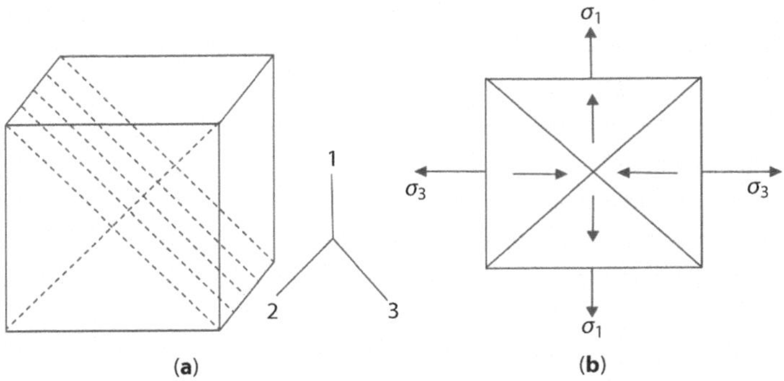


**Figure 2.14** Mode of deformation for the shearing of Tresca material.

If $\delta\varepsilon_1 < \delta\varepsilon_2 < \delta\varepsilon_2$ are the three (principal) plastic strain increments, then we have

$$\begin{aligned} \delta\varepsilon_2 &= 0 \\ 0 &\leq \delta\varepsilon_1 \\ \delta\varepsilon_3 &= -\delta\varepsilon_1 \end{aligned} \tag{2.43}$$

The ambiguous sign indicates that at the yield-point stress, there may or may not be any plastic strain increment. This deformation obviously satisfies the constant-volume requirement as below:

$$\delta\varepsilon_1 + \delta\varepsilon_2 + \delta\varepsilon_3 = 0 \tag{2.44}$$

Another way of writing the result is:

$$(\delta\varepsilon_1, \delta\varepsilon_2, \delta\varepsilon_3) = \lambda(1,0,-1), 0 \leq \lambda \tag{2.45}$$

Similar results apply at yield for the five other possible algebraic orders of principal stress. This sort of relationship is known as a "flow rule."

In the special case where $\sigma_1 > \sigma_2 = \sigma_3$, the situation is more involved because the shearing stress is equal to the critical value $k$ not only on the planes shown in Figure 2.14(a) but also on 45° planes parallel to the 3-axis. We can only suppose that slip can occur in *either* of the two possible modes, and we have, *a priori*, no means of knowing what relative amounts of plastic strain take place on the two possible slip systems. All we can say in these circumstances is

$$(\delta\varepsilon_1, \delta\varepsilon_2, \delta\varepsilon_3) = \lambda(1,0,-1) + \mu(1,-1,0), 0 \le \lambda, 0 \le \mu \quad (2.46)$$

It is instructive to plot plastic strain increments in a space whose axes are parallel to the principal stress axes. Figure 2.15 shows the combined principal stress/principal strain increment space, viewed along the leading diagonal, as before. In order to associate each strain increment vector with the corresponding stress vector, we plot it with the corresponding stress point as a "floating" origin.

For example, anywhere in the region $\sigma_1 > \sigma_2 > \sigma_3$, the plastic strain increment directions are parallel as seen in Equation (2.45), and they are perpendicular to the (Tresca) C-curve locally. Similar relationships apply to the other sides of the hexagon. At the apices of the hexagon, we have the ambiguous case, but the strain increment vector lies between the normal to adjacent sides.

This normality of the strain increment vector to the edge of the hexagon in the π-plane is indeed an aspect of the normality of the strain increment vector to the corresponding face of the three-dimensional hexagonal yield prism; this is easily seen by observing that the constant-volume condition, Equation (2.44), ensures that the strain increment vector is always parallel to the π-plane which in turn is perpendicular to the faces of the

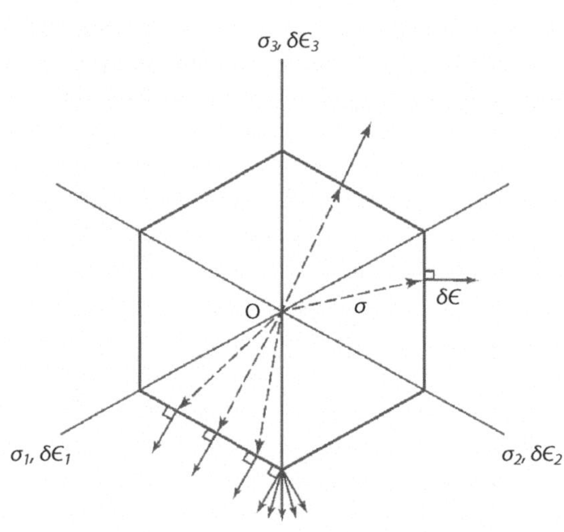


**Figure 2.15** Normality of the strain-increment vector to the Tresca yield surface.

yield surface. The ambiguous direction of the strain increment vector at the edges of the prism does not violate this "normality" rule: the "normal" to a sharp edge is indeterminate to precisely the degree indicated by Equation (2.46).

## 2.9 The Normality Rule

The "normality" of the associated plastic strain increment vector to the yield surface for the Tresca plastic material is indeed not accidental. "Normality" is a general rule which applies not only to plastic *material* but also, as we shall see, in the appropriate load space, to *structures* made of plastic material.

Its generality may be appreciated to some extent by means of the idea of *maximum plastic work.* Suppose that we have a structure made of elastic-plastic material (with a well-defined yield point) which is subject to two (for simplicity) independent loads $X$ and $Y$ acting in specific directions as shown in Figure 2.16(a). In an $X$-$Y$ load space, there will be a "collapse" locus (which corresponds, as we shall see later, to the yield locus for a material) which in general will enclose a simply connected region as illustrated in Figure 2.16(b). Suppose that an external agency now *imposes* an incremental *deflection* at the point of application of $X$ and $Y$, with components *8x, 8y* in the same two directions. The work done in this process is $X\delta x + Y\delta y$, which is the scalar product of vectors $(X, Y)$ and $(\delta x, \delta y)$ in parallel cartesian spaces, so we can conveniently use a geometrical interpretation if we superimpose on $(X, Y)$ space a $(\delta x, \delta y)$ space with parallel axes.

Given $\delta x$, $\delta y$, as displayed in Figure 2.16(c) that since the scalar product is found by multiplying the length of the $(\delta x, \delta y)$ vector by the projection

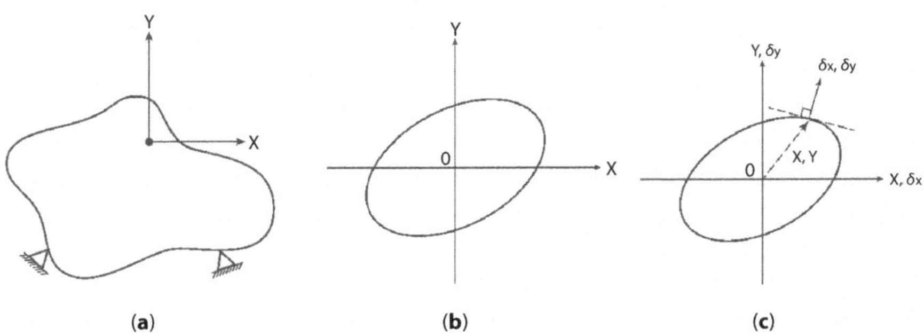


**Figure 2.16** Principle of maximum plastic work.

of the *(X, Y)* vector onto it, the work done by the agency is *maximum* if the collapse load corresponds to the point at which the deflection increment vector $(\delta x, \delta y)$ is *normal* to the collapse load locus. As shown in Figure 2.16, the collapse locus is smooth; the same argument applies equally for collapse loci with corners and flat faces. In the latter case, the load is not uniquely determined for an imposed deflection increment normal to a flat, but the work dissipated is determined uniquely: we shall find that this will have repercussions later on.

Thus, we conclude that the somewhat intuitive principle of "maximum plastic work" does indeed account for the normality rule. Admittedly, the above discussion is sketchy (there has been no reference to elastic deformation, for example), but it does bring out the important point that the displacement increments associated with a collapse locus are related to the loads by means of a *work* relation; the incremental displacement variables are chosen simply so that the scalar product of the load and the displacement increment give the incremental work done. This provides the link between the forces and incremental displacements in the present discussion and the stresses and incremental strains in the discussion of the Tresca yield condition and the corresponding flow rule; it also enables the same ideas to be generalized to other kinds of load which act on structures, such as couples and uniformly distributed loads or pressures.

Finally, we point out again that the normality rule only tells us the *direction* of the incremental strain vector (i.e., the relative proportions of the various components) or incremental deflection vector in the general case; the *magnitude* of the incremental deformation depends on the amount of strain hardening and other factors.

## Bibliography

1. Calladine, C.R., *Plasticity for Engineers: Theory and Applications*, Woodhead, Cambridge, 2011.
2. Chakrabarty, J. (Ed.), *Theory of Plasticity*, Elsevier Butterworth-Heinemann, Oxford, 2006.
3. Chen, W.F., *Limit Analysis and Soil Plasticity*, Elsevier, Amsterdam, 1975.
4. Chen, W.F., *Plasticity in Reinforced Concrete*, McGraw-Hill, New York, 1982.
5. Chen, W.F. and Baladi, G.Y., *Soil Plasticity: Theory and Implementation*, Elsevier, Amsterdam, 1985.
6. Drucker, D.C., On Uniqueness in the Theory of Plasticity. *Q. Appl. Math.*, 14, 35–42, 1956.

7. Hill, R., *The Mathematical Theory of Plasticity*, Oxford University Press, London, 1950.
8. Hoffman, O. and Sachs, G., *Introduction to the Theory of Plasticity*, Mc Graw Hill, New York, 1953.
9. Nielsen, M.P., *Limit Analysis and Concrete Plasticity*, Prentice-Hall, Englewood Cliffs, 1984.
10. Owen, D.R.J. and Hinton, E., *Finite Elements in Plasticity: Theory and Practice*, Pineridge Press Limited, Swansea, 1980.

# 3
# Constitutive Equations

***Abstract***
Chapter 3 delves into the development of constitutive equations essential for modeling plastic deformation in work-hardening materials. Two primary approaches are presented: the total strain theory (e.g., Hencky's model) and the incremental or flow theory. The total strain theory assumes proportional loading and is based on a one-to-one stress-strain relationship, but it struggles with non-proportional loading and unloading paths. The incremental theory, better suited for general loading scenarios, relates plastic strain increments to stress and stress increments, incorporating hardening rules. Associated and non-associated flow rules are introduced, with emphasis on the concept of a plastic potential surface. Drucker's stability postulate ensures material stability through positive work requirements. The chapter further discusses loading criteria, especially distinguishing between elastic and plastic loading, and develops general incremental stress-strain relationships for both perfectly plastic and work-hardening materials. These include expressions for the tangent stiffness tensor, consistency conditions, and plastic flow evolution, which are critical for numerical techniques like finite element analysis. The chapter concludes by highlighting the role of hardening modulus and how associated flow rules ensure solution uniqueness and mathematical robustness.

***Keywords***: Constitutive models, flow rule, Hencky theory, work hardening, incremental theory, yield surface, elastic-plastic modulus

## 3.1 Introduction

Engineering material usually exhibits a work-hardening behavior, increasing the stress beyond the initial yield surface and into the work-hardening range, which produces both plastic and elastic deformation. At each stage of plastic deformation, a new yield surface, called the subsequent loading surface, is established. In developing constitutive equations for work-hardening materials, two basic approaches have been used.

The first type of formulation is the deformation theory in the form of the total stress–strain relation. This theory assumes that the state of stress

Yong Bai. Plasticity and Ductile Fracture in Structural Engineering, (65–82) © 2026 Scrivener Publishing LLC

determines the state of strain uniquely as long as the plastic deformation continues. Thus, the most general form of this theory during loading may be written as

$$\varepsilon_{ij}^{p} = \varepsilon_{ij} - \varepsilon_{ij}^{e} = f\left(\sigma_{ij}\right) \tag{3.1}$$

where $\varepsilon_{ij}^{e}$ and $\varepsilon_{ij}^{p}$ are the elastic and plastic components of the total strain $\varepsilon_{ij}$, respectively. However, this equation cannot adequately describe the phenomenon associated with loading and unloading near the yield surface along a neutral loading path. Nevertheless, such theories have been used extensively in practice for the solution of elastic-plastic problems because of their comparative simplicity. It should be noted that the total strain relation is only valid in the case of proportional loading.

The other type of theory is the incremental theory or flow theory. This type of formulation relates the increment of plastic strain components $d\varepsilon_{ij}^{p}$ to the state of stress, $\sigma_{ij}$, and the stress increment $d\sigma_{ij}$. For work-hardening plasticity, the yield surface is now not fixed in stress space, but rather the stress point $\sigma_{ij}$ is permitted to move outside the yield surface. The response of the material after initial yielding is described by specifying a new yield surface called the subsequent yield surface, and the rule that specifies this post-yield response is called the hardening rule.

## 3.2 The Total Strain Theory

Hencky's stress–strain relation postulates a one-to-one correspondence between stress and strain. Thus, the components of the total plastic strain are taken to be proportional to the corresponding deviatoric stresses. Assuming small strains, the plastic stress–strain relation proposed by Hencky may be written as

$$\varepsilon_{ij}^{p} = \lambda s_{ij} \tag{3.2}$$

where $\lambda$ is positive during loading and zero during unloading. When the material work-hardens, $\lambda$ depends on the equivalent stress $\bar{\sigma}$, which may be regarded as a function of an equivalent total plastic strain defined as

$$\bar{\varepsilon}^{p} = \sqrt{\frac{2}{3}\varepsilon_{ij}^{p}\varepsilon_{ij}^{p}} \tag{3.3}$$

The relationship between $\bar{\sigma}$ and $\bar{\varepsilon}^{p}$ is given by the uniaxial stress-plastic strain curve. Substituting from and using $\bar{\sigma}$, which implies the von Mises yield criterion, we have

$$\sqrt{\frac{3}{2}}\bar{\varepsilon}^{p} = \lambda\sqrt{s_{ij}s_{ij}} = \lambda\sqrt{\frac{2}{3}}\bar{\sigma} \tag{3.4}$$

giving $\lambda = 3\bar{\varepsilon}^{p}/2\bar{\sigma}$. The stress–strain relation may, therefore, be expressed as

$$\varepsilon_{ij}^{p} = \frac{3\bar{\varepsilon}^{p}}{2\bar{\sigma}}s_{ij} = \frac{3}{2}\left(\frac{1}{S} - \frac{1}{E}\right)s_{ij} \tag{3.5}$$

where S is the secant modulus of the uniaxial stress–strain curve at $\sigma = \bar{\sigma}$. For an incompressible material, by Hooke's law, and (4), then furnishes $\varepsilon_{ij} = 3s_{ij}/2S$. The incremental form of (4) is

$$d\varepsilon_{ij}^{p} = \frac{3}{2\bar{\sigma}}\left\{\left(d\bar{\varepsilon}^{p} - \frac{\bar{\varepsilon}^{p}d\bar{\sigma}}{\bar{\sigma}}\right)s_{ij} + \bar{\varepsilon}^{p}ds_{ij}\right\} \tag{3.6}$$

The Hencky equation is equivalent to the Prandtl-Reuss equation when the ratios of the deviatoric stress components are held constant. This may be shown by writing $s_{ij} = (\bar{\sigma}/Y)s_{ij}^{0}$, where $s_{ij}^{0}$ is the deviatoric stress at the initial yielding, and also, we have $ds_{ij} = (d\bar{\sigma}/\bar{\sigma})s_{ij}$.

The quantity $\overline{d\varepsilon}^{p}$, in this case, is identical to $d\bar{\varepsilon}^{p}$. When the stress–plastic–strain relationship in uniaxial tension is represented by a power law, the Hencky equation is initially equivalent to the Prandtl-Reuss equation. Indeed, when $\bar{\sigma}$ varies as the $n$th power of $\bar{\varepsilon}^{p}$, it is easily shown that $d\bar{\sigma}/\bar{\sigma} = n(d\bar{\varepsilon}^{p}/\bar{\varepsilon}^{p})$, while $\bar{\varepsilon}^{p}/\bar{\sigma} \to 0$ as $\bar{\varepsilon}^{p}$ tends to zero. The complete equivalence of the two equations is established by setting $\overline{d\varepsilon}^{p} = (1-n)d\bar{\varepsilon}^{p}$ at the initial stage.

A geometrical illustration of strains and stresses is again possible if the principal axes of the stress do not rotate concerning the element.

As displayed in Figure 3.1, the **PS** and **OP** vectors, representing the plastic strain and the deviatoric stress, respectively, are directed along the same radial line, whatever the stress path. According to the Prandtl-Reuss theory, this happens only when the stress path is a straight line through the source. The **OP** vector also shows the deviatoric elastic strain, and the vector **OS** represents the total deviatoric strain. The magnitudes of the vectors **PS** and **OP** are $2G\sqrt{\frac{3}{2}}\bar{\varepsilon}^p$ and $\sqrt{\frac{2}{3}}\bar{\sigma}$ respectively. Therefore:

$$\frac{PS}{OP} = 3G(\frac{1}{S} - \frac{1}{E}) = \frac{3(E/S-1)}{2(1+\nu)} \tag{3.7}$$

When point $S$ describes a certain curved strain path, the stress vector rotates so that it is always along $OS$. The stress increment **PP′** corresponding to a strain increment **SS′** is obtained by locating $P'$ on $OS'$ by the above relation. The vector **SR**, equal to **PP′**, represents the elastic strain increment, and the vector **RS′** represents the plastic strain increment. This is incompatible with the Prandtl-Reuss flow rule, which requires $RS'$ to be parallel to $OP$. The stress path corresponding to the Hencky equation, therefore, differs from that given by the Prandtl-Reuss equation.

The projection of $RS'$ on a line through $R$ parallel to $OS$ is $RT = 2G\sqrt{\frac{3}{2}}d\bar{\varepsilon}^p$, where $d\varepsilon p$ is the increment of the equivalent total plastic strain. The scalar

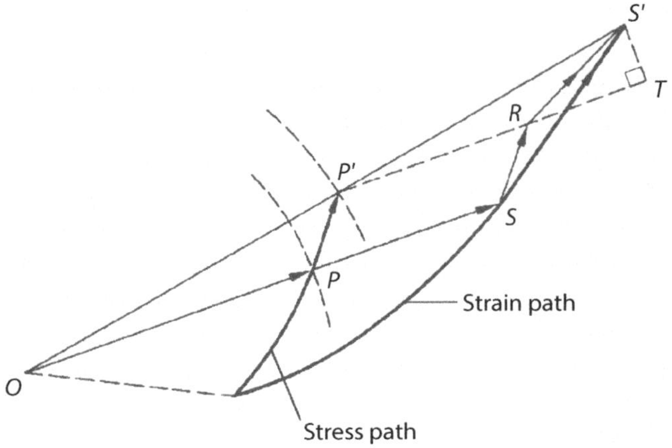


**Figure 3.1** Geometrical representation of the Hencky stress–strain relation in the deviatoric plane.

product of the vectors **OP** and **RS′** gives the increment of plastic work $dW_p = \bar{\sigma} d\bar{\varepsilon}^p$, which agrees with that obtained from the scalar product with the deviatoric stress $s_{ij}$.

The Hencky theory is unsuitable for describing the complete plastic behavior of metals, as can be seen by considering the unloading of the element after a certain amount of plastic deformation. If the element is reloaded to a different stress state on the current yield locus, the ratios of the plastic strain components will be entirely different. This is absurd since only elastic changes in strain can occur while the stress point is inside the yield locus. Nevertheless, when the loading is continuous and the deviatoric stress ratios do not vary significantly during the loading, the Hencky equation should lead to approximately correct results. Unloading can be permitted, provided the element is not reloaded to a plastic state different from that at the time of unloading.

A suitable definition of the strain tensor $\varepsilon_{ij}$ can be used to extend the Hencky theory to large strains; $\varepsilon_{ij} = \int d\varepsilon_{ij}$ is the normal meaning in this context, where the integral is taken along the path of the particle. Once the main axes of strain increment remain fixed in the element, it leads to the logarithmic strains in the main directions.

The total strain theory is, in general, theoretically unacceptable, since it violates the fundamental rule of plastic flow of metals. Except for proportional loading, the Hencky stress–strain relation cannot be associated with a plastic potential continuously turning normal. Consider now the situation in which a corner is formed in the yield locus at the current stress point during the ensuing plastic deformation. The Hencky stress–strain relation is then compatible with the associated flow rule at the corner for a certain range of nonproportional loadings. It is assumed that the yield locus changes in such a way that the stress point continues to be at the corner during the deformation. It is also supposed that the shape of the yield locus in the neighborhood of the corner at each stage is symmetrical about the current deviatoric stress vector.

## 3.3 The Flow Rules

The necessary connection between the loading function $f$ and the stress–strain relation for a work-hardening material will be made here utilizing a flow rule. When the current yield surface f is reached, the material is in a state of plastic flow upon further loading. Introducing the concept of a plastic potential function $g\left(\sigma_{ij}, \varepsilon_{ij}^p, k\right)$, we define the flow rule as below:

$$d\varepsilon_{ij}^{p} = d\lambda \frac{\partial g}{\partial \sigma_{ij}} \tag{3.8}$$

where $d\lambda > 0$ is a scalar function that will vary throughout the history of the straining processes. The gradient of the plastic potential surface $\partial g / \partial \sigma_{ij}$ defines the direction of the plastic strain increment vector $d\varepsilon_{ij}^{p}$, while the length or the magnitude of the vector is determined by the loading parameter $d\lambda$. Here, the flow rule is termed associated if the plastic potential surface has the same shape as the current yield or loading surface.

$$g\left(\sigma_{ij}, \varepsilon_{ij}^{p}, k\right) = f\left(\sigma_{ij}, \varepsilon_{ij}^{p}, k\right) \tag{3.9}$$

And the equation takes the form of

$$d\varepsilon_{ij}^{p} = d\lambda \frac{\partial f}{\partial \sigma_{ij}} \tag{3.10}$$

That is, the plastic flow develops along the normal to the loading surface. The relation is called the associated flow rule because the plastic flow is associated with the current loading surface. Since there is, in general, very little experimental evidence on plastic potential functions for engineering materials, the associated flow rule is applied predominantly to these materials for practical reasons. Apart from its simplicity, the normality condition of the equation ensures a unique solution for a given boundary-value problem using any stress–strain relations developed on this basis.

## 3.4 Drucker's Stability Postulate

Drucker accordingly defines a work-hardening plastic material as one in which the work done during incremental loading is positive, and the work done in the loading–unloading cycle is nonnegative; this definition is generally known as Drucker's postulate.

Figure 3.2 shows stability conditions for three kinds of uniaxial stress–strain curves. In (a), an additional stress $\dot{\sigma} > 0$ gives rise to an additional strain $\dot{\varepsilon} > 0$, with the product $\dot{\sigma}\dot{\varepsilon} > 0$. That is, the additional stress does positive work, which is represented by the shaded section in the diagram. This kind of curve leads to a stable elastoplastic material. For the unstable

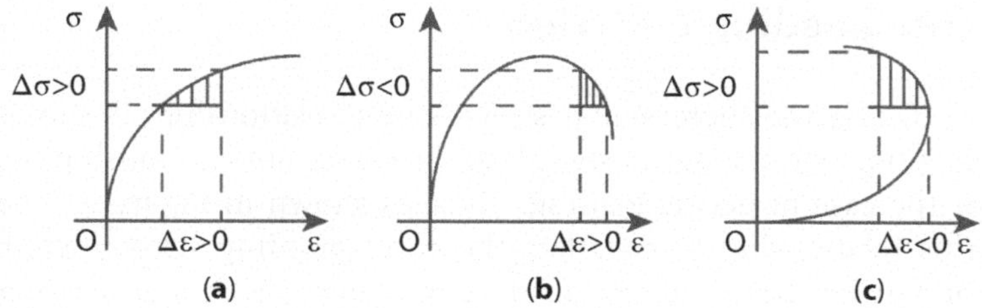


**Figure 3.2** Stable and unstable stress–strain curves: (a) stable and (b) and (c) unstable.

material shown in (b) and (c), however, the work done by the additional stress is always negative.

More strictly, a stable material is defined to be one that satisfies the following conditions:

1. During the application of the added set of forces, the work done by the external agency on the changes in displacements it produces is positive.
2. Over the cycle of application and removal of the added set of forces, the new work performed by the external agency on the changes in displacements it produces is non-negative.

These two conditions can be expressed by mathematical equations in energy form.

$$\oint \sigma_{ij} d\varepsilon_{ij} \geq 0 \tag{3.11}$$

$$\oint \sigma_{ij} d\varepsilon_{ij} \geq 0 \tag{3.12}$$

It should be noted that the stability postulate assures the existence of a unique inverse of the stress–strain relation. For the stable behavior shown in (a), the stress is determined uniquely by a given value of strain. For the unstable material shown in (b) and (c), however, these two strains correspond to a single value of stress.

## 3.5 The Loading Criterion

Before defining the incremental stress–strain relationship, it should be noted that the incremental elastic-plastic stress–strain relationship is valid only for the case of plastic loading. Thus, as shown in Figure 3.3, before using this relationship to calculate the corresponding stress increment, $d\sigma_{ij}$, for a given strain increment, $d\varepsilon_{ij}$, one must check first whether the material is in a plastic loading state corresponding to the given strain increment $d\varepsilon_{ij}$. If not, the following elastic stress–strain relationship should be used instead of the elastic-plastic stress–strain relationship:

$$d\sigma_{ij} = C_{ijkl} d\varepsilon_{kl} \tag{3.13}$$

To this end, a proper loading criterion is needed. We will express the loading criterion in terms of the given strain increment.

The loading criterion for the elastic-perfectly plastic material can be expressed in mathematical form as below:

$$\left.\begin{array}{c} f(\sigma_{ij}) < 0 \quad \textit{elastic state} \\ f(\sigma_{ij}) = 0, df = \dfrac{\partial f}{\partial \sigma_{ij}} \bullet d\sigma_{ij} = 0 \quad \textit{loading} \\ f(\sigma_{ij}) = 0, df < 0 \quad \textit{unloading} \end{array}\right\} \tag{3.14}$$

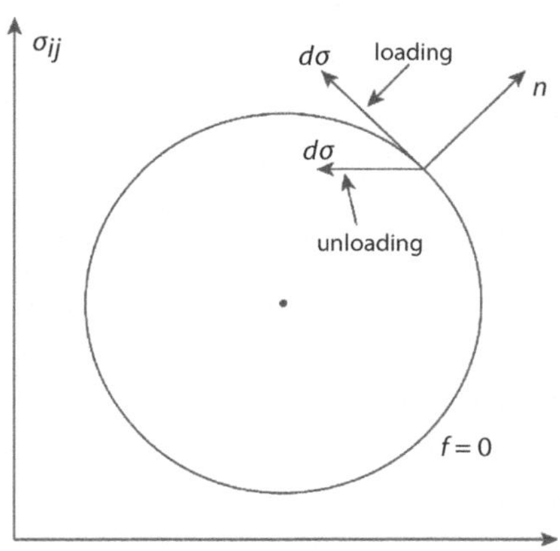


**Figure 3.3** The stress–strain relationship.

## 3.6 Incremental Stress–Strain Relationships

In the elastoplastic analysis by a numerical approach, the most common technique is the incremental method using tangent stiffness. The constitutive relations cannot be directly applied. An incremental relationship between stress and strain is needed in forming the tangent stiffness matrix. These types of constitutive relations are studied in this section.

### 3.6.1 Constitutive Relation for Perfectly Plastic Materials

As described before, a multiaxial perfectly plastic behavior requires that the stress increment vector be tangent to the yield surface and the plastic strain increment vector be normal to the loading surface. According to the concept of perfect plasticity, the magnitude of the plastic strain increment cannot be uniquely determined by the given current stresses and stress increments. However, for given current stresses and a given plastic strain increment satisfying the flow rule, the stress increment can be determined by the so-called consistency condition, which ensures that the stress state remains on the yield surface.

The total strain increment is assumed to be the sum of the elastic strain increment and the plastic strain increment:

$$d\varepsilon_{ij} = d\varepsilon_{ij}^{e} + d\varepsilon_{ij}^{p} \tag{3.15}$$

The elastic strain increment can be obtained from Hooke's law:

$$d\varepsilon_{ij}^{e}=D_{ijkl}d\sigma_{kl} \ or \ d\varepsilon_{ij}^{e} = \frac{d\sigma_{ij}}{2G} - \frac{3v}{E} d\sigma\delta_{ij} \tag{3.16}$$

And the plastic strain increment is obtained from the flow rule:

$$d\varepsilon_{ij}^{p}=d\lambda \frac{\partial f}{\partial \sigma_{ij}} \tag{3.17}$$

Then, the complete strain–stress relations for an elastic-perfectly plastic material are expressed as

$$d\varepsilon_{ij}=D_{ijkl}d\sigma_{kl}+d\lambda\frac{\partial f}{\partial\sigma_{ij}}=\frac{d\sigma_{ij}}{2G}-\frac{3v}{E}d\sigma\delta_{ij}+d\lambda\frac{\partial f}{\partial\sigma_{ij}} \tag{3.18}$$

where $d\lambda$ is an undetermined factor with the value

$$d\lambda=\begin{cases}=0 & \textit{wherever } \text{f<0 or f=0 but df<0} \\ >0 & \textit{wherever } f\text{=0 and df=0}\end{cases} \tag{3.19}$$

We shall determine the form of the factor $d\lambda$ below. This can be accomplished by combining the stress–strain relations with the consistent condition

$$df=\frac{\partial f}{\partial\sigma_{ij}}d\sigma_{ij}=0 \tag{3.20}$$

This ensures that the stress state $(\sigma_{ij}+d\sigma_{ij})$ existing after the incremental change $d\sigma_{ij}$ has taken place still satisfies the yield criterion $f$

$$f\left(\sigma_{ij}+d\sigma_{ij}\right)=f(\sigma_{ij})+df=f\left(\sigma_{ij}\right) \tag{3.21}$$

We can determine the stress increment tensor:

$$d\sigma_{ij}=C_{ijkl}\left(d\varepsilon_{kl}-d\varepsilon_{kl}^{p}\right)=C_{ijkl}d\varepsilon_{kl}-d\lambda C_{ijkl}\frac{\partial f}{\partial\sigma_{kl}} \tag{3.22}$$

Substitute this equation and solve for $d\lambda$:

$$d\lambda=\frac{\dfrac{\partial f}{\partial\sigma_{ij}}C_{ijkl}d\varepsilon_{kl}}{\dfrac{\partial f}{\partial\sigma_{rs}}C_{rstu}\dfrac{\partial f}{\partial\sigma_{tu}}} \tag{3.23}$$

All indices in this equation are dummy indices, indicating the scalar character of $d\lambda$. Therefore, when $f$ is defined for a particular material of interest and strain increments $d\varepsilon_{ij}$ are prescribed, the factor $d\lambda$ is determined uniquely.

The incremental stress–strain relation can be expressed explicitly in the following form:

$$d\sigma_{ij} = \left( C_{ijkl} - \frac{C_{ijmn} \dfrac{\partial f}{\partial \sigma_{mn}} \dfrac{\partial f}{\partial \sigma_{pq}} C_{pqkl}}{\dfrac{\partial f}{\partial \sigma_{rs}} C_{rstu} \dfrac{\partial f}{\partial \sigma_{tu}}} \right) d\varepsilon_{kl} \tag{3.24}$$

in which some dummy subscripts have been properly altered. The coefficient tensor in the parentheses represents the elastic-plastic tensor of tangent moduli for an elastic-perfectly plastic material:

$$C_{ijkl}^{ep} = C_{ijkl} - \frac{C_{ijmn} \dfrac{\partial f}{\partial \sigma_{mn}} \dfrac{\partial f}{\partial \sigma_{pq}} C_{pqkl}}{\dfrac{\partial f}{\partial \sigma_{rs}} C_{rstu} \dfrac{\partial f}{\partial \sigma_{tu}}} \tag{3.25}$$

The equation is the most general formulation of the constitutive relation for the elastic-perfectly plastic material. It is seen that the stress increments can be determined uniquely by the yield function and the strain increments. In other words, if the current stress state is known and the increments of strain are prescribed, the corresponding stress increments can be found. In general, however, if the current stress state is known and the stress increments are prescribed, the corresponding strain increments cannot be uniquely determined because the plastic strain increments can be defined only to within the indeterminate factor $d\lambda$.

### 3.6.2 Constitutive Relation for Work-Hardening Materials

In the previous sections, we have discussed the basic assumptions and equations used in the development of the incremental theory of work-hardening plasticity. Based on these basic equations, a general constitutive equation for an elastic-plastic work-hardening material will be derived in this section in the form of

$$d\sigma_{ij} = C_{ijkl}^{ep} d\varepsilon_{kl} \tag{3.26}$$

where $C_{ijkl}^{ep}$ is the elastic-plastic stiffness tensor of tangent modulus, which is a function of stress state and loading history. For a given stress state and loading history, Equation (3.26) gives the stress increment for a given strain increment, which constitutes a plastic loading. This equation is needed in the numerical analysis of plasticity, such as finite-element analysis.

The general expression of a yield surface or loading surface for a work-hardening material, as discussed before, has the form of

$$f\left(\sigma_{ij}, \varepsilon_{ij}^{p}, k\right) = 0 \tag{3.27}$$

where $k = k\left(\varepsilon_p\right)$ is an isotropic hardening parameter. The strain increment $d\varepsilon_{ij}$ is decomposed into two parts:

$$d\varepsilon_{ij} = d\varepsilon_{ij}^{e} + d\varepsilon_{ij}^{p} \tag{3.28}$$

in which the elastic strain increment, $d\varepsilon_{ij}^{e}$, is related to the stress increment, $d\sigma_{ij}$, by the generalized Hooke's law as

$$d\sigma_{ij} = C_{ijkl} d\varepsilon_{kl}^{e} \tag{3.29}$$

where $C_{ijkl}$ is the tensor of elastic modulus. The plastic strain increment, $d\varepsilon_{ij}^{p}$, can be generally expressed by a non-associated flow rule in the form

$$d\varepsilon_{ij}^{p} = d\lambda \frac{\partial g}{\partial \sigma_{ij}} \tag{3.30}$$

where $g = g\left(\sigma_{ij}, \varepsilon_{ij}^{p}, k\right)$, as for $f\left(\sigma_{ij}, \varepsilon_{ij}^{p}, k\right)$, is a known plastic potential function and $d\lambda$ is a scalar function to be determined later by the consistency condition $df = 0$. Substituting the elastic strain increment and the plastic strain increment into Hooke's law, we have

$$d\sigma_{ij} = C_{ijkl}(d\varepsilon_{kl} - d\lambda \frac{\partial g}{\partial \sigma_{kl}}) \tag{3.31}$$

To obtain the value of $d\lambda$, consider a plastic loading process. At the current state, we know the current stress state $\sigma_{ij}$ and the current plastic deformation state $\varepsilon_{ij}^{p}$ and $k(\varepsilon_{p})$, and they must satisfy the current yield function:

$$f\left(\sigma_{ij}, \varepsilon_{ij}^{p}, k\right) = 0 \tag{3.32}$$

After a small increment in total strain $d\varepsilon_{ij}$, which constitutes a plastic loading, the current state is changed to the new subsequent state, $\sigma_{ij} + d\sigma_{ij}$, $\varepsilon_{ij}^{p} + d\varepsilon_{ij}^{p}$, and $k + dk$, and the new state must satisfy the subsequent yield function, in the mathematical form

$$f\left(\sigma_{ij} + d\sigma_{ij}, \varepsilon_{ij}^{p} + d\varepsilon_{ij}^{p}, k + dk\right) = f\left(\sigma_{ij}, \varepsilon_{ij}^{p}, k\right) + df = 0 \tag{3.33}$$

or

$$df = \frac{\partial f}{\partial \sigma_{ij}} d\sigma_{ij} + \frac{\partial f}{\partial \varepsilon_{ij}^{p}} d\varepsilon_{ij}^{p} + \frac{\partial f}{\partial k} dk = 0 \tag{3.34}$$

This is known as the consistency condition for a general work-hardening material and restricts the increments between $d\sigma_{ij}$, $d\varepsilon_{ij}^{p}$, and $dk$. This condition assures that in a plastic loading process, the subsequent stress and deformation state remain on the subsequent yield surface. The scalar function $d\lambda$ can be determined directly from this condition. This is described in the following:

$$d\varepsilon_{p} = C\sqrt{d\varepsilon_{ij}^{p} d\varepsilon_{ij}^{p}} \tag{3.35}$$

Using the equation of $d\varepsilon_{p}$ and the flow rule, we obtain

$$dk = \frac{dk}{d\varepsilon_p} d\varepsilon_p = \frac{dk}{d\varepsilon_p} C\sqrt{\frac{\partial g}{\partial \sigma_{ij}} \frac{\partial g}{\partial \sigma_{ij}}} d\lambda \tag{3.36}$$

Substituting equations for $d\varepsilon_{ij}^p$, $d\sigma_{ij}$, and $dk$ into the consistency condition, we have

$$df = \frac{\partial f}{\partial \sigma_{ij}} C_{ijkl} d\varepsilon_{kl} - h d\lambda = 0 \tag{3.37}$$

where

$$h = \frac{\partial f}{\partial \sigma_{ij}} C_{ijkl} \frac{\partial g}{\partial \sigma_{kl}} - \frac{\partial f}{\partial \varepsilon_{ij}^p} \frac{\partial g}{\partial \sigma_{ij}} - \frac{\partial f}{\partial k} \frac{dk}{d\varepsilon_p} C\sqrt{\frac{\partial g}{\partial \sigma_{ij}} \frac{\partial g}{\partial \sigma_{ij}}} \tag{3.38}$$

From Equation (3.38), the scalar function $d\lambda$ can be solved as

$$d\lambda = \frac{1}{h} \frac{\partial f}{\partial \sigma_{ij}} C_{ijkl} d\varepsilon_{kl} = \frac{1}{h} H_{kl} d\varepsilon_{kl} \tag{3.39}$$

where the second-order tensor $H_{kl}$ associated with the yield function is defined as

$$H_{kl} = \frac{\partial f}{\partial \sigma_{ij}} C_{ijkl} \tag{3.40}$$

This second-order tensor will be utilized later in connection with the development of the elastic-plastic tangent stiffness tensor $C_{ijkl}^{ep}$ below.

We have expressed the scalar function $d\lambda$ in terms of a given strain increment $d\varepsilon_{ij}$. In the following, we shall show that the scalar function $d\lambda$ can also be expressed in terms of a given stress increment $d\sigma_{ij}$. To this end, we write the no associated flow rule in the following form:

$$d\varepsilon_{ij}^p = d\lambda \frac{\partial g}{\partial \sigma_{ij}} = \bar{G} \frac{\partial g}{\partial \sigma_{ij}} \frac{\partial F}{\partial \sigma_{kl}} d\sigma_{kl} = \frac{1}{\kappa} \frac{\partial g}{\partial \sigma_{ij}} \frac{\partial F}{\partial \sigma_{kl}} d\sigma_{kl} \tag{3.41}$$

in which κ is known as the hardening modulus and is related to the hardening function $\bar{G}$ and the scalar function $d\lambda$ by

$$\kappa = \frac{1}{\bar{G}} \tag{3.42}$$

$$d\lambda = \frac{1}{\kappa}\partial f = \frac{1}{\kappa}\frac{\partial f}{\partial \sigma_{kl}} d\sigma_{kl} \tag{3.43}$$

respectively. The consistency condition can be written as

$$df = \partial f + \frac{\partial f}{\partial \varepsilon_{ij}^{p}} \frac{\partial f}{\kappa} \frac{\partial g}{\partial \sigma_{ij}} + \frac{\partial f}{\partial k} \frac{dk}{d\varepsilon_{p}} \frac{C\partial f}{\kappa} \sqrt{\frac{\partial g}{\partial \sigma_{ij}} \frac{\partial g}{\partial \sigma_{ij}}} = 0 \tag{3.44}$$

From this, the scalar function κ can be solved as

$$\kappa = -\frac{\partial f}{\partial \varepsilon_{ij}^{p}} \frac{\partial g}{\partial \sigma_{ij}} - \frac{\partial f}{\partial k} \frac{dk}{d\varepsilon_{p}} C \sqrt{\frac{\partial g}{\partial \sigma_{ij}} \frac{\partial g}{\partial \sigma_{ij}}} \tag{3.45}$$

As can be seen, the hardening modulus is determined from the isotropic hardening rule. For the special case of elastic-perfectly plastic material, the yield surface does not change during loading, i.e., $\partial f / \partial \varepsilon_{ij}^{p} = 0$ and $dk / d\varepsilon_{p} = 0$; hence, we have $\kappa = 0$. We can also express the scalar function h in terms of the hardening modulus κ by

$$h = \kappa + \frac{\partial f}{\partial \sigma_{ij}} C_{ijkl} \frac{\partial g}{\partial \sigma_{kl}} = \kappa + H_{kl} \frac{\partial g}{\partial \sigma_{kl}} \tag{3.46}$$

Once the scalar function $d\lambda$ is determined, the plastic strain increment, $d\varepsilon_{ij}^{p}$, is known from the flow rule and given by

$$d\varepsilon_{ij}^{p} = d\lambda \frac{\partial g}{\partial \sigma_{ij}} = \frac{1}{h} \frac{\partial f}{\partial \sigma_{mn}} C_{mnst} \frac{\partial g}{\partial \sigma_{ij}} d\varepsilon_{st} = \frac{1}{h} H_{st} \frac{\partial g}{\partial \sigma_{ij}} d\varepsilon_{st} \tag{3.47}$$

and the corresponding stress increment, $d\sigma_{ij}$, can be determined by

$$
\begin{aligned}
d\sigma_{ij} &= C_{ijkl}\left(d\varepsilon_{kl} - \frac{1}{h}\frac{\partial f}{\partial \sigma_{mn}} C_{mnst}\frac{\partial g}{\partial \sigma_{kl}} d\varepsilon_{st}\right) \\
&= C_{ijkl}\left(\delta_{sk}\delta_{tl} - \frac{1}{h}\frac{\partial f}{\partial \sigma_{mn}} C_{mnst}\frac{\partial g}{\partial \sigma_{kl}}\right) d\varepsilon_{st} \\
&= \left(C_{ijkl} - \frac{1}{h}\frac{\partial f}{\partial \sigma_{mn}} C_{mnst}\frac{\partial g}{\partial \sigma_{kl}}\right) d\varepsilon_{st} \\
&= C^{ep}_{ijst} d\varepsilon_{st}
\end{aligned}
\tag{3.48}
$$

Thus, the elastic-plastic tangent stiffness tensor can be written in the form

$$
C^{ep}_{ijkl} = C_{ijkl} + C^{p}_{ijkl} \tag{3.49}
$$

with

$$
C^{p}_{ijkl} = -\frac{1}{h}\frac{\partial f}{\partial \sigma_{mn}} C_{mnkl} C_{ijst}\frac{\partial g}{\partial \sigma_{st}} = -\frac{1}{h} H_{kl} H^{*}_{ij} \tag{3.50}
$$

where $C^{p}_{ijkl}$ is called the plastic tangent stiffness tensor and represents the degradation of the stiffness of the material due to plastic flow. It is obvious that the tensor $C^{p}_{ijkl}$ lacks symmetry, and so does $C^{ep}_{ijkl}$ if a no associated flow rule is used.

Furthermore, in this case, the elastic-plastic tangent stiffness $C^{ep}_{ijkl}$ may not be positive definite. In contrast, if the associated flow rule is adopted, the positive-definiteness of the tensor $C^{ep}_{ijkl}$ can be ensured.

## Bibliography

1. Bathe, K.J., *Finite Element Procedures in Engineering Analysis*, Prentice-Hall, Englewood Cliffs, 1982.
2. Chakrabarty, J. (Ed.), *Theory of Plasticity*, Elsevier Butterworth-Heinemann, Oxford, 2006.
3. Chen, W.F. and Han, D.J., *Plasticity for Structural Engineers*, Springer-Verlag, New York, 1988.

4. Dafalias, Y.F. and Popov, E.P., A Model for Nonlinearly Hardening Materials for Complex Loading. *Acta Mech.*, 21, 3, 173–192, 1975.
5. Hill, R., *The Mathematical Theory of Plasticity*, Oxford University Press, London, 1950.
6. Jackson, L.R., Smith, K.F., Lankford, W.T., Plastic Flow in Anisotropic Sheet Metal. The American Institute of Mining Metallurgical, and Petroleum Engineers, 2440, pp. 1–15, 1948.

# 4
# Cyclic Loading and Shakedown

***Abstract***
Chapter 4 explores the complex response of materials and structures under cyclic loading, emphasizing plasticity and the conditions for shakedown. The chapter begins by contrasting material behavior under strain-controlled and stress-controlled cyclic loading, highlighting cyclic hardening, softening, and ratcheting phenomena. It develops the theoretical basis for cyclic loading analysis in elastic-perfectly plastic materials, using strain decomposition and yield surface concepts. A central contribution is the derivation of the conditions under which a structure's response becomes cyclic and stable after sufficient load repetitions. The concept of shakedown—the transition to elastic behavior after initial plastic deformation—is formally introduced. Various structural failure modes under cyclic loading are reviewed: elastic behavior, alternating plasticity, incremental collapse, and stable shakedown. The chapter concludes with Koiter's shakedown theorem and an intuitive truss example demonstrating how a structure can adapt to repeated loading without accumulating further plastic strain. This chapter offers both theoretical and applied insights, making it vital for assessing long-term structural integrity under repeated or fluctuating loads.

***Keywords***: Cyclic loading, plasticity, shakedown, residual stress, ratcheting, alternating plasticity, incremental collapse

## 4.1 Introduction

The complexity of the plastic behavior of materials under cyclic loading can be appreciated by considering the observed material response under uniaxial cycles of tension and compression. When the process is strain-controlled with a given amplitude of the longitudinal strain, and the material exhibits cyclic hardening, the stress increases in magnitude at the end of each cycle, and the stress–strain curve eventually attains a closed stable loop, as shown diagrammatically in Figure 4.1(a). A cyclically softening material shows a similar behavior, except that the stress decreases in magnitude at the end of each cycle until stabilization occurs. If, on the other

Yong Bai. Plasticity and Ductile Fracture in Structural Engineering, (83–96) © 2026 Scrivener Publishing LLC

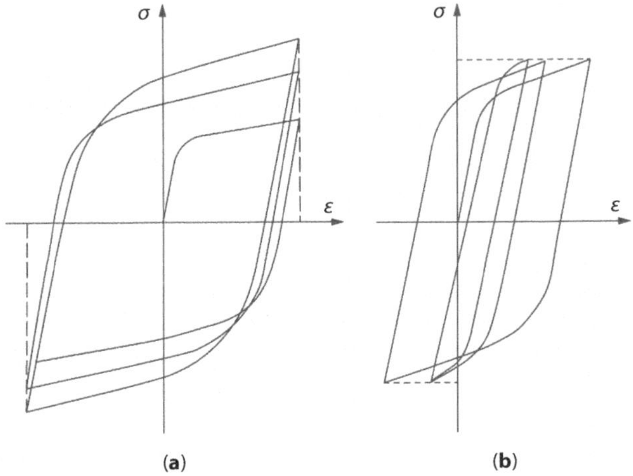


**Figure 4.1** Uniaxial stress and strain cycles in a cyclic hardening material: diagrammatic (a) constant strain cycles and (b) constant stress cycles.

hand, the process is stress-controlled with a constant stress amplitude, a cyclic hardening of the material results in a decrease in the magnitude of the strain at the end of each cycle, until a stable stress–strain loop is finally attained, as indicated in Figure 4.1(b). For a cyclic softening material, the strain increases in magnitude at the end of each stress cycle, leading eventually to a stable closed loop. Such a material may also exhibit ratcheting behavior, in which the stress–strain loop progressively moves outward along the strain axis, causing an excessive accumulation of plastic strain.

## 4.2 Cycling Loading

In many problems of actual engineering, structures or bodies are subjected to a repeated series of loading. For this reason, it is of interest to consider the response of elastic, perfectly plastic structures that are subjected to cyclic loading. That is to say, the hardening is neglected for the material deformation. In this part, the loads are presented by a function of time t as $Q(t)$ with the condition that displacements on the boundary are specified to be zero. Assuming the period of the cyclic loading is T, the loading program can be presented as:

$$Q(t) = Q(t+T) \tag{4.1}$$

For all t.

As discussed in the previous chapters, the stress distribution can be written as the sum of the elastic solution and residual stress distribution.

$$\psi(t) = \psi^{E}(t) + \rho(t) \tag{4.2}$$

It should be noted that $\psi^{E}(t)$ is a cyclic function since it depends on the instantaneous value of the external loads. Thus,

$$\psi^{E}(t) = \psi^{E}(t+T) \tag{4.3}$$

It should be noted that we are dealing with an elastic, perfectly plastic material, so that the following constitutive equations can be obtained:

$$\begin{gathered}\dot{\varepsilon}_j = \dot{\varepsilon}_j^{e} + \dot{\varepsilon}_j^{p}, \\ \dot{\varepsilon}_j^{e} = C_{jk} Q_k, \\ \dot{\varepsilon}_j^{p} = \lambda \frac{\partial \phi}{\partial \psi_j} \; for \; \phi(\psi_j) = 0 \; and \; \frac{\partial \phi}{\partial \psi_j}\dot{\psi}_j = 0 \\ \dot{\varepsilon}_j^{p} = 0 \; for \; \phi(\psi_j) < 0 \; or \; \phi(\psi_j) = 0 \; and \; \frac{\partial \phi}{\partial \psi_j}\dot{\psi}_j < 0\end{gathered} \tag{4.4}$$

where $\phi(\psi_j)$ presents the initial yield surface and limit functions, and $\dot{\varepsilon}_j^{e}$ and $\dot{\varepsilon}_j^{p}$ are the elastic and plastic components of the strain rate. As we are talking about the elastic, perfectly plastic cases, the yield surface will remain unchanged.

Due to the condition, $Q(t) = Q(t+T)$, we note that the difference between the stress fields $Q(t) - Q(t+T)$ is self-equilibrating since each stress field is in equilibrium with the same loads. The principle of virtual velocities thus gives

$$\oiiint \{\psi_j(t) - \psi_j(t+T)\}\{\dot{\varepsilon}_j(t) - \dot{\varepsilon}_j(t+T)\} dV = 0 \tag{4.5}$$

By dividing the total strain rates into elastic and plastic parts and making use of the elastic part of the constitutive relation, we can get one equation:

$$
\begin{aligned}
0 &= \oiiint \{\psi_j(t)-\psi_j(t+T)\}\{\dot{\varepsilon}_j(t)-\dot{\varepsilon}_j(t+T)\}dV = \\
&= \oiiint \{\psi_j(t)-\psi_j(t+T)\}\{\dot{\varepsilon}_j^e(t)-\dot{\varepsilon}_j^e(t+T)\}dV \\
&+ \oiiint \{\psi_j(t)-\psi_j(t+T)\}\{\dot{\varepsilon}_j^p(t)-\dot{\varepsilon}_j^p(t+T)\}dV
\end{aligned}
\tag{4.6}
$$

Making use of the relation between stress and the elastic component of strain in the constitutive relation, the following equation can be obtained:

$$
\begin{aligned}
&\oiiint \{\psi_j(t)-\psi_j(t+T)\}\{\dot{\varepsilon}_j(t)-\dot{\varepsilon}_j(t+T)\}dV \\
&\quad = \oiiint C_{jk}\{\psi_j(t)-\psi_j(t+T)\}\{\dot{\psi}_k(t)-\dot{\psi}_k(t+T)\}dV \\
&\quad = \frac{d}{dt}\oiiint C_{jk}\{\psi_j(t)-\psi_j(t+T)\}\{\psi_k(t)-\psi_k(t+T)\}dV
\end{aligned}
\tag{4.7}
$$

Since $\phi(\psi_j) \le 0$, the fundamental inequality on the convexity of the yield and limit surfaces and the normality rule gives

$$
\begin{aligned}
&(\psi_j(t)-\psi_j(t+T)\dot{\varepsilon}_j^p(t) \ge 0 \\
&(\psi_j(t+T)-\psi_j(t)\dot{\varepsilon}_j^p(t+T) \ge 0
\end{aligned}
\tag{4.8}
$$

Adding the previous two equations, we can get

$$
(\psi_j(t)-\psi_j(t+T)(\dot{\varepsilon}_j^p(t)-\dot{\varepsilon}_j^p(t+T) \ge 0 \tag{4.9}
$$

Derived from the above equations, we can finally get

$$
\begin{aligned}
&\frac{d}{dt}\oiiint C_{jk}\{\psi_j(t)-\psi_j(t+T)\}\{\psi_k(t)-\psi_k(t+T)\}dV \\
&= -\oiiint \{\psi_j(t)-\psi_j(t+T)\}(\dot{\varepsilon}_j^p(t)-\dot{\varepsilon}_j^p(t+T))dV \le 0
\end{aligned}
\tag{4.10}
$$

The matrix of the elastic coefficients $C_{jk}$ is positive definite, and consequently, the expression $A = \oiiint C_{jk}\{\psi_j(t) - \psi_j(t+T)\}\{\psi_k(t) - \psi_k(t+T)\}dV$ is non-negative and zero only when $\psi_j(t) = \psi_j(t+T)$. A may be thought of as a scalar quantity that measures the difference between the stress fields $\psi_j(t)$ and $\psi_j(t+T)$. It should be noted that $\frac{dA}{dt} \leq 0$, i.e., the difference between the stress fields can only decrease with time.

We note also that

$$\begin{aligned} \psi(t) - \psi(t+T) &= \psi^E(t) + \rho(t) - \{\psi^E(t+T) + \rho(t+T)\} \\ &= \rho(t) - \rho(t+T) \end{aligned} \tag{4.11}$$

This is to say that $A(\psi(t), \psi(t+T)) = A(\rho(t), \rho(t+T))$. As the structure responds to the cyclic load, we expect that in general, A will decrease until A = 0. A cannot be negative and hence cannot decrease in every cycle of loads. We expect that there will be a time, $t_0$, denoted by the passage of n cycles, $(n-1)T < t_0 < nT$. Therefore, for $t > nT$,

$$(\psi_j(t) - \psi_j(t+T))(\dot{\varepsilon}_j^p(t) - \dot{\varepsilon}_j^p(t+T)) = 0 \tag{4.12}$$

Thus, for $t > nT$, $\dot{A} = 0$ and A becomes constant. Then, we can get two conditions:

$$\begin{aligned} \psi(t) &= \psi(t+T) \;\; for \; t > nT \\ \dot{\varepsilon}_j^p(t) &= \dot{\varepsilon}_j^p(t+T) \;\; for \; t > nT \end{aligned} \tag{4.13}$$

These conditions imply that the stress field has thus become cyclic and the structure is behaving elastically, and that $\dot{\varepsilon}_j^p(t)$ has been constant, since $\rho_j(t)$ is uniquely determined by $\varepsilon_j^p(t)$. If $\varepsilon_j^p(t)$ is constant for t > nT, it follows that $\rho_j(t)$ is also constant for t > nT, and

$$\psi_j(t) = \psi_j^E(t) + \rho_j(t) \tag{4.14}$$

is cyclic because $\psi_j^E(t)$ is cyclic.

It has been seen that A = 0 implies that the stress field becomes cyclic, and the two conditions discussed above are sufficient for A = 0 for t > nT.

We shall continue our discussion on the assumption that the stress field in a structure subjected to cyclic loading will become cyclic after a sufficiently large number of cycles. It can be seen that no loss is implied in this assumption.

## 4.3 Shakedown Theorem

### 4.3.1 Notion of Shakedown

If the loads are applied in a cyclic manner, without ever reaching the static collapse condition, other forms of collapse may occur. If the strain increments change the sign in every cycle, with yielding on both sides of the cycle, then alternating plasticity is said to occur; the net plastic deformation may remain small, but a weakening of the material may occur nevertheless — a phenomenon called low-cycle fatigue — leading to breaking of the most highly stressed points after a certain number of cycles.

It may also happen that plastic deformation in each cycle accumulates so that after enough cycles, the displacements are large enough to be equivalent to collapse; this is called incremental collapse. On the other hand, it may happen that no further plastic deformation occurs after one or a few cycles — that is, all subsequent unloading–reloading cycles are elastic. In that case, the body is said to have experienced shakedown or adaptation. It is obvious that for bodies subject to repeated loading, the shakedown is more relevant than static collapse, and that the criteria for shakedown are of great importance.

The Shakedown theorem can also be expressed by a physical example. Consider a frame under cyclic loading, as shown in Figure 4.2.

If the material making up the structure is elastic-plastic, the following situations are possible:

(1) If the load intensities remain sufficiently low, the structural response is perfectly elastic (with the exception of stress singularities).
(2) If the load intensities become sufficiently high, the instantaneous load-carrying capacity of the structure becomes exhausted; a plastic, unconstrained flow mechanism develops; and the structure collapses. Obviously, plastic deformations can also develop for loads below the collapse load.

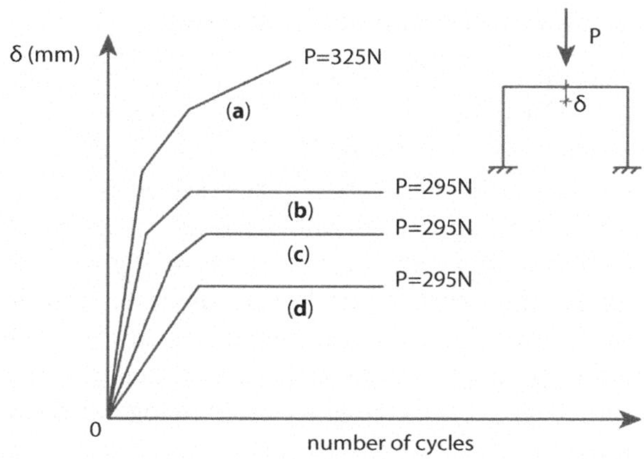


**Figure 4.2** Permanent deflection versus the number of load cycles for various load amplitudes.

(3) If the plastic strain increments in each load cycle are of the same sign, then, after a sufficient number of cycles, the total strains (and therefore displacements) become so large that the structure departs from its original form and becomes unserviceable. Such behavior can be observed in experimental investigations. In Figure 4.2, the development of the permanent side sway of a portal frame subjected to a cycling load is shown. For a sufficiently high-load amplitude (although below the statically load-carrying capacity of the frame), the deflection grows in each cycle. This phenomenon is called incremental collapse.

(4) If the strain increments change the sign in every cycle, they tend to cancel each other out, and the total deformation remains small (this is so-called alternating plasticity). In this case, however, after a number of cycles, the material at the most stressed points begins to break due to low-cycle fatigue.

(5) It may also happen that after some plastic deformation in the initial load cycles, the structural behavior becomes eventually elastic, as can be seen in Figure 1.2 for lower load amplitudes. Such stabilization of plastic deformations is called **shakedown** or **adaptation**.

### 4.3.2 An Intuitive Criterion of Shakedown

An elastic/plastic structure subjected to a cyclic loading program may fail due to excessive deformation, even though the extreme values of the load may be lower than those corresponding to plastic collapse. The plastic strains produced in a critical load cycle progressively increase as the cycle is repeated, and the structure eventually becomes unserviceable. In some cases, cycles of plastic flow may occur in alternating directions, causing early fatigue failure. The structure will be safe, however, if the load varies in such a way that the stress distribution remains within the yield limit after an initial period of plastic flow. The structure is then said to shake down to a state of residual stress, the change in strain being entirely elastic as the load is subsequently varied between the prescribed limits.

The shakedown criterion can be written down as follows:

$$w_p = \int \sigma_{ij} \dot{\varepsilon}_{ij}^{p} \, dtdV < \infty \tag{4.15}$$

However, such a local definition causes substantial mathematical difficulties in formulating the basic shakedown theorems. A less stringent criterion was proposed:

$$w_p = \frac{1}{V} \int_V \int_0^t \sigma_{ij} \dot{\varepsilon}_{ij}^{p} dtdV < w_0 \tag{4.16}$$

where $w_0$ is a material constant suitably measured from low-cycle fatigue tests. Such a formulation enabled him to present the first mathematically rigorous proof of the static shakedown theorem.

Consider a nonhardening elastic/plastic body in which the deformation is small enough for changes in geometry to be negligible. Let $(\sigma_{ij}, \varepsilon_{ij})$ denote the actual stress and strain distributions at any instant, and $(\sigma'_{ij}, \varepsilon'_{ij})$ those which would exist under the same boundary conditions if the body behaved elastically. It is convenient to write

$$\begin{aligned} \sigma_{ij} &= \sigma'_{ij} + \rho_{ij} \\ \dot{\varepsilon}_{ij}^{e} &= \dot{\varepsilon}'_{ij} + \dot{\gamma}_{ij} \end{aligned} \tag{4.17}$$

where $\rho_{ij}$ denotes a self-equilibrating distribution of residual stress and $\gamma_{ij}$ the corresponding elastic strain. A stress field $\rho_{ij}$ that is in equilibrium with zero body force and zero prescribed surface tractions is called self-equilibrated, or, more simply, a field of self-stress. Residual stress is of this kind. The residual stress may be regarded as that remaining in the body on complete removal of the external load by an elastic process. The plastic strain $\varepsilon_{ij}^{p}$ produced by $\rho_{ij}$ is identical to that by $\sigma_{ij}$. It can be shown that not more than one residual stress distribution can exist for a given plastic strain distribution with zero displacements at the constraints.

Suppose it is possible to find a time-independent residual stress $\rho_{ij}^{*}$ such that the superposition of this stress on $\sigma_{ij}'$ gives a safe state of stress $\sigma_{ij}^{*}$ that is always below the yield limit. The elastic strain energy of the body associated with the stress difference $\rho_{ij}-\rho_{ij}^{*}$ may be written as

$$W=\frac{1}{2}\int\left(\rho_{ij}-\rho_{ij}^{*}\right)\left(\gamma_{ij}-\gamma_{ij}^{*}\right)dV=\frac{1}{2}\int c_{ijkl}\left(\rho_{ij}-\rho_{ij}^{*}\right)\left(\rho_{kl}-\rho_{kl}^{*}\right)dV \tag{4.18}$$

Since $\rho_{ij}^{*}$ does not vary with time, and $c_{ijkl}$ is the the fourth-order elastic compliance tensor. The rate of change of $W$ is

$$\dot{W}=\int c_{ijkl}\left(\rho_{ij}-\rho_{ij}^{*}\right)\dot{\rho}_{kl}\,dV=\int\left(\rho_{ij}-\rho_{ij}^{*}\right)\dot{\gamma}_{ij}\,dV \tag{4.19}$$

Let $T_j$ and $T_j^{*}$ denote the surface tractions associated with the stresses $\sigma_{ij}$ and $\sigma_{ij}^{*}$ respectively. If $v_j$ and $v_j'$ are continuous velocity distributions associated with the strain rates $\dot{\varepsilon}_{ij}$ and $\dot{\varepsilon}_{ij}'$, it follows from the virtual work principle that

$$\int\left(\sigma_{ij}-\sigma_{ij}^{*}\right)\left(\dot{\varepsilon}_{ij}-\dot{\varepsilon}_{ij}'\right)dV=\int\left(T_j-T_j^{*}\right)\left(v_j-v_j'\right)dV=0 \tag{4.20}$$

in view of the fact that $T_j=T_j^{*}$ on $S_F$ and $v_j=v_j'$ on $S_v$. Since $\dot{\varepsilon}_{ij}-\dot{\varepsilon}_{ij}'=\gamma_{ij}+\varepsilon_{ij}^{p}$, the above equation becomes

$$\int\left(\sigma_{ij}-\sigma_{ij}^{*}\right)\dot{\gamma}_{ij}dV=-\int\left(\sigma_{ij}-\sigma_{ij}^{*}\right)\dot{\varepsilon}_{ij}^{p}dV=0 \tag{4.21}$$

The integral on the right-hand side is positive by the maximum work inequality whenever plastic flow occurs in the actual loading program. Hence,

$$\dot{W} = \int \left( \rho_{ij} - \rho_{ij}^{*} \right) \dot{\gamma}_{ij} dV = -\int \left( \sigma_{ij} - \sigma_{ij}^{*} \right) \dot{\gamma}_{ij} dV \leq 0 \tag{4.22}$$

with equality only when there is no plastic flow. Since W can never become negative, this equation indicates that plastic flow cannot continue indefinitely, and the body will ultimately shakedown to a steady distribution of residual stress. This completes the proof of the shakedown theorem, which states that an elastic/perfectly plastic structure will shakedown for given extreme values of the load if there exists a time-independent distribution of residual stress that nowhere leads to stresses beyond the yield limit when superimposed on the stress distribution corresponding to the elastic response of the structure. The theorem says nothing about the amount of plastic deformation that may occur before the shakedown state is attained.

### 4.3.3 Example of Structural Behavior in the Case of Cyclic Loads

In order to better understand the shakedown theorem, let us have a look at several simple cases of structures under cyclic loads. In the case of a truss composed of pin-jointed bars, which are indeterminate to degree one and are subjected to a one-parameter loading system, a very convenient model of the behavior structure can be given in the sense that it avoids the necessity of carrying out a complete analysis. Consider the truss, which is produced in Figure 4.3. The bars are uniform along their length and have identical cross-sectional properties. When the material is treated as an elastic, perfectly plastic material, the constitutive equations are

$$\begin{aligned} N &= EA\varepsilon, \;\; \varepsilon \leq \varepsilon_s \\ N &= \sigma_s A, \;\; \varepsilon > \varepsilon_s \end{aligned} \tag{4.23}$$

where the tensile force in the bar is equal to the product of the bar's cross-sectional area, the extension per unit length, and the yield stress.

For the loading system shown in Figure 4.3, the tensile forces in the bars must satisfy an equilibrium relation at joint D:

$$N_1+\sqrt{3}N_2 = P \tag{4.24}$$

The strains in the bars, denoted by $\varepsilon_1$ and $\varepsilon_2$, are related to the displacement $v$ of the joint D by means of the strain-displacement relations:

$$\begin{aligned} \varepsilon_1 &= \frac{v}{h} \\ \varepsilon_2 &= \frac{3}{4}\frac{v}{h} = \frac{3}{4}\varepsilon_1 \end{aligned} \tag{4.25}$$

The elastic solution to this problem is obtained by substituting the bar strain relation, Equation (4.11), into the equilibrium Equation (4.10). This gives

$$\begin{aligned} N_1 &= \frac{P}{1+\frac{3}{4}\sqrt{3}} = 0.435P \\ N_2 &= \frac{\frac{3}{4}P}{1+\frac{3}{4}\sqrt{3}} = 0.326P \end{aligned} \tag{4.26}$$

Before we consider a cyclic loading program, let us treat the simple case of a monotonic increase in $P$. The development history of force in bars is

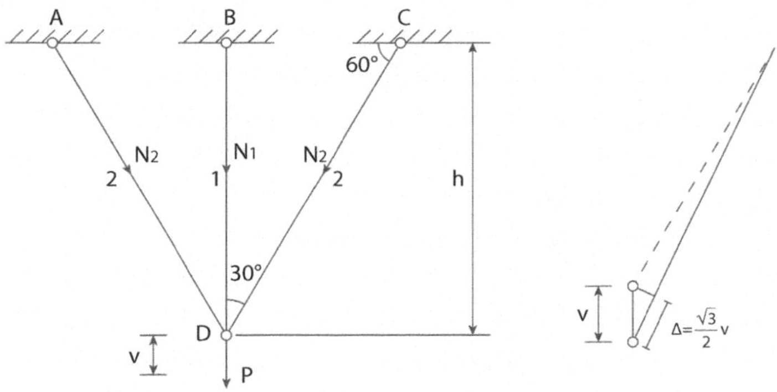


**Figure 4.3** Example of a truss.

shown in Figure 4.4. Based on the result above, we can find that the bar BD will yield firstly.

Assume the structure is initially in its virgin state and the load is also zero, so that the state point is initially at the origin point, $o$. The stress history of the bar BD is shown in the following picture:

$P$ is now increased monotonically, and the behavior will be initially elastic, so that $N_s$ shows a linear growth with strain. During this stage, the forces in bars AD, BD, and CD distribute as in Equation (4.12).

When the state point reaches point $a$, bar BD yields in tension. The load $P$ can be calculated as

$$P_s = N_S + \frac{3\sqrt{3}}{4} N_S = 2.30 N_S \tag{4.27}$$

As the material of the bar is considered perfectly plastic, the force in BD cannot increase and maintain as $N_s$. The tensile forces in the bars will satisfy a new equilibrium relation:

$$\sigma_s A + \sqrt{3} N_2 = P \tag{4.28}$$

Another special point occurs as bars AD and CD yield, which says $N_2 = N_s$. The load in this state can be calculated by

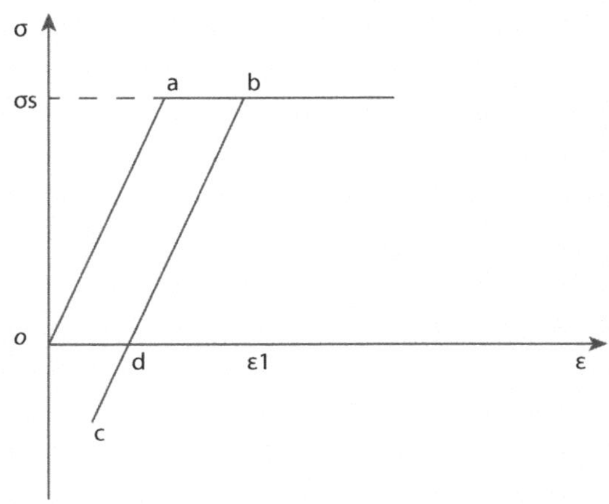


**Figure 4.4** Stress history of BD.

$$P_f = N_s + \sqrt{3}N_s = 2.73N_s \tag{4.29}$$

Thereafter, $P$ cannot be increased further and the whole structure yields.

Now, consider the cyclic loading which goes through the routine $P \to 0 \to P \to 0 \to P$ etc. This process is repeated continuously, and the load is sketched in Figure 4.4. Firstly, let P be in the range of 2.30 *Ns* to 2.73 *Ns*. The application of the load will cause a tensile plastic strain in bar BD and elastic strain in bar AD and CD. The following results can be obtained:

$$\begin{aligned} N_1 &= \sigma_s A \\ N_2 &= \frac{1}{\sqrt{3}}(P - \sigma_s A) \\ \varepsilon_2 &= \frac{N_2}{EA} = \frac{1}{\sqrt{3}EA}(P - \sigma_s A) \le \varepsilon_s \\ \varepsilon_1 &= \frac{4}{3}\varepsilon_2 = \frac{4}{3\sqrt{3}EA}(P - \sigma_s A) > \varepsilon_s \end{aligned} \tag{4.30}$$

Then, let the load change to zero. The removal of the load will take place elastically, and the residual stress and strain can be calculated by

$$\begin{aligned} \Delta N_1 &= 0.435P \\ \Delta N_2 &= 0.326P \\ N_1^r &= N_s - 0.435P \in (-0.005N_s, -0.18755N_s) \\ N_2^r &= \frac{1}{\sqrt{3}EA}(P - \sigma_s A) - 0.435P \in (0.00075N_s, 0.11N_s) \end{aligned} \tag{4.31}$$

From the results, we can find that the removal of the load takes place elastically. Now, let us reapply the load P to point D; the structure will also behave in the elastic stage, and the force distribution will become

$$\begin{aligned} N_1^{(2)} &= N_1^r + \Delta N_1 = N_s \\ N_2^{(2)} &= N_1^r + \Delta N_1 = \frac{1}{\sqrt{3}EA}\left(P - \sigma_s A\right) \end{aligned} \tag{4.32}$$

It can be obtained that the bar AD, CD is still behaving elastically. From now on, the response of the structure has become cyclic after the first cycle, and it will exhibit the same response for further load cycles without any plastic increment. The structure discussed in this case will not exhibit incremental collapse, in other words, the structure's shakedown.

## Bibliography


1. Chakrabarty, J. (Ed.), *Theory of Plasticity*, Elsevier Butterworth-Heinemann, Oxford, 2006.
2. Koiter, W.T., Some Remarks on Plastic Shakedown Theorems. *Proceedings of the Eighth International Congress on Theoretical and Applied Mechanics*, Istanbul, Turkey, 20-28 August 1952, 1952.
3. Koiter, W.T., A new General Theorem on Shakedown of Elastic-Plastic Structures. *Proceedings of the Koninklijke Nederlandse Akademie van Wetenschappen*, North-Holland Publishing Company, Amsterdam, 1956.
4. Martin, J.B., *Plasticity: Fundamentals and General Results*, The MIT Press, Massachusetts, 1975.

# 5

# Some Application Problems


***Abstract***

This chapter discusses a variety of applied problems in plasticity, with an emphasis on real-world structural applications. It begins with the analysis of torsion in thin-walled tubes of arbitrary cross-sections, using equilibrium-based methods to derive shear stress distributions and predict full plastic torque. The chapter extends the analysis to combined tension, torsion, and bending, introducing collapse load loci and safe design regions under multi-axial loading. Bending in thin-walled and rectangular cross-section members is examined, particularly the flattening behavior in curved tubes and non-linear stress distributions. The discussion then shifts to thick-walled cylinders subjected to internal pressure, covering both simplified and complete elastic-plastic solutions under plane axial symmetry. Finally, the chapter explores plane strain indentation problems using both upper-bound and lower-bound approaches, with applications in forming, foundation design, and material testing. Throughout, theoretical developments are supported by simplified assumptions to provide clear insights into complex mechanical behavior in plastic regimes.



***Keywords*:** Plastic torsion, combined loading, thin-walled tubes, bending deformation, thick-walled cylinders, elastic-plastic analysis, indentation mechanics


## 5.1 Torsion in Thin-Walled Tubes

In the analysis and design of the structures, determining the full plastic torque of a given prismatic member has been a challenge for many years, as well as designing the prismatic member to withstand a given torque. It should be noted that in applications, torsion usually concurs with tension or bending or both. Therefore, to obtain a practical analysis, considering this combination is vital. In this chapter, the first pure torsion analysis is performed, and then the combined loading is assessed.

The method of the current study is direct, and it does not reflect elastic and elastic-plastic analysis completely. The advantage of this method is that


Yong Bai. Plasticity and Ductile Fracture in Structural Engineering, (97–132) © 2026 Scrivener Publishing LLC

the preceding familiarity with a specific region of elasticity theory is not necessary.

A thin-walled tube of an *arbitrary cross-section* is shown in Figure 5.1. This tube is loaded by a torque Γ. The load is equal and opposite couples and is applied at the ends of an axis parallel to the tube. The tube is uniform in the sense that it is generated by sliding a cross-section in a direction perpendicular to its plane. The cross-section is simply connected, of any arbitrary shape, and the thickness $t$ is a variable function of the distance, $x$, measured around the periphery from an arbitrary datum, as displayed in Figure 5.2.

As the tube is *thin*, there is no significant difference in $s$ measured around the inside or the outside surface. In specifying torque T, we do not specify the details of how it is supplied; instead, we suppose that end-fittings are provided that "diffuse" the torque into the tube. We, therefore, appeal to the principle of St. Venant and consider a representative cross-section

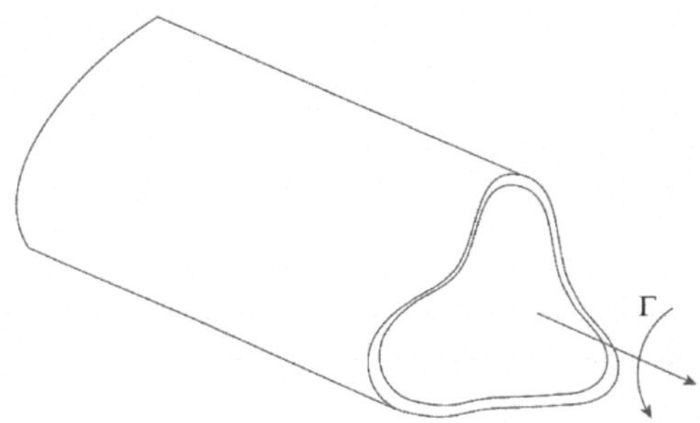


**Figure 5.1** Torsion of a thin-walled tube of irregular cross-section.

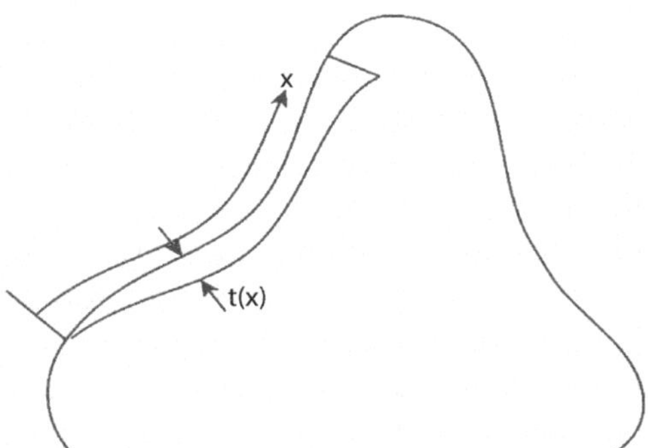


**Figure 5.2** Cross-section of an irregular thin-walled tube.

sufficiently remote from the ends for the stresses to be uniform along the tube. To describe adequately the state of stress in the thin wall of the tube, we need to specify a coordinate system. The obvious one to choose is $(x, z)$ in the thin wall, $z$ being measured axially from an arbitrary datum.

We also use the symbol $y$ to describe the direction of the local normal to the tube wall, and thus, to complete the *local* set of three mutually perpendicular axes. The first steps in the analysis of stress are to observe that it is safe to assume that the stress system does not vary through the (thin) wall of the tube and that several components of the stress on a typical small element, shown in Figure 5.3(a), vanish:

1. $\sigma_x = 0 \rightarrow$ Pressure is equal inside and outside of the tube.
2. $\sigma_z = 0 \rightarrow$ No axial pull exists.
3. $\sigma_y = 0$, $\tau_{yz} = 0$, $\tau_{yx} = 0 \rightarrow$ The tube surfaces are unloaded.

Therefore, $\tau_{zx}$ is the only non-zero stress component. The stress system at any point in the thin wall is thus a very simple one, viz., a state of pure shear referred to the $x$, $z$ directions. Figure 5.3(b) shows the simplified notation for shear stress.

To analyze the stresses in the tube, we first consider the equilibrium of a finite portion of the wall between two arbitrary generators at $x = x_1$ and $x = x_2$, and two arbitrary cross-sections distance L apart, as shown in Figure 5.4. Let the thickness of the wall be $t_1$ and $t_2$, respectively, at $x_1$ and $x_2$ and the corresponding shear stresses $\tau_1$ and $\tau_2$.

The equilibrium equation in the z direction requires

$$L\tau_1 t_1 = L\tau_2 t_2 \tag{5.1}$$

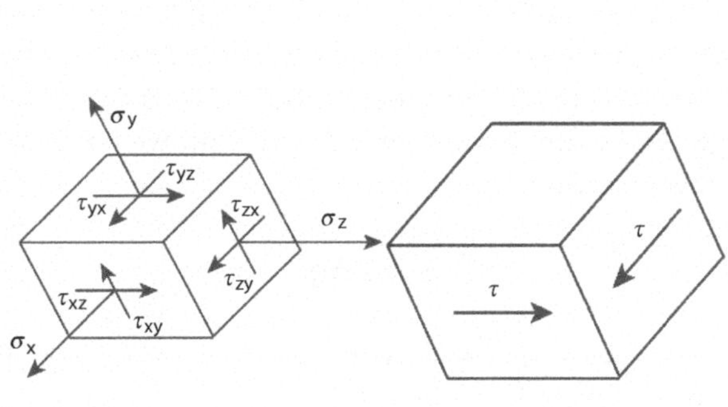


**Figure 5.3** State of stress in an element of the tube wall.

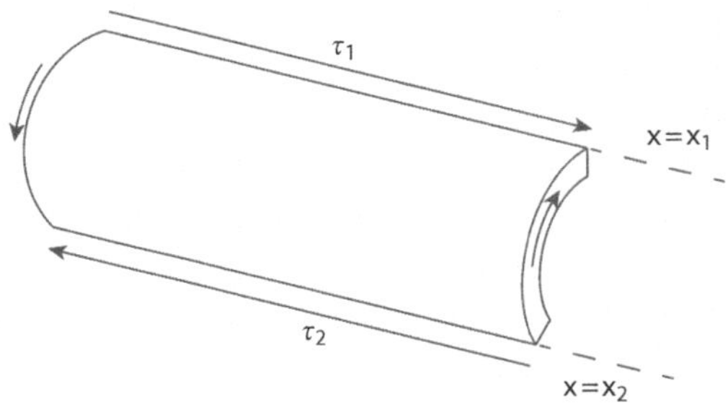


**Figure 5.4** Equilibrium of part of the tube wall.

Since $x_1$ and $x_2$ are arbitrarily:

$$\tau t = q \text{ (constant)} \tag{5.2}$$

around the circumference. The value of *q* depends on the dimensions of the cross-section and the applied torque. The quantity *q* is sometimes known as the "shear flow" in the wall of the tube; it is the shear *force* per unit peripheral length. We can readily check that the other equilibrium conditions for the segment are satisfied.

To find the relationship between Γ and *q*, we work out the resultant of the shear stresses acting on any cross-section and identify it with Γ. In general, the resultant of a set of forces in a plane is a coupled together with a force. In the present case, we expect the resultant to be a pure couple, so we must show that we obtain the *same* quantity by taking moments about an *arbitrary* point in the cross-section.

First, we take moments about an arbitrary point within the tube, such as *D* in Figure 5.5. A small element of wall, *BC*, of length *8s* has an associated force ($\tau t\delta s = q\delta s$) and a moment about *D* of ($DFq\delta s$), when *DF* is the perpendicular from *D* to *BC* (produced). But $DF\delta s$ is twice the area of the narrow triangle *BCD*; therefore, as *q* is a constant, we have on integration around the circumference:

$$T = 2Aq \tag{5.3}$$

where *A* is the total area enclosed by the cross-section. (Note that *A* is *not* the cross-sectional area of the tube wall).

This result is indeed independent of the position of *D* within the tube, and also outside the tube if some areas are reckoned, appropriately, as negative.

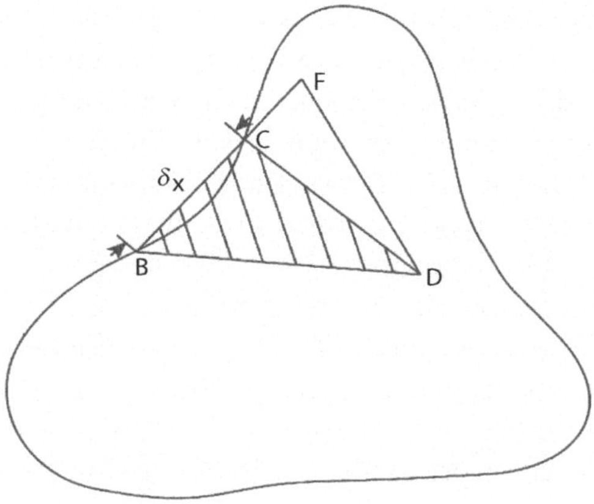


**Figure 5.5** Taking moments about a point.

At any point in the cross-section from Equations (5.2) and (5.3), we have

$$\tau = T/2At \tag{5.4}$$

As we have completed the analysis of stress by the use of equilibrium considerations alone, we note that the tube is *statically determinate.*

From Equation (5.4), we obtain directly a lower bound on the full plastic torque, $\Gamma_p$, for such a tube made of perfectly plastic material with yield stress *k* in pure shear. The shearing stress is largest where the thickness has its smallest value, $t_{min}$; by setting this stress equal to the yield stress, we have

$$\Gamma_p = 2Akt_{min} \tag{5.5}$$

When we consider the plastic *design* of a thin-walled tube, we can see immediately that as far as the economy of material is concerned, it is always best to have a *constant* wall thickness; for then, the shearing stress is constant also. However, even if we decide to have a constant wall thickness, we still have great freedom in design because only the enclosed *area* of the cross-section, and not its shape, is determined by Equation (5.5).

The shape which gives the minimum volume of the material for a given torque is that for which the circumference is least for a given area, i.e., the circle, but a square tube of the same enclosed cross-section area has a perimeter only 13 percent larger.

Considerations of the economy must not be pressed too far, however. In general, the above analysis shows that for a required full plastic torque, the amount of material needed diminishes as the wall thickness decreases and the enclosed cross-sectional area increases. Quite apart from the fact that *compactness* is often desirable in structural components from the point of view of space, there is a tendency for thin-walled structures of all kinds to become unserviceable through *buckling*. The designer must be aware of this possibility when designing unstiffened thin-walled tubes.

However, as our prime concern in the present chapter is the torsion of bars of compact cross-sections, we shall not need to investigate this aspect of the behavior of tubes; as we have already pointed out, our analysis of thin-walled tubes serves primarily as a stepping-stone to the analysis and design of more useful cross-sectional shapes.

## 5.2 Combined Tension and Torsion

The lower-bound approach enables us to make safe estimates of the strength of prismatic members subject simultaneously to tension and torsion, at the expense of very little further effort. The applied torque and longitudinal tension carried by the member can be shown by Γ and T, respectively. A pure torsion analysis furnishes two points in this space, $(\pm \Gamma_0, O)$, while a pure tension analysis furnishes the point $(O, T_0)$. $T_0 = AY$, where $A$ is the total cross-section area of material and $Y$ is the yield stress in pure tension. If we suppose that there is no buckling in compression, and no Bauschinger effect, as usual, we also have a fourth point, $(O, -T_0)$. These four points, Figure 5.6(a), all lie on or within the collapse load locus.

The convexity theorem indicates that the locus formed by joining the four points by straight lines is "safe"; see Figure 5.6(a). This is because the quadrilateral shown is the curve that comes as close to the original as possible without being concave anywhere. The quadrilateral thus represents an "inner" bound on the true Γ and T collapse load locus. Note that this applies to *whatever* the cross-sectional shape of the bar is. It is a simple matter to improve on this lower bound and enlarge the "safe" area of the Γ and T space, as follows.

We know from the "spring onion" analogy that shearing stresses of magnitude $k$ over the cross-section are in equilibrium with torque $\Gamma_0$; therefore, proportional stresses of magnitude $\lambda k$ are in equilibrium with torque $\lambda\Gamma_0$. Similarly, tensile stresses $\mu Y$ are in equilibrium with axial tension $\mu T_0$. Thus, provided the combined stresses $\lambda k$ and $\mu Y$ do not exceed yield, the

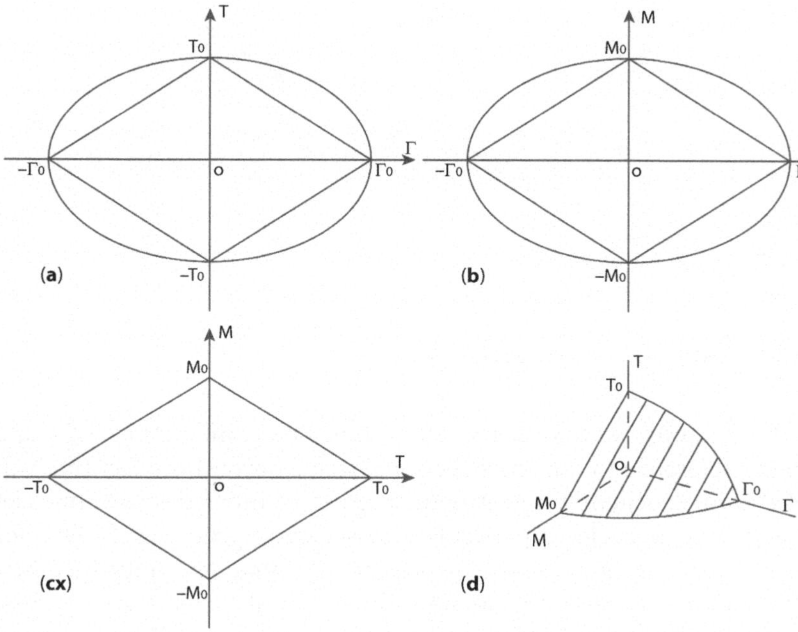


**Figure 5.6** Combinations of tension (T), torsion (Γ), and bending (M).

loads $\lambda\Gamma_o$ and $\mu T_o$ will be "safe." Figure 5.7(a) shows the combined stresses acting on an elementary cube "cut" with one face parallel to the tangent plane to the local thin tube and another perpendicular to the axis.

$$\left(\frac{\tau}{k}\right)^2 + \left(\frac{\sigma}{Y}\right)^2 = 1 \tag{5.6}$$

as shown in Figure 5.7 (b); $k = Y/2$ (Tresca) or $Y/\sqrt{3}$ (Mises). Here, $\tau = \lambda k$ and $\sigma = \mu Y$, so Equation (5.6) represents safe combinations of the load as

$$\left(\frac{\Gamma}{\Gamma_o}\right)^2 + \left(\frac{T}{T_o}\right)^2 = 1 \tag{5.7}$$

The corresponding ellipse is shown in Figure 5.6(a); it encloses considerably more space than the quadrilateral.

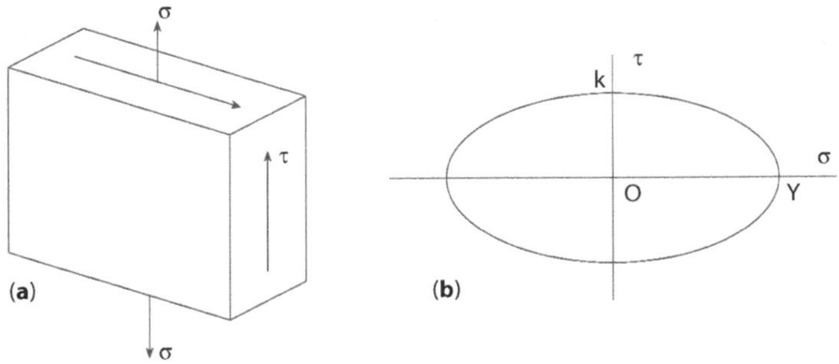


**Figure 5.7** Yield locus for combined tension and shear.

In this analysis, we have been able to find safe combinations of Γ and T without reference to the cross-sectional shape beyond the fact that values of $\Gamma_0$ and $T_0$ depend on it. Any further extension of the safe zone can only be done concerning a particular cross-section geometry. In fact, for the simple cross-sectional shapes which are fully analyzed in the literature under combined tension and torsion, the elliptical locus is a very close lower bound. We conclude, then, that further detailed analysis of this problem would have little practical value.

## 5.3 Combined Tension, Torsion, and Bending

It is easy to show, by a simple extension of the arguments of the previous section, that if $M_0$ is the full-plastic pure bending moment of a prismatic member about a "natural" axis, the ellipse is given by

$$\left(\frac{\Gamma}{\Gamma_o}\right)^2 + \left(\frac{M}{M_o}\right)^2 = 1 \tag{5.8}$$

(Γ, M) space, as shown in Figure 5.6(b), represents safe combinations of torque and bending moment.

The same approach can be applied to combined torsion, tension, and bending. For combined bending and tension (with no torsion), we cannot, unfortunately, improve on the quadrilateral joining the points ($\pm T_0$, O) and (O, $\pm M_0$) in (T, M) space, Figure 5.6(c), as a locus of safe loads without taking into account details of the cross-section geometry; consequently, in (Γ, T, M) space, our best general safe load locus is made up of two intersecting surfaces:

$$\left(\frac{T}{T_o}+\frac{M}{M_o}\right)^2+\left(\frac{\Gamma}{\Gamma_o}\right)^2=1 \tag{5.9}$$

One octant of this surface is shown in Figure 5.6(d). The surface may be derived from the three two-dimensional loci (Figures 5.6(a), (b), and (c)) by application of the *convexity* theorem. It is not difficult to extend the "safe" region of (Γ, M, T) space in specific, simple cases. It should be pointed out that considerable care is necessary for the discussion of plastic bending of members of arbitrary cross-section, due to the presence of "preferred" or "natural" axes of bending within the section, somewhat analogous to principal axes in the theory of elastic bending.

## 5.4 Bending in Thin-Walled Tubes

If a curved tube, such as a steam pipe in a power plant, is subjected to bending moments in its plane at the ends of the tube, the increase or decrease in curvature of the pipe, and its deflections can be calculated by ordinary curved-beam theory, usually by the theory of slightly curved beams because $r/R$ is of the order of *1/10* in a practical case. Bending experiments on such pipes have shown deflections many times greater than the calculation would indicate, so that curved thin-walled pipes are several times (up to 10 times) as flexible as is predicted by simple beam theory.

This was explained by von Karman (1911) based on the theory of least work. Let the pipe of Figure 5.8 be built in at the right end and subjected to a bending moment $M_0$ at the left end. That end will turn through some angle $\varphi_0$, and according to the old beam theory, the cross-sections will not distort, remaining thin-walled circles. This means that the outside fiber of the tube, from $A_0$ to A, will be distorted to run from $A_0$ to $A_1$, but it will still follow the full line, just being extended by an amount $AA_1$. Similarly, the inside fiber $B_0B$ will be distorted to $B_0B_1$, being shortened by $BB_1$. All of this will involve a certain amount of stored energy. However, there exists a means for the fibers to escape being extended and compressed. The fiber $A_0A$ can go from $A_0$ to $A_1$ without any elongation by going down somewhat, following the dashed line, and similarly, the compression fiber $B_0B_1$ can avoid shortening by going up a little. This behavior would certainly

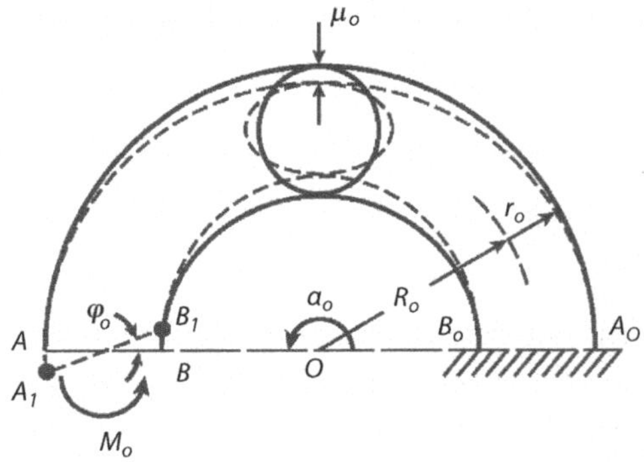


**Figure 5.8** A curved tube.

diminish the energy stored, but the escape of the fibers necessarily means a distortion of the cross-section: a flattening of the circle. Flattening the circular cross-section requires energy: to make that energy as small as possible, we do it by pure bending, leaving the length of the originally circular neutral fiber unchanged so that no hoop compression takes place. This requires the pipe to swell sidewise and acquire an ellipse-like cross-section.

In the analysis, the flattening is expressed by a number $U_0$ shown in Figure 5.8, and the bending deformation is expressed by $\varphi_0$. If $u_0/\varphi_0$ is made zero, then there is much energy in extending the fibers $A_0A$ and $B_0B$ longitudinally and none in distorting the section; on the other hand, if $u_0/\varphi_0$ is made so large that the dotted line $A_0A_1$ is equal in length to the full line $A_0A$, then all the energy is in distortion of the cross-section. We now determine analytically a value of $u_0/\varphi_0$ between the two extremes just mentioned in which the total energy from longitudinal extension and distortion of the cross-section becomes a minimum.

Determining the distortion of the cross-section is complicated, as shown in Figure 5.9; the distortion of the cross-section has to be analyzed first. The original shape of the centerline is a circle of radius $r_0$; the distortion is expressed by the radial displacement $u$, positive outward: $u = f(\theta)$. The slope of the distorted shape concerning the tangent of the dotted circle at point $A$ of Figure 5.9 is

$$\frac{du}{rd\theta} = \frac{u'}{r} \tag{5.10}$$

Now, by developing a Taylor series for $r$, we have

$$\frac{1}{r} = \frac{1}{r_0 + u} = \frac{1}{r_0\left[1 + (u/r_0)\right]} = \frac{1}{r_0}(1 - \frac{u}{r_0} + \frac{u^2}{r_0^2} + ...) \tag{5.11}$$

Then, the slope is given by

$$\frac{u'}{r} = \frac{u'}{r_0}(1 - \frac{u}{r_0} + ...) = \frac{u'}{r_0} - \frac{uu'}{r_0^2} + ... \tag{5.12}$$

The displacement $u$ is considered small concerning $r$, and we shall retain only quantities of the first order of smallness. Hence, $uu'/r_0^2$ is a small second-order, and we neglect it. As can be seen in Figure 5.9, the slope is thus simply $u'/r_0$ at point $A$. At the next point $B$, the slope, relative to the tangent there, is

$$\frac{u'}{r_0} + \frac{d}{d\theta}(\frac{u'}{r_0})d\theta = \frac{u'}{r_0} + \frac{u''}{r_0}d\theta \tag{5.13}$$

The difference in slope between points $B$ and $A$, both concerning a fixed direction, is deduced as

$$d\theta - B_{slope} + A_{slope} = (1 - \frac{u''}{r_0})d\theta \tag{5.14}$$

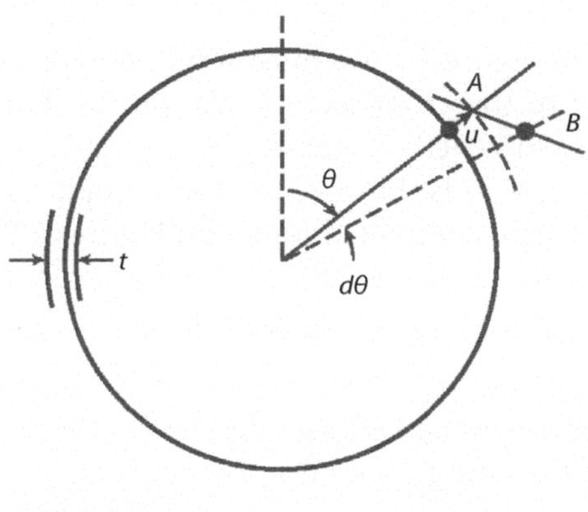


**Figure 5.9** The deformed cross-section.

The local curvature of the distorted shape is

$$\frac{\left[1-(u''/r_0)\right]d\theta}{rd\theta}=\frac{1-(u''/r_0)}{r_0+u}=\frac{1-(u''/r_0)}{r_0\left[1+(u/r_0)\right]}=\frac{1-(u''/r_0)}{r_0}(1-\frac{u}{r_0}+...) \tag{5.15}$$

The original curvature of the non-distorted cross-section was $l/r_0$ so that the increase in curvature due to the deformation is

$$-\frac{u+u''}{r_0^2} \tag{5.16}$$

This, incidentally, is an important general result, which is made the starting point for other problems in the bending of rings. The differential equation of a ring, bent in its plane, is thus

$$M=EI\frac{u+u''}{r_0^2} \tag{5.17}$$

where $M$ is the bending moment, positive when it tends to decrease the curvature of the ring. For long tubes where anticlastic curvature is prevented, the equation is

$$M=\frac{EI}{1-\mu^2}\frac{u+u''}{r_0^2} \tag{5.18}$$

Now, returning to Figure 5.7, we must write an expression for u to simulate the flattening of the cross-section. We do not know the shape the cross-section takes, so strictly speaking, we should write an expression with many parameters. This, however, is too complicated, and we do as best we can with a single parameter. Based on Figures 5.7 and 5.8, we have

$$u=-u_0\cos 2\theta \tag{5.19}$$

which is a reasonable flattened ellipse-like curve. The single parameter $u_0$ tells the amount of flattening. From this assumption, we deduce

$$u'' = 4u_0 \cos 2\theta \tag{5.20}$$

By substituting Equation (5.20) into Equation (5.19), we have

$$-\frac{3u_0 \cos 2\theta}{r_0^2} \tag{5.21}$$

## 5.5 Combined Tension and Torsion in a Thin-Walled Tube

A thin-walled tube is first twisted to the point of yielding and then extended longitudinally while holding the angle of twist constant. The material is pre-strained to obtain a sharp yield point, and the subsequent rate of hardening is assumed negligible compared to the elastic modulus. The loading path is such that the elastic and plastic strain increments are initially of comparable magnitudes, even though the tube is fully plastic. The longitudinal stress $\sigma$ and the shear stress $\tau$ at any stage are approximately uniform through the thickness of the tube, the relevant stress–strain equations being

$$d\varepsilon = \frac{d\sigma}{E} + \frac{2}{3}\sigma d\lambda \tag{5.22}$$

$$d\gamma = \frac{d\tau}{2G} + \tau d\lambda \tag{5.23}$$

where $d\gamma = 0$ during the extension. The elimination of $d\lambda$ between the two equations gives

$$d\varepsilon = \frac{d\sigma}{E} - \frac{\sigma}{3G}\frac{d\tau}{\tau} = \frac{d\sigma}{E} + \frac{\sigma^2}{Y^2 - \sigma^2}\frac{d\sigma}{3G} \tag{5.24}$$

where the last step follows from the yield criterion, the material being assumed nonhardening. Since the shear stress decreases from its initial value $Y/\sqrt{3}$ as the tensile stress increases from zero, the integration of the above equation results in

$$\frac{6G}{Y}\varepsilon = \left(\frac{1-2\nu}{1+\nu}\right)\frac{\sigma}{Y} + \ln\left(\frac{Y+\sigma}{Y-\sigma}\right) \tag{5.25}$$

because of the relation $E = 2(1+\nu)G$. As the longitudinal strain is increased from zero, $\sigma$ rapidly increases to approach the value $Y$ asymptotically. The first term on the right-hand side of Equation (5.25) soon becomes negligible. For $\nu = 0.3$, the longitudinal stress is within one percent of $Y$ when $\varepsilon$ is equal to $3Y/E$. The above theory forms the basis of experimental verification of the Prandtl-Reuss theory. The experiment has been carried out by Hohenemser using pre-strained tubes of mild steel, and reasonable agreement has been found with the theoretical prediction.

Consider the related problem in which a thin-walled tube is brought to the yield point in simple tension, and subsequently twisted in the plastic range, holding the tensile stress constant at the value $Y$. The loading can be continued only if the material work hardens, having an effective stress $\sigma$ at any stage. Then, we have

$$\bar{\sigma}^2 = Y^2 + 3\tau^2 \tag{5.26}$$

$$\bar{\sigma}d\bar{\sigma} = 3\tau d\tau \tag{5.27}$$

The stress-strain Equations (5.22) and (5.23) still hold with $d\sigma = 0$, while $d\lambda$ is given by

$$\frac{2}{3}d\lambda = \frac{d\bar{\sigma}}{H\bar{\sigma}} = \frac{3\tau d\tau}{H(Y^2 + 3\tau^2)} \tag{5.28}$$

where $H$ is the current slope of the stress–plastic strain curve. Hence, Equations (5.22) and (5.23) become respectively

$$d\varepsilon = \frac{3Y\tau d\tau}{H(Y^2 + 3\tau^2)} \tag{5.29}$$

$$d\gamma = \frac{d\tau}{2G} + \frac{9\tau^2 d\tau}{2H(Y^2 + 3\tau^2)} \tag{5.30}$$

It follows from the second of these relations that the torsional rigidity is equal to $G$ when $\tau = 0$, although the tube is plastic. Assuming $H$ to be a constant, the above equations can be integrated to

$$\varepsilon = \frac{Y}{E} + \frac{Y}{2H}\ln(1+\frac{3\tau^2}{Y^2}) \tag{5.31}$$

$$\gamma = \frac{\tau}{2G} + \frac{3}{2H}\left[\tau - \frac{Y}{\sqrt{3}}\tan^{-1}\left(\frac{\sqrt{3}\tau}{Y}\right)\right] \tag{5.32}$$

The incremental shear strain $d\gamma$ at any stage is equal to $rd\varphi/2l$, where $\varphi$ is the total angle of twist, $l$ is the current length of the tube, and $r$ is the current mean radius. Since $r$ and $l$ are varying from stage to stage, $2\gamma$ cannot be interpreted as being the tangent of the angle of the helix formed by an original generator of the tube.

Suppose, now, that a thin-walled tube is subjected to simultaneous torsion and tension following a strain path which is such that the elastic part of the strain is negligible. The Lévy-Mises flow rule furnishes the relations as below:

$$d\varepsilon = \frac{dl}{l} = \frac{\sigma d\bar{\sigma}}{H\bar{\sigma}} \tag{5.33}$$

$$d\gamma = \frac{rd\phi}{2l} = \frac{3\tau d\bar{\sigma}}{2H\bar{\sigma}} \tag{5.34}$$

$$\frac{dr}{r} = \frac{dt}{t} = -\frac{dl}{2l} \tag{5.35}$$

where $l$, $r$, and $t$ denote the current length, mean radius, and thickness of the tube, their initial values being $l_0$, $r_0$, and $t_0$, respectively. The last set of equations immediately furnishes

$$\frac{r}{r_0} = \frac{t}{t_0} = \sqrt{\frac{l_0}{l}} \tag{5.36}$$

Since $r^2l = r_0^2 l_0 = const$, it follows that the internal volume of the tube remains unchanged during the deformation. This is a consequence of the isotropy of the material and the equality of the Lode variables $\mu$ and $\nu$. By using Equation (5.36), we can obtain

$$\frac{3\tau}{\sigma} = r\frac{d\phi}{dl} = r_0\sqrt{\frac{l_0}{l}}\frac{d\phi}{dl} \tag{5.37}$$

$$\bar{\sigma} = \sqrt{\sigma^2 + 3\tau^2} = \sigma\left\{1 + \frac{r_0^2 l_0}{3l}\left(\frac{d\phi}{dl}\right)^2\right\}^{1/2} \tag{5.38}$$

The elimination of $\sigma$ between Equation (5.33) and Equation (5.38) and the integration of the resulting differential equation results in

$$\int_Y^{\bar{\sigma}} \frac{d\bar{\sigma}}{H} = \int_{l_0}^{l}\left\{1 + \frac{r_0^2 l_0}{3l}\left(\frac{d\phi}{dl}\right)^2\right\}^{1/2}\frac{dl}{l} \tag{5.39}$$

If $dl/d\varphi$ is considered as a given function of the length of the tube, the integral on the right-hand side can be evaluated numerically or otherwise to obtain the relationship between $\sigma$ and $l$. Since $\sigma$ and $\tau$ then follow from Equation (5.38), the axial load $2\pi rt\sigma$ and the torque $2\pi r2t\tau$ can be calculated at any stage of the deformation.

## 5.6 Bending in Rectangular Cross-Section

The bending of a curved bar in its plane where the stress distribution is not linear has been displayed in Figure 5.10, where *bh* is the curved beam of rectangular cross-section and $R_0$ is the mean radius of curvature. Besides, the tangential-stress distribution is assumed to be the superposition of a linear one (*a*) and a parabolic one (*b*), which is so adjusted that the total tangential force is zero. The bending stress is given by

$$s_t = C_1 z + (C_0 + C_2 z^2) \tag{5.40}$$

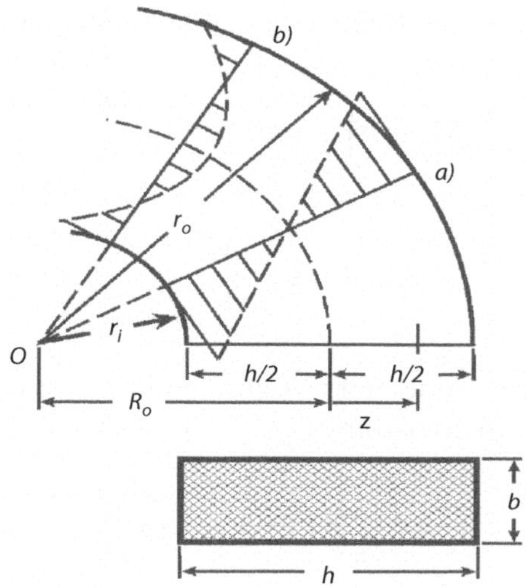


**Figure 5.10** A curved beam with a rectangular cross-section.

The first term is a linear bending distribution, and the term in the parentheses is parabolic stress, symmetrical about the center line of the bar. For pure bending, the total tangential force over a section between $r_i$ and $r_o$ should be zero; the first term $C_1z$ does that automatically, and now, we adjust $C_o$ and $C_2$ so that the net force of the parabolic distribution also is zero, so we have

$$\int_{-h/2}^{+h/2} (C_0 + C_2 z^2) dz = 0 = C_0 h + C_2 \frac{h^3}{12} \tag{5.41}$$

$$C_2 = -\frac{12C_0}{h^2} \tag{5.42}$$

And the stress distribution is

$$s_t = C_1[z + \frac{C_0}{C_1}(1 - 12\frac{z^2}{h^2})] \tag{5.43}$$

The size constant C determines the bending moment $M_o$ as

$$M_o = b\int_{-h/2}^{+h/2} s_t z dz = C_1 \frac{bh^3}{12} \tag{5.44}$$

$$C_1 = \frac{12M_o}{bh^3} \tag{5.45}$$

The term $1\text{-}12(z^2/h^2)$ does not give any bending, as is obvious from looking at $b$ in Figure 5.10. Renaming $C_0/C_1 = A$ for short, the bending stress is

$$s_t = \frac{12M_o}{bh^3}[z + A(1 - 12\frac{z^2}{h^2})] \tag{5.46}$$

This contains one parameter $A$, which makes the stress distribution vary from type a (Figure 5.10), for $A = 0$ to type $b$ for $A = \infty$. Now, this set of stresses $s_t$ by itself is not in equilibrium, which we recognize by considering a thin curved string element about $O$ as the center. The ends are pulled and are not in line, so that radial stress is necessary to hold the string. Mathematically, this is expressed by the equilibrium equation (for $\omega^2 = 0$), from which $s_r$ can be calculated. But when we do that and specify that $s_r = 0$ on $r = r_i$ and $r = r_o$, we find that parameter A in Equation (5.46) becomes fixed instead of floating. We should have started with one extra parameter in Equation (5.45) to get a good solution. This is a great labor, so that here, as previously, we assume the comparatively unimportant radial stress to be zero, so that Equation (5.46) represents the complete stress picture, even if it is somewhat out of equilibrium. Then, we have

$$U = \frac{1}{2E}\int s_t^2 dvol = \frac{1}{2E}\int s_t^2 r d\theta b dr = \frac{b\theta}{2E}(\frac{12M_o}{bh^3})^2 \int [z + A(1 - 12\frac{z^2}{h^2})]^2 r dr \tag{5.47}$$

However, in Figure 5.10, $r = R_o + z$, so we have

$$U = const\int_{-h/2}^{+h/2}\left[z + A(1 - 12\frac{z^2}{h^2})\right]^2 (R_o + z)dz \tag{5.48}$$

Besides, for the least work, we have

$$\frac{\partial U}{\partial A} = 0 = \int_{-h/2}^{+h/2} \frac{\partial}{\partial A}\left(s_t^2\right) dvol \tag{5.49}$$

$$\int_{-h/2}^{+h/2} \left[ z + A\left(1 - 12\frac{z^2}{h^2}\right) \right] \left(1 - 12\frac{z^2}{h^2}\right)(R_o + z)dz \tag{5.50}$$

Splitting this into terms containing $A$ and independent of $A$, we find

$$\frac{I_1}{I_2} = \frac{\int_{-h/2}^{+h/2} z(1-12\frac{z^2}{h^2})(R_o + z)dz}{\int_{-h/2}^{+h/2} (1-12\frac{z^2}{h^2})^2 (R_o + z)dz} = -A \tag{5.51}$$

These integrals are easily evaluated. We recognize that

$$\int_{-h/2}^{+h/2} z^{2n+1} dz = 0 \tag{5.52}$$

For when the power of z in the integrand is odd, the integrand curve is symmetrical about the center of the $z$ system, and the integral is zero. Therefore, we calculate only even powers of $z$ in the integrands. Worked out, we find for $I$ the numerator integral as below:

$$I_1 = \int z^2(1-12\frac{z^2}{h^2})dz = \frac{z^3}{3} - \frac{12}{5}\frac{z^5}{h^2}\Big|_{-h/2}^{+h/2} = h^3(\frac{1}{12} - \frac{3}{20}) = -\frac{h^3}{15} \tag{5.53}$$

And the denominator integral is

$$I_2 = \int (1-12\frac{z^2}{h^2})^2 R_o dz = R_o \left[ z - 8\frac{z^3}{h^2} + \frac{144}{5}\frac{z^5}{h^4} \right]_{-h/2}^{+h/2} = R_o h(1 - 2 + \frac{9}{5}) = \frac{4}{5} R_o h \tag{5.54}$$

Then, we have

$$A = -\frac{I_1}{I_2} = \frac{h^2}{12R_o} \tag{5.55}$$

So, by calculating Equation (5.55), the stress distribution is

$$s_t = \frac{12M_o}{bh^3}(z + \frac{h^2}{12R_o} - \frac{z^2}{R_o}) \tag{5.56}$$

## 5.7 Internal Pressure in Thick-Walled Tubes

In this part, we consider the stress distribution in a thick-walled ($r_o$-$r_i$) tube under internal pressure ($p_i$) as illustrated in Figure 5.11. To have a free parameter, we must start with three and assume that for the tangential stress, we have

$$s_t = C_1 + C_2 r + C_3 r^2 \tag{5.57}$$

The tangential stress in a thick cylinder must keep equilibrium with the prescribed internal pressure $p_i$ as

$$p_i r_i = \int_{r_i}^{r_o} s_t dr \tag{5.58}$$

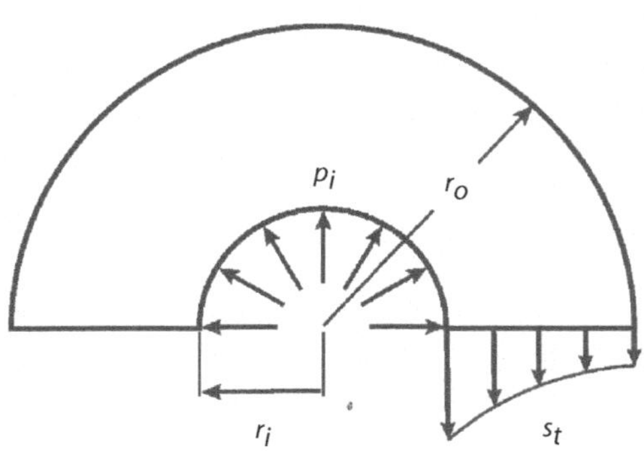


**Figure 5.11** A thick-walled tube under internal pressure.

Then, satisfying the two equilibrium conditions, we can derive expressions for the stresses $s_t$ and $s_r$ in terms of a single parameter, and we can proceed as in the previous example. The algebra involved in this process is very large in amount, although not difficult, and the process leads to a very good answer. We shall not apply it now, but shall proceed in a simpler but much cruder manner. We say that of the two stresses, $s_t$ and $s_r$, the tangential one, $s_t$, is more important, especially for reasonably thin-walled tubes. Now, we take $s_r = 0$ for simplicity and work with $s_t$ only. Then, we need one parameterless, and we write

$$s_t = C_1 + C_2 r \tag{5.59}$$

Figure 5.12 demonstrates a thick-walled tube with an assumed linear tangential stress distribution. The radial stress is arbitrarily assumed to be zero. Now, by applying the equilibrium criterion of Equation (5.59), we have

$$p_i r_i = \int_{r_i}^{r_o} (C_1 + C_2 r)dr = C_1(r_o - r_i) + \frac{C_2}{2}(r_0^2 - r_i^2) \tag{5.60}$$

We can determine $C_1$ by

$$C_1 = \frac{p_i r_i}{r_o - r_i} - \frac{C_2}{2}(r_o + r_i) \tag{5.61}$$

And the stress distribution can be computed by

$$s_t = \frac{p_i r_i}{r_o - r_i} + C_2\left(r - \frac{r_o + r_i}{2}\right), s_r = 0 \tag{5.62}$$

Now, this contains an arbitrary parameter $C_2$; it does satisfy the equilibrium condition of Figure 5.12. Therefore, the theorem of least work should not be applied, and any answer we may get will be out of equilibrium radially. But since $s_t$ is the more important stress, we proceed anyhow. The energy $U$ is as below:

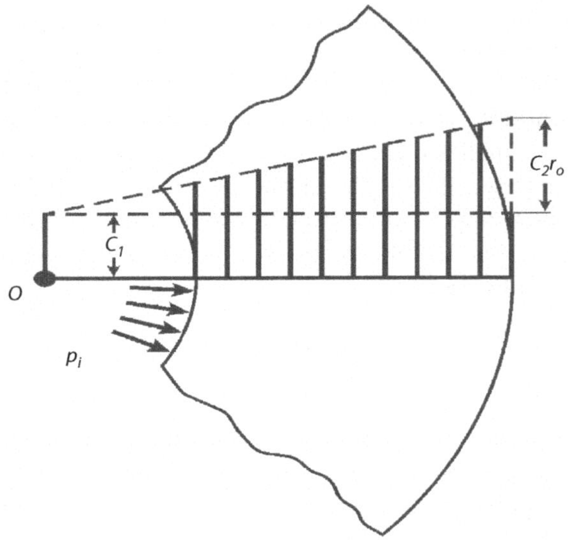


**Figure 5.12** A thick-walled tube with tangential stress distribution.

$$U = \frac{1}{2E}\int s_t^2 dvol = \frac{1}{2E}\int s_t^2 r d\theta dr dl = \frac{2\pi l}{2E}\int_{r_i}^{r_o} s_t^2 r dr \tag{5.63}$$

A slight simplification can be obtained by differentiating before integrating

$$0 = \frac{\partial U}{\partial C_2} = \int_{r_i}^{r_o} s_t \frac{\partial s_t}{\partial C_2} r dr \tag{5.64}$$

$$0 = \int_{r_i}^{r_o}\left[\frac{p_i r_i}{r_o - r_i} + C_2\left(r - \frac{r_o + r_i}{2}\right)\right]\left(r - \frac{r_o + r_i}{2}\right) r dr \tag{5.65}$$

Splitting this up into terms with and without $C_2$, we have

$$C_2 = -\frac{\frac{p_i r_i}{r_o - r_i}\int \left(r - \frac{r_o + r_i}{2}\right) r dr}{\int \left(r - \frac{r_o + r_i}{2}\right)^2 r dr} \tag{5.66}$$

$$C_2 = -\frac{p_i r_i}{r_o - r_i} \frac{\frac{r_o^3 - r_i^3}{3} - \frac{(r_o + r_i)(r_o^2 - r_i^2)}{4}}{\frac{r_o^4 - r_i^4}{4} - \frac{(r_o + r_i)(r_o^3 - r_i^3)}{3} + \frac{(r_o + r_i)(r_o^2 - r_i^2)}{8}} \tag{5.67}$$

### 5.7.1 Elastic Analysis of Thick-Walled Cylinders

When the ratio of the outer radius (b) of the cylinder to the inner radius (a) is b/a > 1.2, it is called a thick-walled cylinder. Its geometric shape is symmetrical to the central axis and has no change along the axial direction of the cylinder. If the load of the cylinder the distribution is also symmetrical about the central axis and the same along the axial direction, then it is a plane axis symmetry problem. In this type of problem, the stress, strain, and displacement components are independent of the tangential coordinate $\theta$, but are only functions of the radial coordinate. In engineering, the cylinder, collar, die for cold extrusion, underground or underwater culverts, and artillery barrels of cylindrical pressure vessels can all be simplified into thick-walled cylinders.

In a thick-walled sphere, if the load distribution is symmetrical to the center point of the sphere, it is a spherical symmetry problem, and all components are only functions of $r$. High-pressure spherical tanks in the chemical industry are typically thick-walled spheres.

When a thick-walled cylinder or thick-walled sphere is in an elastic state, the stress method or the displacement method can be used to solve it. The solution should satisfy the basic equations of the elastic boundary value problem and the corresponding boundary conditions. When part of the component region enters the plastic state, all components should satisfy the basic equations of the elastic-plastic boundary value problem, that is, the equilibrium equation, geometric relations, and plastic constitutive relations should be satisfied in the plastic zone, while the elastic zone is solved as the corresponding elastic problem, and the corresponding boundary conditions must be met on the boundaries of the given force and given displacement. On the elastic-plastic interface, the continuity conditions of normal displacement and stress should be met. For some components of ideal elastic-plastic materials, the stress field can be determined by the equilibrium equation, yield condition, and stress boundary condition, without involving the material's constitutive equation. Under the conditions of simple loading, full theoretical solutions can be used. Even for the structure of reinforced materials, only relatively simple mathematical methods are used to determine the stress field and displacement field. When the yield condition is linear, the solution is also very simple. Since elastic analysis is easy to solve, using the obtained results for design can make the structure have a sufficient safety margin, so it has been widely used. Designing and checking structures based on elastic-plastic analysis can better utilize the potential of material plastic properties, achieve the purpose of saving materials, or improve load-bearing capacity. Accurately determine the safety margin of the structure. Elastic analysis is the basis of elastic-plastic analysis, and when the results of elastic-plastic analysis are taken into account, the allowable value of stress can be appropriately increased in the design based on elastic analysis. With the development of science, technology, and production, the impact of plastic analysis methods and their results on structural design and verification will also have received increasing attention from people. The discussion of thick-walled spheres will begin with elastic analysis, then elastoplastic analysis, and finally plastic limit analysis. When performing elastoplastic analysis, unless otherwise stated, it is assumed that the material is ideal elastoplastic, and the material in the plastic zone is volume incompressible. In order to apply the analysis results to engineering practice, the application of combined thick-walled cylinders and self-tightening technology in thick-walled cylinders was also discussed.

### 5.7.2 Plain Axial Symmetry Problem

In the analysis, the polar coordinates $(r, \theta)$ are used to indicate the weight. Due to axial symmetry, $\tau_{r\theta} = 0$, radial stress, and tangential stress are only functions of r, independent of $\theta$, that is, $\sigma_r(r)$, $\sigma_\theta(r)$. Similarly, strain components $\varepsilon_r(r)$, $\varepsilon_\theta(r)$. Due to axial symmetry, it produces uniform expansion or contraction along the radius direction, that is, only the radial displacement $u(r)$, while the axial displacement is only related to $z$, that is, $\omega(z)$.

The unknown amount in the plane axial symmetry problem is $\sigma_r$, $\sigma_\theta$, $\varepsilon_r$, $\varepsilon_\theta$, and $u$. They should meet the basic equation and the corresponding boundary conditions, of which the equilibrium equation (not counting physical strength) is

$$\frac{d\sigma_r}{dr} + \frac{\sigma_r - \sigma_\theta}{r} = 0 \tag{5.68}$$

Geometric equation of force:

$$\varepsilon_r = \frac{du}{dr}, \varepsilon_\theta = \frac{u}{r} \tag{5.69}$$

The constitutive equation:

$$\begin{aligned} \varepsilon_r &= \frac{1}{E}\left(\sigma_r - \nu\sigma_\theta\right) \\ \varepsilon_\theta &= \frac{1}{E}\left(\sigma_\theta - \nu\sigma_r\right) \end{aligned} \tag{5.70}$$

or write as

$$\begin{aligned} \sigma_r &= \frac{E}{1-\nu^2}\left(\varepsilon_r - \nu\varepsilon_\theta\right) \\ \sigma_\theta &= \frac{E}{1-\nu^2}\left(\varepsilon_\theta - \nu\varepsilon_r\right) \end{aligned} \tag{5.71}$$

Boundary conditions:

$$\sigma_r|s_\sigma = F_r \tag{5.72}$$

$$u|s_u = \bar{u} \tag{5.73}$$

If the elastic constants $E$ and $\nu$ in the plane stress constitutive equations (5.51) and (5.52) are changed to

$$E^* = \frac{E}{1-\nu^2}, \nu^* = \frac{\nu}{1-\nu} \tag{5.74}$$

then it becomes the instructive process of the plane strain. If you solve it according to the intrinsic square of the plane strain, and transform the sum of $E$ in the inner component of the surface as above, the corresponding value in the in-plane component can be obtained in the plane strain state. In the plane force ($\sigma_z = 0, \varepsilon_z \neq 0$) and the plane strain change ($\varepsilon_z = 0, \sigma_z \neq 0$), non-zeroes of $\sigma_z$ and $\varepsilon_z$ can be expressed by the in-plane component of the phase, and they are not independent.

## 5.8 Identification Problems

There are numerous practical applications of the indentation problem, and four notable examples are outlined below:

(i) The act of indenting a surface may be viewed as a fundamental forming operation. Several real-world forming processes, although more intricate, are closely related to the indentation in terms of their mechanical behavior.
(ii) The problem holds broad relevance in the context of foundation design within the field of civil engineering.
(iii) It serves as the underlying principle for hardness testing, wherein the yield stress of a material can be determined by creating and measuring a microscopic indentation on its surface.
(iv) The analysis is also pertinent to investigations into friction and wear phenomena on a microscopic scale.

If we had the freedom to select a single indenter shape for our study, a circular indenter in plain view would likely be the preferred choice. This form provides a general resemblance to most practical cases mentioned above, while also offering a high degree of symmetry. Somewhat unexpectedly, however, this kind of axial symmetry does not yield a particularly straightforward mathematical treatment. Conversely, the mathematical analysis becomes significantly more manageable when the deformation occurs under plane strain conditions, that is, when the plastic flow is constrained between smooth, rigid, parallel plates, effectively preventing deformation in one specific direction. Due to these mathematical advantages, most research on indentation problems has been conducted within the framework of plane strain, and we will adopt the same approach in this chapter. Physically, this setup is analogous to indentation by an infinitely long strip-shaped punch similar in form to a long, blunt-edged knife. Although our focus on this somewhat atypical scenario is primarily driven by mathematical considerations, it is reasonable to assume that the indentation pressures observed in plane strain cases will be generally comparable to those found in corresponding axisymmetric problems.

Plane plastic flow has become one of the most extensively developed areas within plasticity theory, largely due to Hill's influential work in formulating and applying the theory of the "slip-line field." The elegance of this theory is largely attributable to the simplifying constraints imposed by plane strain conditions. While we will provide a brief overview of slip-line theory in the following chapter, our immediate attention will be directed toward applying the lower- and upper-bound theorems: the lower-bound theorem is suitable for the conservative design of foundations and similar applications, whereas the upper-bound theorem is more applicable to forming processes.

### 5.8.1 Upper-Bound Approach

We begin by attempting an upper-bound solution to the plane-strain indentation problem, depicted schematically in Figure 5.13. The illustration shows a cross-sectional view of a long, rigid indenter A being forced into a large block of perfectly plastic material by a load FFF per unit length. The initial step in our analysis involves proposing a geometrically consistent mode of deformation for the plastic material. Given our assumption of plane strain, which eliminates any out-of-plane motion, the deformation pattern must lie entirely within the plane. Clearly, if the indenter is to move

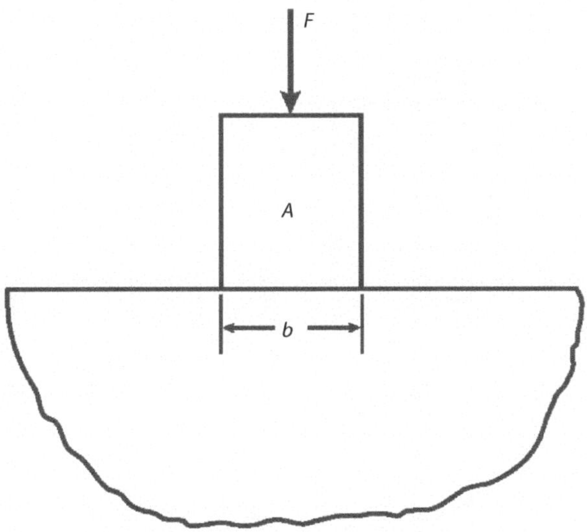


**Figure 5.13** Fundamental setup of the indentation problem.

downward and the material is incompressible, the material must experience significant lateral and upward displacement.

The simplest admissible deformation mode is shown in Figure 5.14. This configuration is geometrically feasible if no external constraints restrict the vertical motion of the indenter. In this case, the block of material *B* rotates as a rigid body with angular velocity $\theta$, while a semi-circular zone of intense shear separates it from the rest of the material. Although the mode is mechanically straightforward, one may question whether it is valid to assume a mechanism involving a discontinuity in tangential velocity across an interface.

To address this, we consider the unit mechanism illustrated in Figure 5.14. Here, a rigid upper block moves to the right with velocity $v$ relative to a rigid lower block. These are separated by a thin zone of plastic deformation with a unit area and thickness $h$, where the shear strain rate is uniform. We assume the material has unit thickness in the out-of-plane direction, which is perpendicular to the diagram. We now calculate the rate of energy dissipation $D$ within this plastic zone. Since the deformation is one of pure shear, the shear strain rate $\gamma$ is $v/h$. Therefore, the energy dissipation rate per unit volume is $kv/h$, where $k$ is the yield stress in pure shear. Given that the zone's volume is numerically equal to $h$, we have

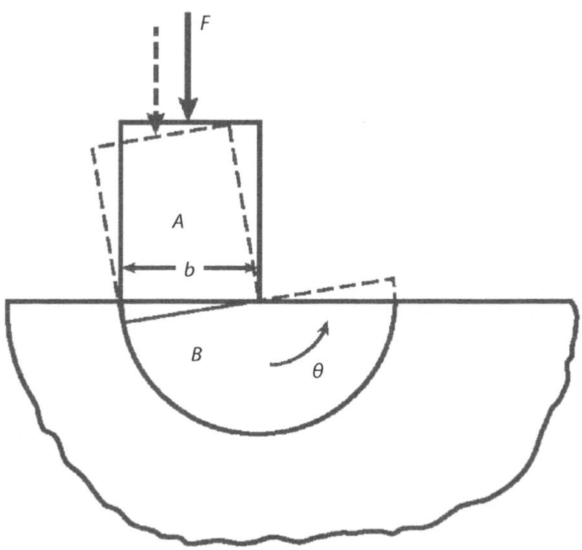


**Figure 5.14** A basic assumed mode of deformation.

$$D = kv \tag{5.75}$$

This equation shows that the energy dissipation rate per unit area of the interface is the product of the yield stress in shear and the relative velocity between the two rigid blocks. Importantly, this result is independent of the thickness $h$, which means $h$ can be made arbitrarily small (even zero). While an infinitesimal $h$ leads to an infinite shear rate $\gamma$, this is permissible for perfectly plastic, non-hardening materials, which are assumed under the bound theorem approach. The conceptual benefit of taking $h = 0$ is that it allows the modeling of simple mechanisms in which rigid blocks slide over each other. Whether such velocity discontinuities exist in real physical systems or even in exact analytical solutions is not our concern here. We are merely using a postulated mechanism that is geometrically valid, albeit potentially unrealistic, to obtain an upper bound on the indentation load. One might also object to the mode in Figure 5.14 on the grounds that it introduces surface discontinuities in the plastic material. However, the upper-bound load derived is independent of the angular velocity $\theta$, so we may treat $\theta$ as sufficiently small to avoid distorting the overall geometry. Nevertheless, assuming a flat surface confines our analysis to the initial indentation phase. If we wanted to study the continued advance of the indenter into a region with considerable lateral material pile-up, surface profile changes would need to be accounted for.

The upper-bound work calculation yields

$$F^u b\theta / 2 = kb\theta\pi b \tag{5.76}$$

where $b$ is the width of the indenter as shown in Figure 5.13. The superscript notation indicates that this is an upper bound. Solving for Equation (5.57), we obtain

$$F^u = 2\pi kb = 6.29kb \tag{5.77}$$

This provides a straightforward upper bound for the indentation force per unit length of the indenter. The mechanism in Figure 5.14 can easily be generalized by allowing variation in the radius and center of the circular shear zone. By exploring multiple configurations, we might obtain a smaller and therefore better upper bound.

Before proceeding to modes that do not involve rotation of the indenter, it is worth noting that slip-circle failure mechanisms, similar to the one

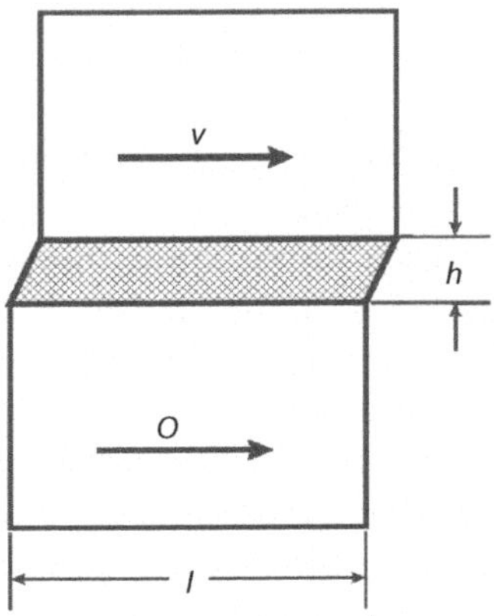


**Figure 5.15** Localized zone of intense shear.

in Figure 5.15, where entire structures such as buildings, docks, embankments, or hillsides rotate and settle, have been observed in soft clays worldwide. When applying upper-bound methods to such problems, the self-weight of the soil must often be considered. We typically neglect the material's self-weight compared to the indentation forces, which is a reasonable simplification for most mechanical engineering applications.

The next deformation mode we consider is shown in Figure 5.16. Like the previous case, all internal energy dissipation occurs along interfaces between rigid blocks, but now, there are multiple blocks, and all interfaces

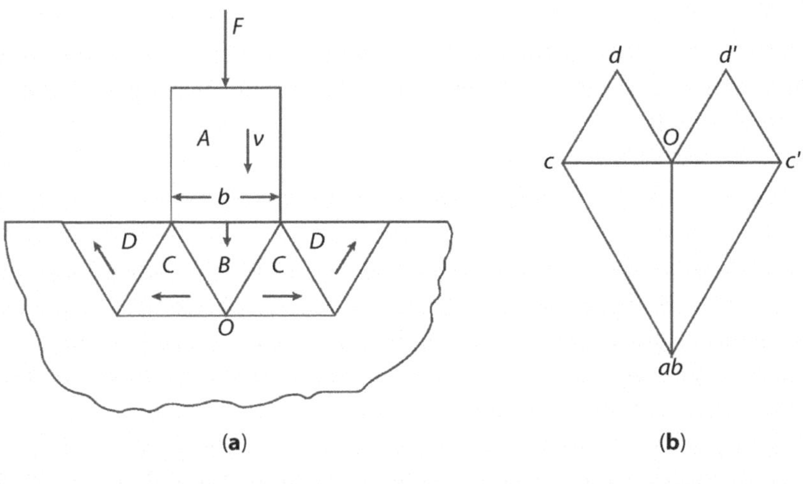


**Figure 5.16** Basic rigid block deformation mode with corresponding velocity diagram.

are planar. In Figure 5.16(a), the blocks are drawn as equilateral for simplicity in calculation, although various geometries could be used. To evaluate energy dissipation, we need the relative sliding velocities between adjacent blocks. This is most directly done using a velocity diagram (also called a hodograph), as shown in Figure 5.16(b). Each block in Figure 5.16(a) is labeled with a capital letter, and the stationary zone is marked as *O*. In Figure 5.16(b), each block's velocity is represented as a point, labeled with the corresponding lowercase letter. Since the blocks do not rotate, their velocities are single vectors. In the velocity diagram, the line between two points represents the relative velocity between the corresponding blocks.

To construct this diagram, we first draw vector oa to represent the indenter's velocity, using an arbitrary scale and direction appropriate to Figure 5.16(a). While the indenter's motion need not be vertical, choosing a vertical orientation introduces convenient symmetry. Relative to block *A*, block *B* can only move horizontally; otherwise, a gap would form. For simplicity, we assume block *B* moves vertically as an extension of the indenter, making points *a* and *b* coincide in the velocity diagram. Point *c* is then fixed by noting that the velocity directions of block *C* relative to blocks *B* and *O* align with their interface directions. This allows ccc to be located between *b* and *o*. Similarly, *d* is constructed from *c* and *o* and *c′* and *d′* follow by symmetry or similar construction.

Letting $l_{BC}$ denote the length of the interface between blocks *B* and *C*, and similarly for others, the work equation for unit thickness becomes

$$F^u oa = 2k\left\{bc.l_{BC} + co.l_{co} + cd.l_{cd} + do.l_{DO}\right\} \tag{5.78}$$

Since all terms on the right-hand side are positive, there is no ambiguity with respect to sign. The overall size of the velocity diagram and the indentation velocity magnitude do not influence the result.

In this example, the trigonometry of the velocity diagram is particularly simple. By inspection,

$$oc\,/\,oa = 1\sqrt{3}, etc. \tag{5.79}$$

$$l_{BC} = l_{CO} = \ldots = b \tag{5.80}$$

$$F^u = 2kb\left(\frac{2}{\sqrt{3}} + \frac{1}{\sqrt{3}} + \frac{1}{\sqrt{3}} + \frac{1}{\sqrt{3}}\right) = \left(\frac{10}{\sqrt{3}}\right)kb = 5.76kb \quad (5.81)$$

This result provides an upper bound about 8% lower than the one found using the simpler slip circle mechanism.

A potential concern with the Figure 5.16(a) mechanism is that block $B$ appears unable to descend, as it is already in contact with region $O$. This issue is resolved if the interfaces are assigned a small but finite thickness. Regardless, the magnitude of the incremental deformation does not affect the energy calculation.

If similar calculations are repeated (preferably with the aid of a drawing board) for other mechanisms in the same family, but with different interface inclinations, it is theoretically possible to find a smaller $F^u$. One form of partial minimization can be done analytically by maintaining congruent triangles and varying their depth. Although this only reduces $F^u$ by about 2%, the triangle depth in the optimal case is roughly 20% smaller than in the equilateral configuration. In other words, the minimum is relatively flat, and various geometries in this family are likely to yield similar upper-bound estimates. Ultimately, the smallest value of $F^u$ across all admissible mechanisms would equal the exact value of $F$, and the corresponding stress distribution would satisfy equilibrium. However, it is clear that equilibrium cannot be maintained at the junction where three slip planes intersect in Figure 5.16(a), since this would require the maximum shear stress to occur along planes that are not mutually orthogonal, something the Mohr circle construction deems impossible. Therefore, no matter how refined the geometry, the exact solution cannot be obtained from this mechanism family. For engineering applications, though, this limitation is generally acceptable since exact precision is not always required.

### 5.8.2 Lower-Bound Approach

We now proceed to develop a lower-bound solution for the same indentation problem by constructing simple stress fields that satisfy equilibrium and remain within the yield condition. Our approach involves defining regions or blocks within the material where the stress is assumed to be constant. These regions are, by default, in internal equilibrium, assuming the material's self-weight is negligible. Additionally, we take advantage of the possibility that the stress may discontinuously change across boundaries between adjacent blocks. To begin, we examine whether the stress states on either side of a planar interface can differ while still maintaining overall equilibrium. This scenario is depicted in Figure 5.17(a), which illustrates a boundary plane along with the stress states on both sides. These stress components are expressed in terms of a local coordinate system, with axes $t$ (tangential) and $n$ (normal) relative to the interface. Under plane strain conditions, the direction perpendicular to the diagram (normal to the plane of the paper) serves as a principal stress direction. Therefore, equilibrium

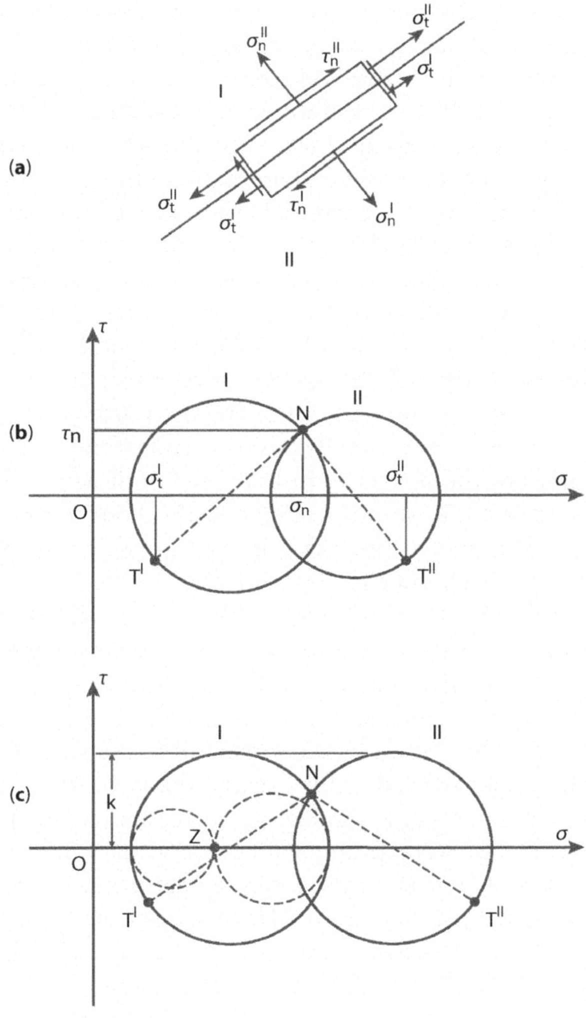


**Figure 5.17** Stress discontinuity across an interface.

analysis can be limited to the in-plane components of stress. We resolve the forces in the $n$ and $t$ directions for a small, representative element straddling the interface to assess compatibility across the boundary:

$$\sigma_n^I = \sigma_n^{II} \tag{5.82}$$

$$\tau_n^I = \tau_n^{II} \tag{5.83}$$

In this context, the superscripts indicate the respective stress zones. By taking moments about any point across the interface, we arrive at the so-called "complementary shear" relationship. It is important to observe that the $\sigma_t$ component of stress cannot be incorporated into these equilibrium equations in any meaningful way. Therefore, we must conclude that from the standpoint of equilibrium alone, a discontinuity in the tangential normal stress component $\sigma_t$ is permissible across the boundary. However, the normal component $\sigma_n$ and the shear component $\tau_n$ must remain continuous across the interface, as dictated by Equations (5.63) and (5.64). This behavior is clearly illustrated in the Mohr circle diagram for stress shown in Figure 5.17(b). Although the values of $\sigma_n$ and $\tau_n$ match on both sides of the interface, the full stress states in regions I and II, represented by distinct Mohr circles, can still differ. For the purpose of applying the lower-bound theorem, it is generally advantageous if the material is yielding on both sides of the interface. In that case, both Mohr circles would have the same radius $k$, as shown in Figure 5.17(c). From basic geometric construction, it then follows that the principal stress directions in the two regions are mirror images of each other relative to the boundary plane. It is crucial to emphasize that this interpretation applies strictly to the plane strain condition under discussion. Since deformation in the z-direction (perpendicular to the diagram) is not allowed, the stress $\sigma_z$ must always act as the intermediate principal stress according to Tresca's criterion. As a result, the Mohr circles shown here correspond to the largest of the three principal Mohr circles needed to fully define the stress state. The smaller, dotted circles shown for region I in Figure 5.17(c) are included to reinforce this point. If instead we were dealing with plane stress, where $\sigma_z = 0$, then $\sigma_t$ would not necessarily represent the intermediate principal stress. Finally, it is worth noting that the diagram in Figure 5.17(c) and, by extension, the analysis that follows throughout this chapter, remain valid for the Mises yield criterion as well, albeit with minor modifications.

## Bibliography

1. Calladine, C.R., *Plasticity for Engineers: Theory and Applications*, Woodhead, Cambridge, 2011.
2. Johnson, W. and Mellor, P.B., *Engineering Plasticity*, Ellis Horwood, Hertfordshire, 1983.
3. Martin, J.B., *Plasticity: Fundamentals and general results*, MIT Press, Massachusetts, 1975.
4. Prager, W. and Hodge, P.G., *Theory of Perfectly Plastic Solids*, John Wiley, New York, 1951.
5. Timoshenko, S., *Strength of Materials, Volume II: Advanced Theory and Problems*, CBS Publishers & Distributors, New Delhi, 2002.
6. Timoshenko, S. (Ed.), *Strength of Materials, Volume I: Elementary Theory and Problems*, CBS Publishers & Distributors, New Delhi, 2004.

# 6
# Plasticity in Finite Element Analysis

***Abstract***
This chapter focuses on the integration of plasticity theory into finite element analysis (FEA) for structural and mechanical simulations. It begins with a review of the fundamental principles of FEA and proceeds to describe how elastic-plastic material behavior is incorporated into numerical models. Yield criteria such as Tresca and von Mises are discussed alongside associated flow rules and hardening laws, including isotropic and kinematic hardening. The chapter explains the computational strategies required for implementing plasticity within FEA, including the treatment of incremental loading, stress return mapping algorithms, and the construction of consistent tangent moduli. Special attention is given to convergence and stability issues in plastic analyses. Through practical examples, the chapter demonstrates how plasticity models enhance the accuracy of stress-strain predictions in components subjected to complex loading paths. Overall, this chapter provides a comprehensive framework for engineers and researchers to perform realistic and reliable plastic deformation simulations using finite element techniques.

***Keywords*:** Finite element method, yield criteria, flow rule, hardening, elastic-plastic analysis, numerical implementation, material modeling

## 6.1 Introduction

The constitutive equation is a relation between stresses and strains. For instance, elasticity, plasticity, viscoelasticity, viscoelasticity, and creep are examples of constitutive equations. On the other hand, the constitutive equation explains the response of the material, whereas the stress is a function of the deformation. In order to differentiate between steel and rubber or polymer, we need to use a constitutive equation. The constitutive equation is also referred to as the stress–strain law for the material. Truesdell and Noll [1] studied the thermodynamic foundations of constitutive equations at finite strains. Lately, Marsden and Hughes [2] presented the mathematical foundations of elasticity, and computational aspects of plasticity

Yong Bai. Plasticity and Ductile Fracture in Structural Engineering, (133–172) © 2026 Scrivener Publishing LLC

are presented by Simo and Hughes [3]. The plastic constitutive relations have a nonlinear nature; therefore, determining the analytic solutions of boundary-value problems is not simple. The number of available exact solutions of elastic–plastic boundary-value problems is small. For the case

**Table 6.1** Overview of material models.

| **Material model** | **Characteristics** | **Examples** |
|---|---|---|
| Elastic | Stress is only a function of strain. | Steels, cast iron, glass, rock, wood, and so on, before yielding or fracture |
| Thermoelastic | Involves one variable that is temperature more than the elasticity | The loads involve variations in temperature |
| Hypoelastic | Stress increments are derived from strain increments. | Concrete models |
| Hyperelastic | Stress is derived from the function of strain energy. | Rubbers |
| Viscoelastic | Exhibit a rate and time-dependent behavior. | Metals in high temperature |
| Elastic–plastic | Elastic behavior is linear until yield, by using the yield condition and also flow and hardening rules, stress, and plastic strain increments are derived; plastic strain increments are prompt. | Metals, Soils, and Rocks when subjected to high stresses |
| Hypoelastic–plastic | Elastic strains are small compared to the plastic strains. | Rock models |
| Hyperelastic–plastic | Elastic strains are large compared to the plastic strains. | Rubbers in high temperature |
| Viscoplastic | Inelastic strains are time-dependent. | Polymers |
| Plastic | There is no hardening material. | Alloy steel |

of 3D procedures where either redundant strain is significant or geometric complexity is the typical feature of the problem, future prospects for precise solutions are far less bright. Due to the fast development of supercomputers and modern numerical systems, there is a powerful finite element method that can carry out the incremental inelastic analysis of virtually any structural problem. The field of plasticity has significantly benefited from this development. Hence, the classical theory of plasticity with the newer concepts and wider applications is provided.

The use of the material models requires knowing how the complete nonlinear analysis is achieved incrementally. The comprehensive process is given by Bathe [4] and is shown in Table 6.1. Based upon this table, the material relationships are used at two points of the solution process: one is the evaluation of the stresses, and the other is the evaluation of the tangent stress–strain matrices. The stresses are used in the computation of the nonlinear strain stiffness matrices and the nodal point force vectors, and the tangent stress–strain matrices are used in the computation of the linear strain stiffness matrices. The stresses must be evaluated with high precision; otherwise, the solution result is not correct, and also the stiffness matrices must be truly tangent matrices; otherwise, more repetitions to convergence are needed than are essential. The most common constitutive models which have been used in solid mechanics applications are characterized in this chapter. In fact, the objective of this chapter is to assess the applications of the constitutive relations in the plasticity of structures by the nonlinear finite-element method. In the following, the constitutive equations for different types of materials are described. The special emphasis of this study is placed on rubbers and polymer constitutive equations.

## 6.2 Elastic Materials

All solid materials possess a domain in the stress space within which a load variation results only in a variety of elastic strains. The hypothesis of elastic isotropy has been amply verified for metallic materials and a little less so for concrete. For composites and wood, however, it is necessary to use special anisotropic theories of elasticity. There are different constitutive equations for multiaxial elasticity. Besides, the same constitutive equation can be written in several different ways. We have to consider the difference between different material models and the same material model with different mathematical expressions. It is always possible to transform from one form of the constitutive equation to the other. The constitutive models for large-strain elasticity are explained for the special case of a Kirchhoff material, which is a straightforward generalization of linear elasticity to large deformations.

### 6.2.1 Kirchhoff Material

In engineering applications that involve small strains and large rotations, the effects of large deformation mainly depend on rotations, i.e., in the bending of an offshore riser. Generalizing the linear elastic relations is the simplest explanation of elastic material for large deformation analysis. The extension of the linear elastic laws can model the response of the material as below:

$$S_{ij} = C_{ijkl}.E_{kl} \tag{6.1}$$

where the S and E are the components of the second Piola-Kirchhoff stress and Green-Lagrange strain tensors, and the C is the constant elasticity fourth-order tensor. Meanwhile, the stress–strain law, which is often referred to as Hooke's law for linear elastic materials, is:

$$\sigma_x = E\varepsilon_x \tag{6.2}$$

Likewise, the strain energy density (w) that is a quadratic function of strains can be written as

$$dw(\varepsilon_x) = \sigma_x d\varepsilon_x \tag{6.3}$$

$$w = \int_0^{\varepsilon_x} \sigma_x d\varepsilon_x \tag{6.4}$$

$$w = \frac{1}{2} E\varepsilon_x^2 \tag{6.5}$$

Indeed, the generalization of Equation (6.22) is applicable for multiaxial states of stress and strain. It could include an entirely anisotropic material response. In the following, the properties of C are described. These are mostly the symmetry properties that enable the number of material constants to be reduced to a controllable number. Several of these properties apply to material tangent moduli, as well. The material coefficients in Equation (6.1) have the minor symmetries from the symmetries of the stress and strain tensors, which means

$$C_{ijkl} = C_{jikl} = C_{ijlk} \tag{6.6}$$

The Saint-Venant-Kirchhoff material is path-independent and possesses an elastic strain energy potential. The strain energy per unit volume was calculated by Equation (6.4). This equation was in one dimension, so by generalization, it is applicable for multiaxial states, which can be computed as

$$w = \int S_{ij} dE_{ij} = \int C_{ijkl} E_{kl} dE_{ij} = \frac{1}{2} C_{ijkl} E_{ij} E_{kl} \tag{6.7}$$

After that, the stress is given by

$$S_{ij} = \frac{\partial w}{\partial \mathrm{E}_{ij}} \tag{6.8}$$

It is assumed that the strain energy is positive-definite, so this is

$$w = \frac{1}{2} C_{ijkl} E_{ij} E_{kl} \geq 0 \tag{6.9}$$

Once E = 0, C is a positive-definite fourth-order tensor. From the existence of a strain energy potential, Equation (6.7), it is derived that

$$C_{ijkl} = \frac{\partial^2 w}{\partial E_{ij} \partial E_{kl}} \tag{6.10}$$

Then, the smoothness of the potential indicates that

$$\frac{\partial^2 w}{\partial E_{ij} \partial E_{kl}} = \frac{\partial^2 w}{\partial E_{kl} \partial E_{ij}} \tag{6.11}$$

Consequently, $w$ and $C$ have major symmetry if the material possesses a smooth potential. It means

$$C_{ijkl} = C_{klij} \tag{6.12}$$

The major symmetry of the material response matrix is an essential condition for the symmetry of the tangent stiffness matrix. The tangent stiffness is not symmetric if the tangent moduli do not display major symmetry. Note that for symmetry of the tangent stiffness, major symmetry of the tangent moduli is not an adequate condition. For some objective rates, the tangent stiffness is not symmetric even when the tangent moduli have major symmetry. We will inspect the major symmetry of the tangent moduli for the material laws that follow as they are advanced.

### 6.2.2 Continuum Mechanics

We consider the motion of a general body in a stationary Cartesian coordinate system, as shown in Figure 6.1, and assume that the body can experience large displacements, large strains, and nonlinear constitutive responses.

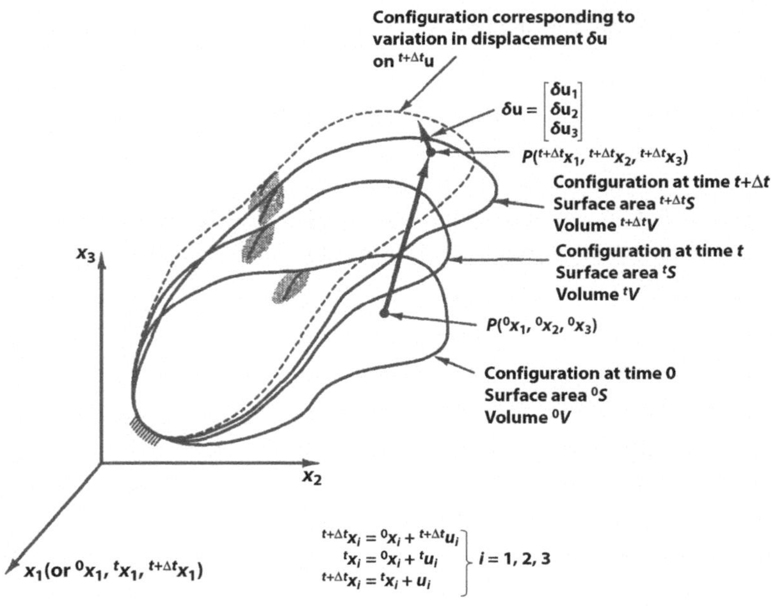


**Figure 6.1** The motion of the body in the Cartesian coordinate frame [5].

The aim is to evaluate the equilibrium positions of the complete body at the discrete-time points *0*, $\Delta t$, $2\Delta t$, $3\Delta t$, ..., where $\Delta t$ is an increment in time. To develop the solution strategy, assume that the solutions for the static and kinematic variables for all time steps from time 0 to time t, inclusive, have been obtained. Then, the solution process for the next required equilibrium position corresponding to time t + Δt is typical and is applied repetitively until the complete solution path has been solved. Hence, in the analysis, we follow all particles of the body in their motion, from the original to the final configuration of the body, which means that we adopt a Lagrangian or material formulation of the problem. This approach stands in contrast to a Eulerian formulation, which is usually used in the analysis of fluid mechanics problems, in which attention is focused on the motion of the material through a stationary control volume. Considering the analysis of solids and structures, a Lagrangian formulation usually represents a more natural and effective analysis approach than a Eulerian formulation. For example, using a Eulerian formulation of a structural problem with large displacements, new control volumes have to be created (because the boundaries of the solid change continuously), and the nonlinearities in the convective acceleration terms are difficult to deal with.

### 6.2.3 Lagrangian Formulations

In the Lagrangian incremental analysis approach, we express the equilibrium of the body at time $t +$ at using the principle of virtual displacements. Using tensor notation, this principle requires that

$$\int_{^{t+\Delta t}V} {}^{t+\Delta t}\tau_{ij}\delta_{t+\Delta t}e_{ij}d^{t+\Delta t}V = {}^{t+\Delta t}\Re \tag{6.13}$$

where

${}^{t+\Delta t}\tau_{ij}$ = Cartesian components of the *Cauchy stress* tensor (forces per unit area in the deformed geometry)

$\delta_{t+\Delta t}e_{ij} = \frac{1}{2}(\frac{\partial \delta u_i}{\partial^{t+\Delta t}x_j} + \frac{\partial \delta u_i}{\partial^{t+\Delta t}x_i})$ = strain tensor corresponding to virtual displacements

$\delta u_i$ = components of virtual displacement vector imposed on the configuration at time $t + \Delta t$, a function of ${}^{t+\Delta t}x_j, j=1,2,3$

${}^{t+\Delta t}x_i$ = Cartesian coordinates of the material point at time $t + \Delta t$

${}^{t+\Delta t}V$ = volume at time $t + \Delta t$

and

$$ {}^{t+\Delta t}\Re = \int_{{}^{t+\Delta t}V} {}^{t+\Delta t}f_i^B \delta u_i d^{t+\Delta t}V + \int_{{}^{t+\Delta t}S_f} {}^{t+\Delta t}f_i^S \delta u_i^S d^{t+\Delta t}S \quad (6.14) $$

where

${}^{t+\Delta t}f_i^B$ = components of externally applied forces per unit volume at time $t + \Delta t$

${}^{t+\Delta t}f_i^S$ = components of externally applied surface tractions per unit surface area at time $t + \Delta t$

${}^{t+\Delta t}S_f$ = surface at time $t + \Delta t$ on which external tractions are applied

$\delta u_i^S = \delta u_i$ evaluated on the surface ${}^{t+\Delta t}S_f$ (the $\delta u_i$ components are zero at and corresponding to the prescribed displacements on the surface ${}^{t+\Delta t}S_u$)

For an effective incremental analysis, appropriate stress and strain measures need to be employed. We consider in this section only the continuum mechanics equations without reference to a particular finite element solution scheme.

The equilibrium and compatibility requirements of the general body are considered in the configuration corresponding to time $t + \Delta t$. Since, in general, the body can undergo large displacements and large strains, and the constitutive relations are nonlinear, an approximate solution can be obtained by referring to all variables. This solution can then be improved by iteration.

In order to develop a governing linearized equation, the solutions for times $0, \Delta t, 2\Delta t, \ldots, t$ can be calculated by

$$ \int_{{}^tV} {}^t\tau_{kl}\delta_t e_{kl} d^tV = \int_{{}^tV} \left( \frac{{}^tp}{{}^0p} {}_0^tS_{ij}\, {}_0^tx_{k,i}\, {}_0^tx_{l,j} \right) \left( {}_t^0x_{m,k}\, {}_t^0x_{n,l} \delta_0^t\varepsilon_{mn} \right) d^tV = $$

$$ \int_{{}^tV} \frac{{}^tp}{{}^0p} {}_0^tS_{ij}\delta_{mi}\delta_{nj}\delta_0^t\varepsilon_{mn} d^tV = \int_{{}^0V} {}_0^tS_{ij}\delta_0^t\varepsilon_{ij} d^0V $$

$$ (6.15) $$

Then, by applying the second Piola-Kirchhoff stresses ${}_\tau^tS_{ij}$ and Green-Lagrange strains ${}_\tau^t\varepsilon_{ij}$, we have

$$ \int_{{}^tV} {}^t\tau_{mn}\delta_t e_{mn} d^tV = \int_{{}^\tau V} {}_\tau^tS_{ij}\delta_\tau^t\varepsilon_{ij} d^tV \quad (6.16) $$

Therefore, refer to the stresses and strains to one of these known equilibrium configurations, the choice lies essentially between two formulations, which have been termed total Lagrangian (TL) and updated Lagrangian (UL) formulations. The TL formulation has also been referred to as the Lagrangian formulation. In this solution scheme, all static and kinematic variables are referred to as the initial configuration at time 0. The UL formulation is based on the same procedures that are used in the TL formulation, but in the solution, all static and kinematic variables are referred to as the last calculated configuration. Both the TL and UL formulations include all kinematic nonlinear effects due to large displacements, large rotations, and large strains. But whether the large strain behavior is modeled appropriately depends on the constitutive relations specified. The only advantage of using one formulation rather than the other lies in its greater numerical efficiency.

Using Equation (6.17) in the TL formulation, we consider the basic equation

$$\int_{{}^{0}V} {}^{t+\Delta t}_{0}S_{ij}\,\delta_{0}^{t+\Delta t}\varepsilon_{ij}\,d^{0}V = {}^{t+\Delta t}\Re \tag{6.17}$$

whereas in the UL formulation, we consider

$$\int_{{}^{t}V} {}^{t+\Delta t}_{t}S_{ij}\,\delta_{t}^{t+\Delta t}\varepsilon_{ij}\,d^{t}V = {}^{t+\Delta t}\Re \tag{6.18}$$

in which ${}^{t+\Delta t}\Re$ is the external virtual work and can be calculated by Equation (6.14). This expression also depends, in general, on the surface area and the volume of the body under consideration. However, for simplicity of discussion, we assume for the moment that the loading is deformation independent, a very important form of such loading being concentrated forces whose directions and intensities are independent of the structural response.

The linearized equilibrium equations in the TL formulation are

$$\int_{{}^{0}V} {}_{0}C_{ijrs}\,{}_{0}e_{rs}\,\delta_{0}e_{ij}\,d^{0}V + \int_{{}^{0}V} {}^{t}_{0}S_{ij}\,\delta_{0}\eta_{ij}\,d^{0}V = {}^{t+\Delta t}\Re - \int_{{}^{0}V} {}^{t}_{0}S_{ij}\,\delta_{0}e_{ij}\,d^{0}V \tag{6.19}$$

The linearized equilibrium equations in the UL formulation are

$$\int_{{}^tV} {}_tC_{ijrs}\, {}_te_{rs}\delta_t e_{ij} d^tV + \int_{{}^tV} {}^t\tau_{ij}\delta_t\eta_{ij} d^tV = {}^{t+\Delta t}\Re - \int_{{}^tV} {}^t\tau_{ij}\delta_t e_{ij} d^tV \tag{6.20}$$

where ${}_0C_{ijrs}$ and ${}_tC_{ijrs}$ are the incremental stress–strain tensors at time $t$ referred to the configurations at times 0 and $t$, respectively. It should be noted that in Equations (6.19) and (6.20) ${}^t_0S_{ij}$ and ${}^t\tau_{ij}$ are the known second Piola-Kirchhoff and Cauchy stresses at time $t$, and ${}_0e_{ij}$, ${}_0\eta_{ij}$ and ${}_te_{ij}$, ${}_t\eta_{ij}$ are the linear and nonlinear incremental strains which are referred to the configurations at times 0 and $t$, respectively.

### 6.2.4 Hook's Law

The fundamental observation that "the second Piola-Kirchhoff stress and Green-Lagrange strain components do not change when measured in a fixed coordinate system when the material is subjected to rigid body motions" is important not only for elastic analysis. Indeed, this observation implies that any material description which has been developed for infinitesimal displacement analysis using engineering stress and strain measures can be directly employed in large displacement and large rotation but small strain analysis, provided second Piola-Kirchhoff stresses and Green-Lagrange strains are used. Figure 6.2 illustrates this fundamental fact.

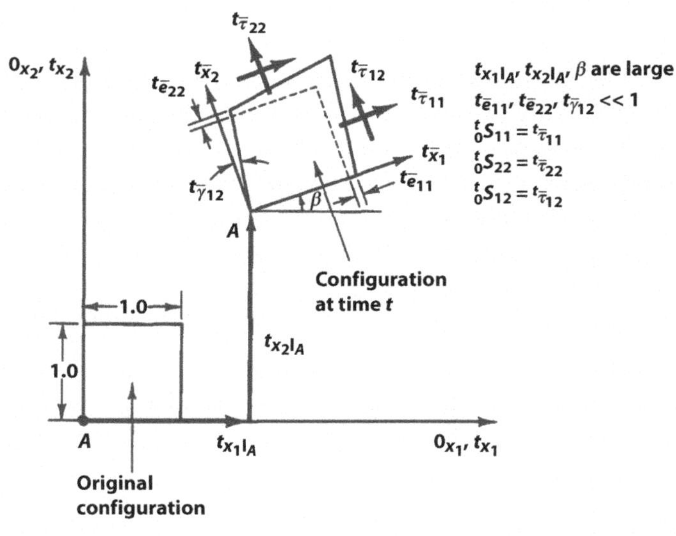


**Figure 6.2** The large displacement of large rotation but small strain conditions [5].

The preceding observations are of special importance because, in practice, Hooke's law applies only to small strains and because there are many engineering problems in which large displacements, large rotations, but only small strain conditions, are encountered.

The stress–strain description given in Equation (6.1) implicitly assumes that a formulation is used to analyze the physical problem. Let us now assume that we want to employ a formulation but that we are given the constitutive relationship in Equation (6.7). If we define a new constitutive tensor,

$$ {}_{t}^{t}C_{mnpq} = \frac{{}^{t}p}{{}_{0}p}\, {}_{0}^{t}x_{m,i}\, {}_{0}^{t}x_{n,j}\, {}_{0}^{t}C_{ijrs}\, {}_{0}^{t}x_{p,r}\, {}_{0}^{t}x_{q,s} \tag{6.21} $$

This equation can change to

$$ {}_{0}^{t}C_{ijrs} = \frac{{}^{0}p}{{}_{t}p}\, {}_{t}^{0}x_{i,m}\, {}_{t}^{0}x_{j,n}\, {}_{t}^{t}C_{mnpq}\, {}_{t}^{0}x_{r,p}\, {}_{t}^{0}x_{s,q} \tag{6.22} $$

Then, by using

$$ \delta_{t}e_{mn} = {}_{t}^{0}x_{i,m}\, {}_{t}^{0}x_{j,n} \delta_{0}^{t}\varepsilon_{ij} \tag{6.23} $$

Equation (6.7) can be written as

$$ \int_{{}^{t}V} {}_{t}^{t}C_{mnpq}\, {}_{t}^{t}\varepsilon_{pq}^{A} \delta_{t}e_{mn} d^{t}V = {}^{t}\Re \tag{6.24} $$

and

$$ {}^{t}\tau_{mn} = {}_{t}^{t}C_{mnpq}\, {}_{t}^{t}\varepsilon_{pq}^{A} \tag{6.25} $$

where ${}_{t}^{t}\varepsilon_{pq}^{A}$ are the components of the Almansi strain tensor:

$$ {}_{t}^{t}\varepsilon_{pq}^{A} = {}_{t}^{0}x_{r,p}\, {}_{t}^{0}x_{s,q}\, {}_{0}^{t}\varepsilon_{rs} \tag{6.26} $$

Like the Green-Lagrange strain tensor, the Almansi strain tensor can also be defined in several different but completely equivalent ways.

However, in contrast, the Green-Lagrange strain tensor, the components of the Almansi strain tensor, are not invariant under a rigid body rotation of the material. Therefore, we have

$$ {}_t^t\varepsilon_{ij}^A = \frac{1}{2}\left({}_t^t u_{i,j} + {}_t^t u_{j,i} - {}_t^t u_{k,i}\,{}_t^t u_{k,j}\right) \tag{6.27} $$

Then

$$ {}_t^t\varepsilon_{ij}^A d^t x_i d^t x_j = \frac{1}{2}\{(d^t s)^2 - d^0 s)^2\} \tag{6.28} $$

Applying Equation (6.24) with the Almansi strain and the constitutive tensor ${}_t^t C_{mnpq}$ is quite equivalent to transforming the second Piola-Kirchhoff stress ${}_0^t S_{ij}$ (obtained using ${}_0^t S_{ij} = {}_0^t C_{ijrs}\,{}_0^t \varepsilon_{rs}$) to the Cauchy stress and then evaluating ${}^t\mathfrak{R}$. Indeed, ${}_0^t C_{ijrs}$ is known; this procedure is computationally more efficient, and the definition and use of the Almansi strain with Equation (6.24) may be regarded as only of theoretical interest.

By considering TL and UL Equations (6.19) and (6.20), the corresponding integral terms in the formulations are identical, provided the transformations for the stresses and the constitutive tensors given in Equation (6.21) are used. Hence, whether we choose the TL or the UL continuum formulation is decided merely by considerations of numerical efficiency.

## 6.3 Thermo-Elastic Materials

The temperature induces the additional phenomenon of dilatation. The intrinsic dissipation is always zero, but thermal dissipation does take place. In the rest, concerning aspects of the constitutive law itself and the determination of its characteristic coefficients as functions of temperature are discussed.

Assume $T_0$ is a reference temperature of a volume element of an elastic material, and $T$ is the temperature at a subsequent instant so that the temperature difference, $\theta = T\text{-}T_0$, is small compared to $T_0$. Under such conditions, the elastic properties and the density $\rho$ can be considered to be constant.

The law of classical linear thermoelasticity is obtained by again selecting as the thermodynamic potential a positive definite quadratic form which, in addition to strain components, has temperature as one of the variables.

Assuming isotropy of responses in both the deformation and dilatation phenomena and taking cognizance of the classically used law, we are led to write

$$\Psi = \frac{1}{\rho}\left[\frac{1}{2}\left(\lambda\varepsilon_I^2 + 4\mu\varepsilon_{II}\right) - \left(3\lambda + 2\mu\right)\alpha\theta\varepsilon_1\right] - \frac{C_\varepsilon}{2T_o}\theta^2 \tag{6.29}$$

where $\alpha$ and $C_\varepsilon$ are two coefficients with a meaning given by the state laws derived from $\Psi$:

$$\sigma = \rho\left(\partial\Psi / \partial\varepsilon\right) = \lambda Tr(\varepsilon)1 + 2\mu\varepsilon - (3\lambda + 2\mu)\alpha\theta 1 \tag{6.30}$$

$$\sigma_{ij} = \lambda\varepsilon_{kk}\delta_{ij} + 2\mu\varepsilon_{ij} - (3\lambda + 2\mu)\alpha\theta\delta_{ij} \tag{6.31}$$

$$s = -\frac{\partial\Psi}{\partial T} = -\frac{\partial\Psi}{\partial\theta} = \frac{1}{\rho}(3\lambda + 2\mu)\alpha Tr(\varepsilon) + \frac{C_\varepsilon}{T_o}\theta \tag{6.32}$$

By using Equation (6.32), we have

$$\left(\partial s / \partial T\right)_{\varepsilon=cnt} = \left(\partial s / \partial\theta\right)_{\varepsilon=cnt} = C_\varepsilon / T_o \tag{6.33}$$

which shows that $C_\varepsilon = T_0(\partial s/\partial T)$ is the specific heat at constant strain. By inverting the first law of the state or by employing the dual potential $\Psi^*(\sigma,\theta)$, we may express strains as functions of stresses and temperature:

$$\varepsilon = \frac{1+\nu}{E}\sigma - \frac{\nu}{E}Tr(\sigma)1 + \alpha\theta 1 \tag{6.34}$$

$$\varepsilon_{ij} = \frac{1+\nu}{E}\sigma_{ij} - \frac{\nu}{E}\sigma_{kk}\delta_{ij} + \alpha\theta\delta_{ij} \tag{6.35}$$

which constitutes the law of linear isotropic thermoelasticity. By considering a variation of temperature with zero stress, we obtain

$$\varepsilon_{ij} = \alpha\theta\delta_{ij} \tag{6.36}$$

which defines the isotropic dilatation coefficient $\alpha$.

The coefficients $\lambda$, $\mu$, $\alpha$, and $C_\varepsilon$ are taken to be constant for small temperature variations ($\theta/T_0 \ll 1$). As illustrated by Figure 6.3, they vary with the temperature but "weakly."

The thermoelastic constitutive equation enables us to calculate the strains and stresses due to the thermal effect as soon as the temperature field is known. For the treatment of a complete thermoelastic problem in which the flux (and not the temperature itself) is given, it is necessary to use Fourier's law and the heat equation.

The thermoelastic constitutive equation is written under the assumption of small temperature variations. If such is not the case, it is used step

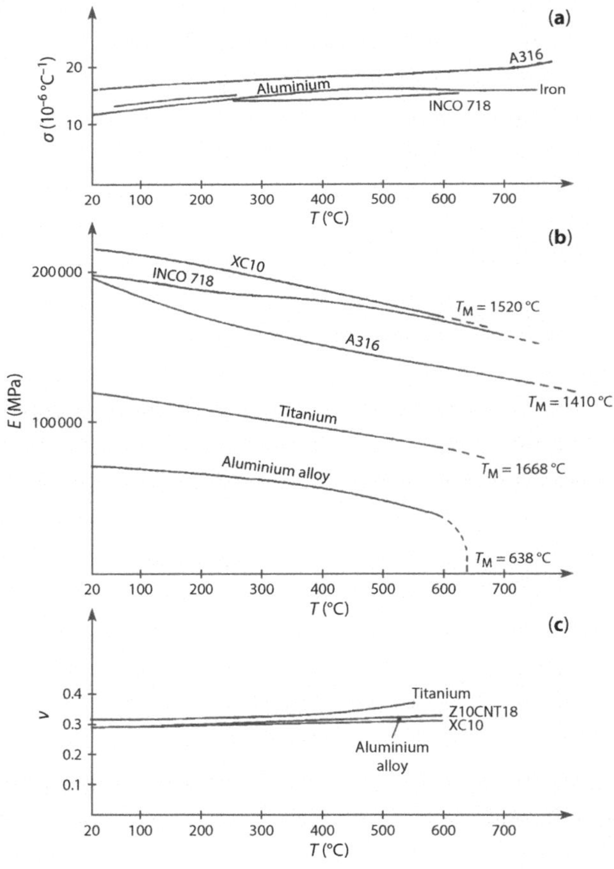


**Figure 6.3** Evolution of (*a*) the dilatation coefficient α, (*b*) Young's modulus *E*, and (*c*) Poisson's ratio *ν*, as functions of the temperature *T* [5].

by step with the variable coefficients $E$, v, and a; this implies that their evolution as a function of temperature is known.

We are therefore confronted with a major experimental difficulty: the almost impossible task of measuring, with sufficient precision, the local deformation as soon as the temperature exceeds the limit for using electric resistance strain gauges. This limit varies from 200 to 300°C. On the other hand, it is easy to obtain the dilatation coefficient by measuring the elongation of long wires heated in ovens; some values are given in Figure 6.3.

For the elastic modulus, we may also use rather long wires with an extensometer attached to the part of the wire outside the oven. However, this method raises the problem of the reliability of the fixture used for applying the forces (slipping, the wrong definition of the useful length...).

## 6.4 Hypoelastic Materials

The hypoelastic material laws relate the rate of stress to the rate of deformation. The stress rate should be objective and should be related to an objective measure of the rate of deformation to satisfy the principle of material frame indifference. A general form of the hypoelastic relation is given by

$$\sigma^{\nabla} = f(\sigma, D) \tag{6.37}$$

In the above equation, all the parameters are objective. $\sigma^{\nabla}$ shows any rate of the Cauchy stress, and D represents the rate of deformation, and f is a function of the stress and rate of deformation. In the form of a linear relationship between the objective measure of stress and deformation rate, a large class of hypoelastic constitutive relations can be written. Therefore,

$$\sigma^{\nabla} = C : D \tag{6.38}$$

where C is the fourth-order tensor of elastic moduli, which is an objective function of the stress state. The above equation is rate-independent and incrementally linear, and reversible (Prager) [7]. This means that for small increases about a finitely deformed state, the increases in strain and stress are recovered upon unloading and are linearly related. Nevertheless, energy is not necessarily conserved for huge deformations, and the work done in a closed deformation path is not necessarily zero. Hypoelastic laws are used primarily for representing the elastic response in phenomenological elastic–plastic laws where the elastic deformations are small, and the dissipative effects are also small. Some commonly used forms of hypoelastic constitutive relations are

$$\sigma^{\nabla J} = C^{\sigma J} : D\sigma^{\nabla T} = C^{\sigma T} D\sigma^{\nabla G} = C^{\sigma G} : D \tag{6.39}$$

In the above relation, Jaumann's rate of Cauchy stress, the Truesdell rate, and the Green–Naghdi rate of Cauchy stress are shown by $\sigma^{\nabla J}$, $\sigma^{\nabla G}$, and $\sigma^{\nabla T}$, respectively (Belytschko) [8]. The superscripts on the moduli identify the objective rate of stress given by the moduli. These tangent moduli possess minor symmetries since the objective stress rate and the rate of deformation are symmetric. In hypoelastic models, the tangent moduli are generally assumed to also have major symmetries.

For an isotropic material, the tangent modulus tensor is a fourth-order isotropic tensor as

$$C_{ijkl} = \lambda\delta_{ij}\delta_{kl} + \mu(\delta_{ik}\delta_{il} + \delta_{il}\delta_{jk}), C = \lambda I \otimes I + 2\mu I \tag{6.40}$$

Therefore, for an isotropic material, the tangent moduli of the Jaumann rate could be derived easily. Besides, for a general anisotropic material, the Voigt form of the tangent moduli is

$$\begin{pmatrix} S_{11} \\ S_{22} \\ S_{33} \\ S_{23} \\ S_{13} \\ S_{12} \end{pmatrix} = \begin{pmatrix} C_{11} & C_{12} & C_{13} & C_{14} & C_{15} & C_{16} \\ C_{21} & C_{22} & C_{23} & C_{24} & C_{25} & C_{26} \\ C_{31} & C_{32} & C_{33} & C_{34} & C_{35} & C_{36} \\ C_{41} & C_{42} & C_{43} & C_{44} & C_{45} & C_{46} \\ C_{51} & C_{52} & C_{53} & C_{54} & C_{55} & C_{56} \\ C_{61} & C_{62} & C_{63} & C_{64} & C_{65} & C_{66} \end{pmatrix} \begin{pmatrix} E_{11} \\ E_{22} \\ E_{33} \\ 2E_{23} \\ 2E_{13} \\ 2E_{12} \end{pmatrix} \tag{6.41}$$

The Voigt form is obtained from the tensor components by mapping the first and second pairs of indices according to the Voigt rule. The [C] matrix above has 21 independent entries in three dimensions, and the major symmetry of the fourth-order tensor manifests itself as asymmetry of the Voigt matrix form. This law is the same for all the fully anisotropic Kirchhoff materials. Nye [9] proved that by consideration of material symmetry, the number of independent material constants can be reduced more. This theory has been applied to the Kirchhoff material. Isotropy is a significant sample of material symmetry. In fact, anisotropic material has no preferred orientations or directions, so that the stress–strain relation is identical when expressed in any rectangular Cartesian coordinate system. Many materials (such as metals and ceramics) can be modeled as isotropic for small strains. In addition, for an orthotropic material (e.g., wood or composites with aligned fibers), the symmetries yield a tangent modulus of the form given in Voigt notation as

$$\begin{pmatrix} S_{11} \\ S_{22} \\ S_{33} \\ S_{23} \\ S_{13} \\ S_{12} \end{pmatrix} = \begin{pmatrix} C_{11} & C_{12} & C_{13} & 0 & 0 & 0 \\ C_{21} & C_{22} & C_{23} & 0 & 0 & 0 \\ C_{31} & C_{32} & C_{33} & 0 & 0 & 0 \\ 0 & 0 & 0 & C_{44} & 0 & 0 \\ 0 & 0 & 0 & 0 & C_{55} & 0 \\ 0 & 0 & 0 & 0 & 0 & C_{66} \end{pmatrix} \begin{pmatrix} E_{11} \\ E_{22} \\ E_{33} \\ 2E_{23} \\ 2E_{13} \\ 2E_{12} \end{pmatrix} \tag{6.42}$$

The above Kirchhoff stress–strain relation shows that there are only nine independent elastic constants for an orthotropic material. In an isotropic Kirchhoff material, the tensor C is isotropic. An anisotropic tensor has the same components in any (rectangular Cartesian) coordinate system. It is important to observe that, for anisotropic materials, the tangent moduli depend on the deformation. Therefore, they have to be updated as the material deforms.

### 6.4.1 Cauchy Material

A Cauchy material is considered as a material that is independent of the history of the motion. The constitutive relation for a Cauchy elastic material is derived as

$$\sigma = G(F) \tag{6.43}$$

where G is the material response function, and the obvious dependence on position X and time t has been suppressed for notational convenience. The response function depends only on the current value of the deformation gradient and not on its history. For a Cauchy material, the restriction is given by

$$\sigma = R.G(U).R^T \tag{6.44}$$

It should be noted that a constitutive equation cannot be expressed in terms of F except the F dependence has a special form. The dependence can be on $F^T.F = 2E+I$ or $U = (F^T.F)^{1/2}$. Alternative forms of the same constitutive equation for other representations of stress and strain followed from the stress transformation relations can be the nominal stress for a Cauchy material, which is given by $P = JU^{-1}.G(U).\ R^T$. The relationship for the second Piola–Kirchhoff stress is $S = JU^{-1}.G(U)$. In $U^{-1} = h(U) = h(C)$, C is the right Cauchy–Green deformation tensor. The stress can be calculated for a Cauchy material according to Equation (6.35) and is independent of the deformation history. Nevertheless, the work done depends on the deformation history or load path. Therefore, the Cauchy material has some elastic properties but not all. For example, the stress is path-independent, but the energy may not be.

## 6.5 Hyperelastic Materials

Hyperelastic materials are elastic materials for which the work is independent of the load path. Hyperelastic materials are characterized by the existence of a strain energy function that is a potential for the stress as below:

$$S = 2\frac{\partial \psi(C)}{\partial C} = \frac{\partial w(E)}{\partial E} \tag{6.45}$$

The deposited energy potential is shown by $\psi$. When the potential is demonstrated as a function of the Green strain E, the notation w is the relation between the two scalar functions, which is given by $w(E) = \psi(2E+1)$. Rivlin and Saunders [16] developed a hyperelastic constitutive model for large deformations of rubber. The model is incompressible and initially isotropic. Hyperelastic materials provide a natural framework for the frame-invariant formulation of anisotropic material response by simply embodying the anisotropy in the potential. Expressions for different stress measures are obtained through the appropriate transformations. Hence, the Kirchhoff stress is given by

$$\tau = J\sigma = F.S.F^T \tag{6.46}$$

A consequence of the existence of a deposited energy function is that the effort done on a hyperelastic material is independent of the deformation path. This behavior is approximately observed in rubbers. In order to demonstrate the independence of work on the deformation path, consider the deposited energy per unit reference volume in going from deformation state $C_1$ to $C_2$. Since the second Piola–Kirchhoff stress tensor S and the Green strain E = (C – I)/2 are work conjugates,

$$\int_{E_1}^{E_2} S : dE = w(E_2) - w(E_1), or \frac{1}{2}\int_{C_1}^{C_2} S : dC = \psi(C_2) - \psi(C_1) \tag{6.47}$$

It should be mentioned that deposited energy in the material only depends on the initial and final states of deformation and is independent of the deformation or load path. In order to determine an expression for the nominal stress tensor P as a derivative of the potential, we recall that P is conjugate in power to $F^T$. Then, in terms of the potential, the nominal stress is

$$\frac{\partial w}{\partial F^T} = \frac{\partial \psi}{\partial C} : \frac{\partial C}{\partial F^T} = S.F^T = P, or P_{ij} = \frac{\partial w}{\partial F_{ji}} \tag{6.48}$$

The deformation gradient tensor F is not symmetric, so the nine components of the nominal stress tensor P are not symmetric, as well. Malvern [20] showed that the potential strain energy for a hyperelastic material which is isotropic concerning the initial, unstressed configuration can be written as a function of the principal invariants ($I_1$, $I_2$, $I_3$) of the right Cauchy–Green deformation tensor, that is, $\psi = \psi(I_1, I_2, I_3)$. The second Piola–Kirchhoff stress tensor for a hyperelastic material is derived. Accordingly, for an isotropic material, we have

$$S = 2\frac{\partial\psi}{\partial C} = 2(\frac{\partial\psi}{\partial I_1}\frac{\partial I_1}{\partial C} + \frac{\partial\psi}{\partial I_2}\frac{\partial I_2}{\partial C} + \frac{\partial\psi}{\partial I_3}\frac{\partial I_3}{\partial C}) = 2(\frac{\partial\psi}{\partial I_1} + I_1\frac{\partial\psi}{\partial I_2})I - 2\frac{\partial\psi}{\partial I_2}C + 2I_3\frac{\partial\psi}{\partial I_3}C^{-1} \tag{6.49}$$

Therefore, the Kirchhoff stress tensor is computed by

$$\tau = F.S.F^T = 2(\frac{\partial\psi}{\partial I_1} + I_1\frac{\partial\psi}{\partial I_2})B - 2\frac{\partial\psi}{\partial I_2}B^2 + 2I_3\frac{\partial\psi}{\partial I_3}I \tag{6.50}$$

The left Cauchy–Green deformation tensor is B = F.F$^T$. It should be noticed that while B is coaxial (has the same principle directions) with τ, C is coaxial with the S parameter in Equations (6.49) and (6.50).

### 6.5.1 Neo-Hookean Material

Hooke's law is an isotropic linear law, so by extending it to large deformations, the Neo-Hookean material model is derived. The stored energy function for a compressible Neo-Hookean material, which is isotropic due to the initial state and has an unstressed configuration, is deduced as

$$\psi(C) = \frac{1}{2}\lambda_0(\ln J)^2 - \mu_0 \ln J + \frac{1}{2}\mu_0(traceC - 3) \tag{6.51}$$

where $\lambda_0$ and $\mu_0$ are the Lame constants of the linearized theory and $J = det\ F$. Besides, the stresses are

$$S = \lambda_0 \ln JC^{-1} + \mu_0(I - C^{-1}) \tag{6.52}$$

$$\tau = \lambda_0 \ln JI + \mu_0(B - I) \tag{6.53}$$

By considering $\lambda = \lambda_0$ and $\mu = \mu_0 - \lambda \ln J$, the elasticity tensors (tangent moduli) in the component form are

$$C^{SE}_{ijkl} = \lambda C^{-1}_{ij}C^{-1}_{kl} + \mu(C^{-1}_{ik}C^{-1}_{jl} + C^{-1}_{il}C^{-1}_{kj}) \tag{6.54}$$

$$C^{\tau}_{ijkl} = \lambda\delta_{ij}\delta_{kl} + \mu(\delta_{ik}\delta_{jl} + \delta_{il}\delta_{kj}) \tag{6.55}$$

The tangent moduli in Equation (6.55) have the same form as Hooke's law for small-strain elasticity, except for the dependence of the shear modulus (μ) on the deformation. Nearly incompressible behavior is obtained for $\lambda_0 >> \mu_0$. In Voigt matrix notation, the spatial elastic moduli for a Neo-Hookean material for plane strain are given by:

$$\left[C^{\tau}_{ab}\right] = \begin{bmatrix} \lambda + 2\mu & \lambda & 0 \\ \lambda & \lambda + 2\mu & 0 \\ 0 & 0 & \mu \end{bmatrix} \tag{6.56}$$

### 6.5.2 Modified Mooney-Rivlin Material

The hyperelastic constitutive model for large deformations of rubber was developed by Rivlin and Saunders [16]. The model is incompressible and initially isotropic. The stored energy function of this model is in the form of $\Psi = \Psi(I_1, I_2, I_3)$. For an incompressible material $J = det\ F = 1$, and then $I_3 = det\ C = J^2 = 1$. The potential can be written as a series expansion in $I_1$ and $I_2$:

$$\psi = \psi(I_1, I_2) = \sum_{i=0}^{\infty}\sum_{j=0}^{\infty} \overline{c}_{ij}(I_1 - 3)^i (I_2 - 3)^j, \overline{c}_{00} = 0 \tag{6.57}$$

where $c_{ij}$ is constant. Mooney and Rivlin showed that the simple form is

$$\psi = \psi(I_1, I_2) = c_1(I_1 - 3) + c_2(I_2 - 3) \tag{6.58}$$

which closely matches the experimental results. The Mooney–Rivlin material is an example of a Neo-Hookean material. The components of the second Piola–Kirchhoff stress are obtained by differentiating Equation (6.58) concerning the right Cauchy–Green deformation tensor. However, since the material is incompressible, the deformation is constrained such that $I_3 = 1$, which is equivalent to $J = 1$. One way in which this constraint can be enforced is through the use of a constrained potential. Alternatively, a penalty function formulation can be used. The corresponding penalty function is expressed as $ln\ I_3 = 0$. In this case, the modified strain energy function and the constitutive equation are respectively as below:

$$\hat{\psi} = \psi + p_0 \ln I_3 + \frac{1}{2}\beta(\ln I_3)^2 \tag{6.59}$$

$$S = 2\frac{\partial \hat{\psi}}{\partial C} = 2\frac{\partial \psi}{\partial C} + 2(p_0 + \beta(\ln I_3))C^{-1} \tag{6.60}$$

The penalty parameter $b$ must be large enough so that the compressibility error is negligible ($I_3 = 1$), yet not so large that numerical ill-conditioning occurs. Numerical experiments reveal that $\beta = 10^3 * max(c_1, c_2)$ to $\beta = 10^7 * max(c_1, c_2)$ is adequate for a floating-point word length of 64 bits. The constant $p_0$ is chosen so that the components of S are all zero in the initial configuration, that is, $p_0 = -(c_1 + 2c_2)$.

## 6.6 Viscoelastic Materials

The materials that have a rate- and time-dependent behavior are known as viscoelastic, such as metals at high temperatures. The specified properties of viscoelastic material response are a fading memory. Figure 6.4 shows that the model comprises a linear spring in series with a dashpot. The behaviors of the Maxwell model are both fluid-like and solid-like. The spring models the elastic response and is assigned a stiffness $E$. The dashpot models the viscous response and is assigned a viscosity $\eta$. The total strain of the spring and the dashpot combined is taken to be the sum of the elastic and viscous strains:

$$\varepsilon = \varepsilon^e + \varepsilon^v \tag{6.61}$$

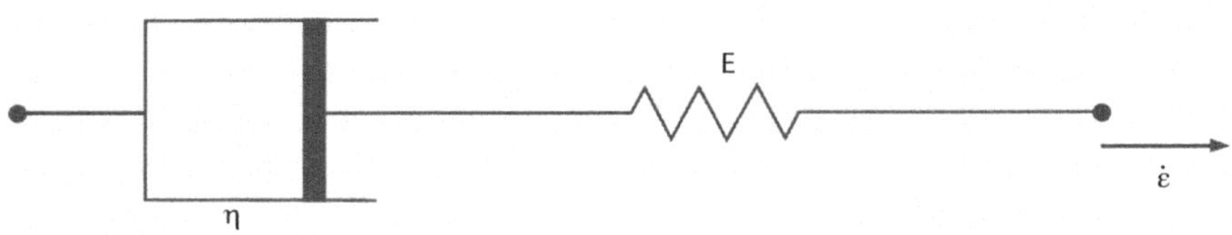


**Figure 6.4** A simple schematic of linear viscoelastic material behavior (Maxwell Model).

By considering the material time derivative, we have

$$\dot{\varepsilon} = \dot{\varepsilon}^e + \dot{\varepsilon}^v \tag{6.62}$$

The above equation can be changed to

$$\dot{\varepsilon} = \frac{\dot{\sigma}}{E} + \frac{\sigma}{\eta} \tag{6.63}$$

This is an ordinary differential equation for stress $\sigma$ with constant coefficients. A solution is deduced by the convolution integral as

$$\sigma(t) = \int_{-\infty}^{t} E \exp[(t' - t)/\tau] \frac{d\varepsilon(t')}{dt'} dt' \tag{6.64}$$

where $\tau$ is the relaxation time and is calculated by dividing $\eta$ by E. For more general one-dimensional models, the stress can be calculated as

$$\sigma(t) = \int_{-\infty}^{t} E \exp(-t/\tau) \frac{d\varepsilon(t')}{dt'} dt' \tag{6.65}$$

The above convolution integral is extended to the multiaxial case by

$$\sigma_{ij} = \int_{-\infty}^{t} \hat{C}_{ijkl}(t - t') \frac{d\varepsilon_{ij}(t')}{\partial t'} dt' \tag{6.66}$$

where $C_{ijkl}$ is the relaxation modulus. The relaxation moduli have the same minor symmetries as the linear elastic moduli. Major symmetry is assumed. For an isotropic material, Equation (6.50) can be written in terms of two relaxation functions relating the deviatoric and hydrostatic parts of stress and strain as below:

$$\sigma_{ij}^{dev} = 2\int_{-\infty}^{t} \hat{\mu}(t-t')\frac{\partial \varepsilon_{ij}^{dev}(t')}{\partial t'}dt' \tag{6.67}$$

$$\sigma_{kk} = 3\int_{-\infty}^{t} \hat{K}(t-t')\frac{\partial \varepsilon_{kk}(t')}{\partial t'}dt' \tag{6.68}$$

To fit polymer behavior, the relaxation moduli may be represented by a finite Dirichlet series of exponential relaxation functions (generalized Maxwell model) as

$$K(t) = \sum_{i=1}^{N_b} K^i \exp(-t/\tau_i^b) \tag{6.69}$$

$$\mu(t) = \sum_{i=1}^{N_s} \mu^i \exp(-t/\tau_i^s) \tag{6.70}$$

where $K^i$ and $m^i$ are the bulk and shear moduli of the $N_b$ and $N_s$ elements in the series, and $\tau^b_i$ and $\tau^s_i$ are the corresponding relaxation times. For a viscoelastic solid, one of the relaxation times will be infinite in each of the previous expressions (so that the functions $K(t)$ and $m(t)$ relax to constant values, called the long-term or rubbery moduli, and not to zero).

## 6.7 Elastic–Plastic Materials

The elastoplastic response is governed by the classical incremental theory of plasticity based on the Prandtl-Reuss equations. Using the additive decomposition of strains, $de = de^E + de^P$, the stress increment is given by the basic relationship.

$$d\sigma = C^E\left(de - de^P\right) \tag{6.71}$$

where the $C^E$ is the elastic constitutive tensor and the de, $de^E$, and $de^P$ are the components of the total strain increment, the elastic strain increment, and the plastic strain increment, respectively. This incremental relationship holds throughout the inelastic response. The elastic–plastic tangent stiffness matrix $[C^{ep}]$ is introduced to describe the constitutive response of the material during elastic–plastic deformation. The elements of this matrix are expressed in terms of the derivatives of the yield function and the plastic potential function. Based on these definitions, the incremental stress–strain relationship can be written as follows.

$$d\sigma_{ij} = C^{ep}_{ijkl} d\varepsilon_{kl} = (C_{ijkl} + C^{p}_{ijkl}) d\varepsilon_{kl} \tag{6.72}$$

where $C^{ep}_{ijkl}$, $C_{ijkl}$, and $C^{P}_{ijkl}$ are the elastic–plastic tangent stiffness tensor, the elastic tangent stiffness tensor, and the plastic tangent stiffness tensor, respectively. Besides, the incremental stress and strain tensors $d\sigma_{ij}$ and $d\varepsilon_{kl}$ are generally expressed in vector forms:

$$\{d\sigma\}^T = \{d\sigma_x d\sigma_y d\sigma_z d\tau_{yz} d\tau_{zx} d\tau_{xy}\} \tag{6.73}$$

$$\{d\varepsilon\}^T = \{d\varepsilon_x d\varepsilon_y d\varepsilon_z d\gamma_{yz} d\gamma_{zx} d\gamma_{xy}\} \tag{6.74}$$

The matrix form of the elastic tangent stiffness tensor $C_{ijkl}$ is expressed by

$$[C] = \begin{pmatrix} K+\frac{4}{3}G & K-\frac{2}{3}G & K-\frac{2}{3}G & 0 & 0 & 0 \\ K-\frac{2}{3}G & K+\frac{4}{3}G & K-\frac{2}{3}G & 0 & 0 & 0 \\ K-\frac{2}{3}G & K-\frac{2}{3}G & K+\frac{4}{3}G & 0 & 0 & 0 \\ 0 & 0 & 0 & G & 0 & 0 \\ 0 & 0 & 0 & 0 & G & 0 \\ 0 & 0 & 0 & 0 & 0 & G \end{pmatrix} \tag{6.75}$$

in which K and G are the bulk and shear moduli, respectively. In addition, the plastic tangent stiffness tensor $C^p_{ijkl}$ is deduced as

$$C^p_{ijkl} = -\frac{36G^2}{h}\bar{s}_{ij}\bar{s}_{kl} \tag{6.76}$$

The scalar function is shown by $h$ and is given by

$$h = 36G^2.H \tag{6.77}$$

$$h = 4(3G + H_p)\bar{\sigma}_e^2 \tag{6.78}$$

$$H_p = \frac{d\sigma_e}{d\varepsilon_p} \tag{6.79}$$

where $H_p$ is the definition of the plastic modulus, and it is associated with the rate of expansion of the yield or loading surface. It should be noted that $\bar{s}_{kl}$ is the stress value and reduced stress values are demonstrated by $\bar{\sigma}_e$ and $\bar{s}_{ij}$ parameters. These two parameters referred to the stress space with its origin at the current center of the kinematically translated yield surface. The reduced stress deviator tensor $\bar{s}_{ij}$ could be expressed in vector form as

$$[\bar{S}]^T = \{\bar{s}_x\,\bar{s}_y\,\bar{s}_z\,\bar{s}_{yz}\,\bar{s}_{zx}\,\bar{s}_{xy}\} \tag{6.80}$$

Consequently, the plastic stiffness tensor can be calculated as a 6 × 6 matrix:

$$[C^p] = -\frac{1}{H}\begin{pmatrix} \bar{s}_x^2 & \bar{s}_y\bar{s}_x & \bar{s}_z\bar{s}_x & \bar{s}_{yz}\bar{s}_x & \bar{s}_{zx}\bar{s}_x & \bar{s}_{xy}\bar{s}_x \\ \bar{s}_y\bar{s}_x & \bar{s}_y^2 & \bar{s}_z\bar{s}_y & \bar{s}_{yz}\bar{s}_y & \bar{s}_{zx}\bar{s}_y & \bar{s}_{xy}\bar{s}_y \\ \bar{s}_z\bar{s}_x & \bar{s}_z\bar{s}_y & \bar{s}_z^2 & \bar{s}_{yz}\bar{s}_z & \bar{s}_{zx}\bar{s}_z & \bar{s}_{xy}\bar{s}_z \\ \bar{s}_{yz}\bar{s}_x & \bar{s}_{yz}\bar{s}_y & \bar{s}_{yz}\bar{s}_z & \bar{s}_{yz}^2 & \bar{s}_{zx}\bar{s}_{yz} & \bar{s}_{xy}\bar{s}_{yz} \\ \bar{s}_{zx}\bar{s}_x & \bar{s}_{zx}\bar{s}_y & \bar{s}_{zx}\bar{s}_z & \bar{s}_{zx}\bar{s}_{yz} & \bar{s}_{zx}^2 & \bar{s}_{xy}\bar{s}_{zx} \\ \bar{s}_{xy}\bar{s}_x & \bar{s}_{xy}\bar{s}_y & \bar{s}_{xy}\bar{s}_z & \bar{s}_{xy}\bar{s}_{yz} & \bar{s}_{xy}\bar{s}_{zx} & \bar{s}_{xy}^2 \end{pmatrix} \tag{6.81}$$

In addition, under the condition of plane strain, $d\varepsilon_z = d\gamma_{zx} = d\gamma_{yz} = 0$. Equations (6.75) and (6.81) can be reduced, and the elastic–plastic stiffness matrix is expressed in the form

$$[C^{ep}] = \begin{pmatrix} K + \frac{4}{3}G - \frac{\overline{s}_x^2}{H} & K - \frac{2}{3}G - \frac{\overline{s}_y\overline{s}_x}{H} & -\frac{\overline{s}_y\overline{s}_x}{H} \\ K - \frac{2}{3}G - \frac{\overline{s}_y\overline{s}_x}{H} & K + \frac{4}{3}G - \frac{\overline{s}_y^2}{H} & -\frac{\overline{s}_{xy}\overline{s}_y}{H} \\ -\frac{\overline{s}_y\overline{s}_x}{H} & -\frac{\overline{s}_{xy}\overline{s}_y}{H} & G - \frac{\overline{s}_{xy}^2}{H} \end{pmatrix} \quad (6.82)$$

The nonzero stress components in the axisymmetric case are $\sigma_r$, $\sigma_z$, $\sigma_\theta$, and $\tau_{rz}$, and the corresponding strains are $\varepsilon_r$, $\varepsilon_z$, $\varepsilon_\theta$, and $\gamma_{rz}$. Equations (6.75) and (6.81) can be reduced, and the corresponding matrix $[C^{ep}]$ is given by

$$[C^{ep}] = \begin{pmatrix} K + \frac{4G}{3} - \frac{\overline{s}_r^2}{H} & K - \frac{2G}{3} - \frac{\overline{s}_z\overline{s}_r}{H} & K - \frac{2G}{3} - \frac{\overline{s}_\theta\overline{s}_r}{H} & -\frac{\overline{s}_{rz}\overline{s}_r}{H} \\ K - \frac{2G}{3} - \frac{\overline{s}_z\overline{s}_r}{H} & K + \frac{4G}{3} - \frac{\overline{s}_z^2}{H} & K - \frac{2G}{3} - \frac{\overline{s}_\theta\overline{s}_z}{H} & -\frac{\overline{s}_{rz}\overline{s}_z}{H} \\ K - \frac{2G}{3} - \frac{\overline{s}_\theta\overline{s}_r}{H} & K - \frac{2G}{3} - \frac{\overline{s}_\theta\overline{s}_z}{H} & K + \frac{4G}{3} - \frac{\overline{s}_\theta^2}{H} & -\frac{\overline{s}_{rz}\overline{s}_\theta}{H} \\ -\frac{\overline{s}_{rz}\overline{s}_r}{H} & -\frac{\overline{s}_{rz}\overline{s}_z}{H} & -\frac{\overline{s}_{rz}\overline{s}_\theta}{H} & G - \frac{\overline{s}_{rz}^2}{H} \end{pmatrix}$$

(6.83)

In the plane stress case, $d\sigma_z = d\tau_{yz} = d\tau_{xz} = 0$, but the strain component $d\varepsilon_z$ is not zero; only the shear strain components $d\gamma_{yz}$ and $d\gamma_{zx}$ are zero. $d\varepsilon_z$ must be solved from the condition $d\sigma_z = 0$ and substitute the dEz so obtained into the first and second equations of Equation (6.66). Then the elastic–plastic stiffness matrix for the plane stress case can be obtained. However, the expression of $[C^{ep}]$ so obtained is quite involved. An alternative form proposed by Yamada [19] is relatively simple. It has the form

$$[C^{ep}] = \begin{pmatrix} \dfrac{E}{1-\nu^2} - \dfrac{\bar{s}_1^2}{\bar{s}} & \dfrac{\nu E}{1-\nu^2} - \dfrac{\bar{s}_1 \bar{s}_2}{\bar{s}} & -\dfrac{\bar{s}_1 \bar{s}_6}{\bar{s}} \\ \dfrac{\nu E}{1-\nu^2} - \dfrac{\bar{s}_1 \bar{s}_2}{\bar{s}} & \dfrac{E}{1-\nu^2} - \dfrac{\bar{s}_2^2}{\bar{s}} & -\dfrac{\bar{s}_2 \bar{s}_6}{\bar{s}} \\ -\dfrac{\bar{s}_1 \bar{s}_6}{\bar{s}} & -\dfrac{\bar{s}_2 \bar{s}_6}{\bar{s}} & \dfrac{E}{1-\nu^2} - \dfrac{\bar{s}_6^2}{\bar{s}} \end{pmatrix} \tag{6.84}$$

which in the above equation we have

$$\bar{s}_1 = \frac{E}{(1-\nu^2)}(\bar{s}_x + \nu \bar{s}_y), \quad \bar{s}_2 = \frac{E}{(1-\nu^2)}(\nu \bar{s}_x + \bar{s}_y), \quad \bar{s}_6 = \frac{E}{(1+\nu)} \bar{s}_{xy} \tag{6.85}$$

$$\bar{s} = \frac{4}{9} \bar{\sigma}_e H_p + \bar{s}_2 \bar{s}_x + \bar{s}_2 \bar{s}_y + 2 \bar{s}_6 \bar{s}_{xy} \tag{6.86}$$

Equation (6.84) is derived directly from the component form of the flow rule. The elastic–plastic tangent stiffness matrix [$C^{ep}$] for an orthotropic material in the plane stress case takes a similar form.

## 6.8 Hypoelastic–Plastic Materials

Hypoelastic–plastic models are typically used when elastic strains are small compared to plastic strains. As discussed in previous sections, energy is not conserved in a closed deformation cycle for hypoelastic materials. However, for small elastic strains, the energy error is insignificant, and hypoelastic descriptions of the elastic response are often adequate. In these constitutive models, the additive decomposition of the rate-of-deformation tensor, D, into elastic and plastic parts is assumed:

$$D = D^e + D^p \tag{6.87}$$

The elastic response is hypoelastic: a suitable objective rate of stress is related to the elastic part of the rate-of-deformation tensor (in pure hypoelasticity, the stress rate is related to the total rate-of-deformation tensor). The choice of objective stress rate in the constitutive response depends on several factors. The Jaumann rate of the Cauchy (true) stress leads to nonsymmetric tangent stiffness matrices, while the Jaumann rate of the Kirchhoff stress $\tau = J\sigma$ can lead to symmetric stiffness matrices. We will identify the circumstances under which these choices are appropriate and what advantages are to be gained by the choice of one rate over another. The choice of objective stress rate should not be confused with the expression of a given constitutive relation in terms of different stress rates. The latter is accomplished by simply using the appropriate transformation between rates.

We first present a model based on the Cauchy stress with elastic response specified in terms of the Jaumann rate. The elastic response is specified by applying the hypoelastic law to the elastic part of the rate of deformation and is derived as

$$\sigma^{\nabla J} = C_{el}^{\sigma J} : D^e = C_{el}^{\sigma J} : (D - D^p) \tag{6.88}$$

If the elastic moduli, $C^{\sigma J}_{el}$, are taken to be constant, they must be isotropic to satisfy the principle of material frame indifference. The rate of plastic flow is given by

$$D^p = \dot{\lambda} r(\sigma, q) \tag{6.89}$$

$$D_{ij}^p = \dot{\lambda} r_{ij}(\sigma, q) \tag{6.90}$$

where λ is a scalar plastic flow rate and r(σ, q) is the plastic flow direction. The plastic flow direction is often specified as *r* where *y* is called the plastic flow potential. To avoid confusion of the plasticity parameter with the Lame constant, the Lame constant will subsequently be denoted as *le*. The plastic flow direction depends on the Cauchy stress, s, and on a set of internal variables denoted collectively as q. Examples of scalar internal variables are the accumulated effective plastic strain and the void volume fraction. The back stress in kinematic hardening models is an example of an internal variable which is a second-order tensor.

Evolution equations for the internal variables are required and, for most plasticity models, can be specified as below:

$$\dot{q} = \dot{\lambda} h(\sigma, q) \tag{6.91}$$

$$\dot{q}_a = \dot{\lambda} h_a(\alpha, q) \tag{6.92}$$

where $a$ ranges over the number of internal variables. Here, the internal variables are a collection of scalars and the material time derivative is an objective rate. Note that the plastic parameter $\lambda$ or some function of it may be one of the internal variables. The evolution equation for the plastic parameter is obtained through the consistency condition below. The yield condition is

$$f(\sigma, q) = 0 \tag{6.93}$$

As in the one-dimensional case, the loading-unloading conditions are given by

$$\begin{cases} \dot{\lambda} \geq 0 \\ f \leq 0 \\ \dot{\lambda} f = 0 \end{cases} \tag{6.94}$$

During plastic loading ($\lambda > 0$), the stress is required to remain on the yield surface $f = 0$. This can be also be stated in terms of the consistency condition $f = 0$ which can be expanded by the chain rule to give

$$\dot{f} = f_\sigma : \dot{\sigma} + f_q . \dot{q} = 0 \tag{6.95}$$

$$\dot{f} = (f_\sigma)_{ij} \dot{\sigma}_{ij} + (f_q)_a \dot{q}_a = 0 \tag{6.96}$$

where we have adopted the notation

$$\begin{cases} f_\sigma = \partial f / \partial \sigma \\ f_q = \partial f / \partial q \end{cases} \tag{6.97}$$

The consistency condition involves the normal, $f\sigma$, to the yield surface. If the plastic flow direction is proportional to the normal to the yield surface, plastic flow is said to be associative; otherwise, it is said to be nonassociative. In associative plasticity, the flow direction is determined by the derivative of the plastic flow potential. For many materials, an appropriate choice for the plastic potential is $\Psi = f$, which gives rise to an associative flow rule. Drucker has shown that when the yield surface is convex, associative plasticity models are stable for small strains if the strain hardening is positive. From Prager [7], it follows that if $f\sigma$ and $\sigma$ commute, that is

$$f_\sigma.\sigma = \sigma.f_\sigma \tag{6.98}$$

$$f_\sigma : \sigma = f_\sigma : \sigma^{\nabla J} \tag{6.99}$$

For the derivatives of the principal invariants of a second-order tensor, it can be seen that when $f$ is a function of the invariants of the stress, $f\sigma$ and $\sigma$ commute, and thus, Equations (6.98) and (6.99) hold. Besides, objectivity requires yield functions of Equation (6.93) to be isotropic functions of the stress and hence functions of the principal invariants of stress. By submitting Equation (6.93) into Equation (6.96), we have

$$\dot{f} = f_\sigma.\sigma^{\nabla J} + f_q.\dot{q} = 0 \tag{6.100}$$

Using the hypoelastic relation (6.80), the plastic flow relation (6.81), and the evolution Equations (6.81) and (6.82) in (6.92) gives

$$0 = f_\sigma : C_{el}^{\sigma J} : (D - D^p) + f_q.\dot{q} = f_\sigma : C_{el}^{\sigma J} : (D - \dot{\lambda} r) + f_q.\dot{\lambda} h \tag{6.101}$$

which can be solved for λ to give

$$\dot{\lambda} = \frac{f_\sigma : C_{el}^{\sigma J} : D}{-f_q.h + f_\sigma : C_{el}^{\sigma J} : r} \tag{6.102}$$

Equation (6.99) also holds for other spin-based rates of stress but does not hold for the Truesdell rate. When the elastic response is specified in terms of a spin-based rate, simplification of Equation (6.101) occurs. With the Truesdell rate, additional terms need to be accounted for.

Substituting Equation (6.102) together with the plastic flow rule Equation (6.98) into Equation (6.99), we obtain a relation between the Jaumann rate of Cauchy stress and the total rate-of-deformation tensor:

$$\sigma^{\nabla J} = C_{el}^{\sigma J} : (D - \dot{\lambda} r) = C_{el}^{\sigma J} : (D - \frac{f_\sigma : C_{el}^{\sigma J} : D}{-f_q . h + f_\sigma : C_{el}^{\sigma J} : r} r) = C^{\sigma J} : D \tag{6.103}$$

The fourth-order tensor $C^{\sigma J}$ is called the continuum elastic–plastic tangent modulus and is obtained by rearrangement of the expression in Equation (6.103):

$$C^{\sigma J} = C_{el}^{\sigma J} - \frac{(C_{el}^{\sigma J} : r) \otimes (f_\sigma : C_{el}^{\sigma J})}{-f_q . h + f_\sigma : C_{el}^{\sigma J} : r}, C_{ijkl}^{\sigma J} = (C_{el}^{\sigma J})_{ijkl} - \frac{(C_{el}^{\sigma J})_{ijmn} r_{mn} (f_\sigma)_{pq} (C_{el}^{\sigma J})_{pqkl}}{-(f_q)_\alpha h_\alpha + (f_\sigma)_{rs} (C_{el}^{\sigma J})_{rstu} r_{tu}} \tag{6.104}$$

The symbol ⊗ denotes the tensor or open product and is defined in the Glossary. The elastoplastic tangent modulus consists of the elastic tangent modulus and is a term due to plastic flow. When written in Voigt matrix form, the plastic flow contribution is a matrix of rank 1 and is often referred to as a rank 1 correction (to the elastic moduli). From the symmetry of the stress rate and the rate-of-deformation, the elastic–plastic tangent modulus $C^{\sigma J}$ has both minor symmetries.

## 6.9 Hyperelastic–Plastic Materials

There are several drawbacks to the hypoelastic–plastic constitutive relations described in Section 6.7:

1. The elastic response is hypoelastic, and therefore, the work done in a closed cycle of deformation is not exactly zero.
2. If the elastic moduli are assumed to be constant, frame invariance restricts the moduli to be isotropic.
3. The yield function is required to be an isotropic function of stress.
4. The hypoelastic relation must be integrated in time to compute the stress. Hypoelastic formulations require incrementally objective stress update schemes to ensure that finite rotations do not induce unacceptably large erroneous stresses.

For the corotational formulation, items 2 and 3 do not apply, that is, the elastic and plastic response is not restricted to be isotropic. Hyperelastic–plastic constitutive models were developed to eliminate the above drawbacks of the hypoelastic–plastic formulations [14, 15, 18]. Since the elastic response is derived from a hyperelastic potential, the work done in a closed elastic deformation path vanishes exactly. Furthermore, no stress rate equations are integrated to compute the stress, and the need for incrementally objective stress update algorithms is eliminated. In addition, the hyperelastic–plastic formulation provides a natural framework for frame-invariant formulations of anisotropic elasticity *and* anisotropic plastic yield.

For a thorough understanding of this material, the reader should be familiar with pull-back and push-forward operations and the Lie derivative. Two key concepts distinguish hyperelastic–plastic materials from hypoelastic–plastic materials:

1. Multiplicative decomposition of the deformation into elastic and plastic parts by

$$F = F^e . F^p \tag{6.105}$$

   where $F^e$ and $F^P$ are the elastic and plastic parts of the deformation gradient, respectively.

2. The calculation of the stress in terms of the elastic strain through a hyperelastic potential:

$$\bar{S} = \frac{\partial \bar{w}(\bar{E}^e)}{\partial \bar{E}^e} \tag{6.106}$$

The stress S here is not in the reference configuration, and so, it differs from S and is defined with $E^e$.

## 6.10 Viscoplastic Materials

The material with the rate-dependent plasticity is called a viscoplastic material, such as a polymer. The plasticity model considered above does not model time effects that physically occur in the material. Such effects may be important, and a viscoplastic material model may be more appropriate to characterize the material response. Let us consider a quite widely used viscoplastic model proposed by Perzyna [10]. The model uses the concepts of the von Mises plasticity model but introduces time-rate effects. An important aspect of the theory of viscoplasticity is that there is no yield condition, but instead, the rate of the inelastic response is determined by the instantaneous difference between the effective stress and the material effective stress. In this paper, we use the Perzyna method without temperature effects. Therefore, this model postulates that the increments in strain occur at any time:

$$de_{ij} = de_{ij}^{E} + de_{ij}^{VP} \tag{6.107}$$

In the above equation, the elastic and viscoplastic strain increments are shown by E and VP. The calculation of elastic strain increments is regular, and the viscoplastic strain increments at time t are calculated in this way:

$$de_{ij}^{VP} = \begin{Bmatrix} \beta\phi({}^{t}\bar{\sigma})\dfrac{3}{2\,{}^{t}\bar{\sigma}}\,{}^{t}S_{ij}dt, if\ {}^{t}\bar{\sigma}\ \text{f}\ {}^{t}\bar{\sigma}_0 \\ \\ 0, if\ {}^{t}\bar{\sigma} \leq {}^{t}\bar{\sigma}_0 \end{Bmatrix} \tag{6.108}$$

where $\beta$ is a material constant, and the current effective stress ${}^{t}\bar{\sigma}$ is given by

$$\phi({}^{t}\bar{\sigma}) = \left(\frac{{}^{t}\bar{\sigma} - {}^{t}\bar{\sigma}_0}{{}^{t}\bar{\sigma}_0}\right)^{N} \tag{6.109}$$

In these relations, ${}^{t}\bar{\sigma}_0$ is the material effective stress, that is, the effective stress consistent with the accumulated effective viscoplastic strain ${}^{t}\bar{e}^{VP}$ (Figure 6.4), and $N$ is another material constant. The model requires the elastic constants (Young's modulus E and Poisson's ratio υ), the material constants which obtain the rate behavior of the material (β (l/time), and N), and the curvature in Figure 6.5. That is, viscoplastic strains are accumulated as long as the material's effective stress is smaller than the effective stress, and β and N specify the rate of such accumulation.

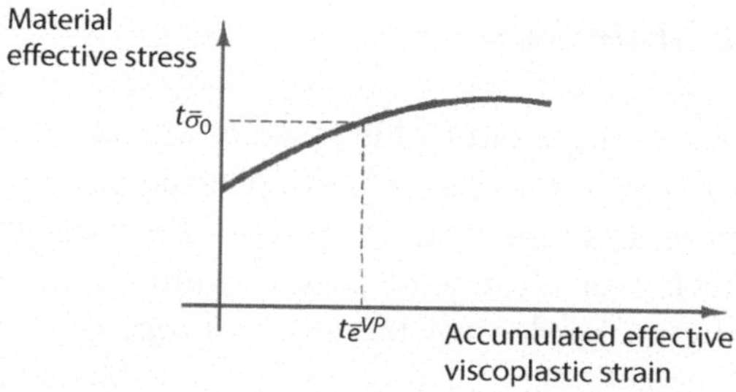


**Figure 6.5** Schematic representation of material effective stress versus accumulated effective viscoplastic strain [5].

In addition, the strain increment of viscoplastic is deduced by

$$\Delta e^{VP} = \Delta t^{\tau}\gamma^{\tau}S \tag{6.110}$$

Then, the α-method of integration has been used:

$${}^{\tau}S = (1-\alpha)^{t}S + \alpha^{t+\Delta t}S \tag{6.111}$$

and the parameter ${}^{\tau}\gamma$ is given by

$${}^{\tau}\gamma = \begin{Bmatrix} \beta^{\tau}\phi\dfrac{3}{2^{\tau}\bar{\sigma}}, if\ {}^{\tau}\bar{\sigma}\ \mathrm{f}\ {}^{\tau}\bar{\sigma}_0 \\ 0, if\ {}^{\tau}\bar{\sigma} \le {}^{\tau}\bar{\sigma}_0 \end{Bmatrix} \tag{6.112}$$

We note that ${}^{\tau}\gamma$ depends on ${}^{\tau}\bar{\sigma}$ and ${}^{\tau}\bar{\sigma}_0$, where ${}^{\tau}\bar{\sigma}_0$ depends on the accumulated effective viscoplastic strain. In point of fact, in the use of viscoplastic models, it is important to consider the various functional dependencies which can be directly, and with relative ease, involved in the calculations, since there is no explicit yield condition. In its place, the solution is determined by integrating the inelastic strains until the effective stress is equivalent to the material's effective stress (obtained by the effective viscoplastic strain). This integration is efficiently achieved with the α-method, for the reason that with α ≥ 1/2, the integration can proceed with moderately large time steps.

## 6.11 Plastic Materials

An important point brought out by Figures 6.6 (a) and (b) is that the *shape* of the plastic stress–strain curve for a particular material may appear quite different over different ranges of strain. Perhaps the most striking instance of this is mild steel: over a range of plastic strain of, say, 0–1 percent, it exhibits practically no hardening, but over a range of 0–20 percent, the hardening is appreciable. Similar remarks apply to the rolled aluminum-magnesium alloy and—to a lesser extent—to other materials.

It follows that in choosing a conceptual model for a real material, we must be aware of the *range of strain* in which we are interested. In general, in the design of structures, we shall be concerned with behavior over a strain range of the order of 1 percent in tension, while in the analysis of forming processes, we shall be concerned with much larger ranges of strain. It may seem surprising that conspicuous deformation of structures is possible at such small strains, but it is not difficult to see that if a structure consists of slender members like beams or plates, appreciable transverse deflections may occur at the expense of relatively small strain.

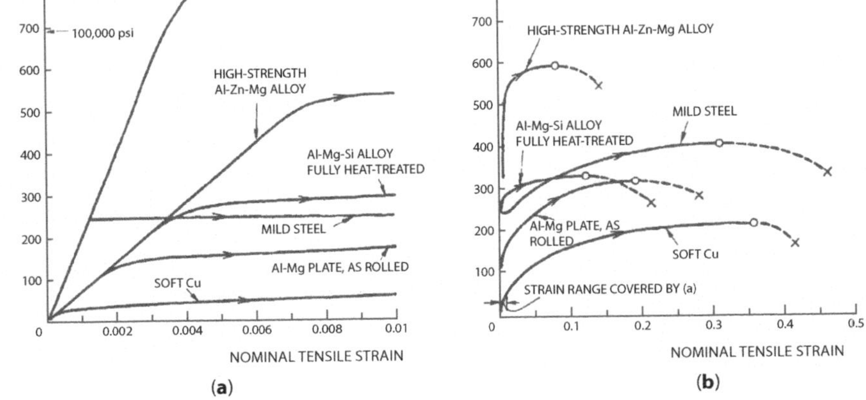


**Figure 6.6** (a) and (b) Tension tests on different materials.

There is a very large volume of experimental work on structures that support the use of an ideal perfectly plastic material in the development of the theory of plasticity. One of the important requirements for the applicability of the simple plastic theory we shall be developing in this book is that the elastic strain range of the material should be sufficiently small for deformation of the structure within the elastic range to be inconspicuous and sufficiently small *in comparison with the plastic range* for such deformations to be eventually overshadowed by the deformations of the structure due to plastic straining. For some materials, therefore, we should require data over a larger range of strain than that given in Figure 6.6(a). We shall return to these points in several subsequent chapters. Having established, then, a *prima facie* case for the adoption of perfectly plastic material to represent at least several real metals and alloys in the analysis and design of structures, we turn to the question of suitable idealizations for materials undergoing large plastic strain in forming processes, for which a non-hardening model is not obvious, to say the least, from the curves of Figure 6.6(b).

Here, we must take account of the consequences, as far as calculations are concerned, of the adoption of various possible idealizations of material. It turns out, not altogether surprisingly, that a non-hardening material leads to a *simple* analysis of structures and forming processes at least, simple in comparison with the analysis which would be necessary to deal with more complicated and "realistic" idealizations of the material. One consequence of the choice of a non-hardening material is that powerful *theorems* are available, which in many cases—as we shall see—lead to particularly simple analyses. If then, there are enormous special conceptual and computational advantages to be gained from the use of a perfectly plastic ideal material, it is worthwhile to investigate whether even a fairly strongly hardening material may be represented sufficiently well for some purposes by a non-hardening ideal material with yield stress chosen to be "representative" in some sense over the appropriate range of strain. Practically, all theoretical studies that have been made of forming processes have embodied a perfectly plastic ideal material, and, in general, agreement with the experiment has been remarkably good when the value of the "yield stress" has been assigned intelligently.

Table 6.2 illustrates typical and approximate values of ultimate tensile strength and specific gravity of a few metals and alloys. There are very many different iron-, aluminum-, and nickel-based alloys, and detailed information on their composition and mechanical properties should be sought in manufacturers' handbooks, etc.

**Table 6.2** Material tensile tests.

| Material | Specific gravity | Ultimate tensile strength | |
|---|---|---|---|
| | | MN/m² | psi |
| Carbon steel, hot rolled: | 7.8 | | |
| 0.2%C (mild steel) | | 420 | 60000 |
| 0.5%C | | 660 | 95000 |
| 0.8%C | | 830 | 120000 |
| Alloy steel | 7.9 | 1120 | 160000 |
| Structural aluminum: | 2.8 | | |
| Al-Mg alloy plate, as rolled | | 315 | 45000 |
| Al-Mg-Si alloy, fully heat-treated | | 315 | 45000 |
| High-strength Al-Zn-Mg alloy | | 600 | 85000 |
| Nickel-based alloy (Nimonic) | 7.9 | 1260 | 180000 |
| Commercially pure: | | | |
| Aluminum | 2.7 | 112 | 16000 |
| Copper | 8.9 | 225 | 32000 |
| Iron | 7.9 | 295 | 42000 |
| Nickel | 8.9 | 485 | 69000 |

# Bibliography


1. Truesdell, C. and Noll, W., *The non-linear field theories of mechanics, in Encyclopaedia of Physics*, Springer-Verlag, Berlin, 1965.
2. Marsden, J. and Hughes, T., *Mathematical Foundations of Elasticity*, Englewood Cliffs, New Jersey, 1983.
3. Simo, J. and Hughes, T., *Computational Inelasticity*, Springer-Verlag, New York, 1998.
4. Bathe, K.J., *Finite Element Procedures*, Prentice-Hall, New Jersey, 1996.
5. Bathe, K.J., Ramm, E., Wilson, E.L., Finite Element Formulations for Large Deformation Dynamic Analysis. *Int. J. Numer. Methods Eng.*, 9, 353–386, 1975.
6. Lemaitre, J. and Chaboche, J.L., *Mechanics of Solid Materials*, Cambridge University Press, Cambridge, 1990.
7. Prager, W., *Introduction to Mechanics of Continua*, Ginn & Co, New York, 1961.

8. Belytschko, T., Liu, W.K., Moran, B., *et al.* (Eds.), *Nonlinear Finite Elements for Continua and Structures*, Wiley, New Jersey, 2013.
9. Nye, J.F., *Physical Properties of Crystals: Their Representation by Tensors and Matrices*, Oxford University Press, Oxford, 1985.
10. Perzyna, P., Fundamental Problems in Viscoplasticity. *Adv. Appl. Mech.*, 9, 243–377, 1966.
11. Chen, W.F. and Han, D.J., *Plasticity for Structural Engineers*, Springer-Verlag, New York, 1988.
12. Dunne, P.C., Constitutive equations for engineering materials: Vol. 1-elasticity and modelling, W.F. Chen and A.F. Saleeb (Wiley, New York, 1982). *Comput. Methods Appl. Mech. Eng.*, *36*, 373–374, 1983.
13. Fish, J. and Belytschko, T., *A First Course in Finite Elements*, John Wiley & Sons, Chichester, 2007.
14. Miehe, C., Aspects of the Formulation and Finite Element Implementation of Large Strain Isotropic Elasticity. *Int. J. Numer. Methods Eng.*, 37, 1981–2004, 1994.
15. Moran, B., Ortiz, M., Shih, C.F., Formulation of Implicit Finite Element Methods for Multiplicative Finite Deformation Plasticity. *Int. J. Numer. Methods Eng.*, 29, 483–514, 1990.
16. Rivlin, R.S. and Saunders, D.W., Large Elastic Deformations of Isotropic Materials. VII. Experiments on the Deformation of Rubber. *Philos. Trans. R. Soc. Lond. Ser. A. Math. Phys. Sci.*, 243, 865, 251–288, 1951.
17. Salcido, A., A Phenomenological Gradient Approach to Generalized Constitutive Equations for Isotropic Fluids. *J. Appl. Math. Phys.*, 6, 7, 1494–1506, 2018.
18. Simo, J.C. and Ortiz, M., A Unified Approach to Finite Deformation Plasticity Based on the Use of Hyperelastic Constitutive Equations. *Comput. Methods Appl. Mech. Eng.*, 49, 221, 1985.
19. Yamada, Y., Recent Japanese Developments in Matrix Displacement Method for Elasto-Plastic Problems. *Japan-U.S. Seminar on Matrix Methods of Structural Analysis and Design*, Tokyo, Japan, 1969.
20. Malvern, L.E., Introduction to the Mechanics of a Continuous Medium. Prentice-Hall, Inc., Englewood Cliffs, New Jersey, 1969.

# 7
# Plasticity in Total Force Model


***Abstract***
Chapter 7 introduces a total force-based plasticity model tailored for structural systems subjected to large dynamic loads that induce permanent deformation. Unlike classical rigid-plastic methods, this chapter presents a more realistic approach by incorporating elastic-plastic behavior, strain hardening, and strain-rate effects using the plastic node method. Beam-column elements are developed with both linear and nonlinear stiffness matrices, derived from Timoshenko theory and geometric deformation contributions. The formulation includes a consistent treatment of isotropic and kinematic hardening, as well as strain-rate hardening using constitutive relations like the Cowper-Symonds equation. Plastic deformation is concentrated at nodes using a hinge mechanism, enabling efficient numerical implementation without full integration. A transformation matrix is introduced to accommodate large rotations and displacements between global and local coordinate systems. The chapter also revisits fundamental plasticity principles—such as yield criteria, hardening models, and flow rules—to support the formulation. Finally, a stress-based elastic–plastic constitutive model is derived for dynamic beam analysis, culminating in a robust elastic-plastic stiffness equation capable of handling unloading and reversals. This chapter provides a comprehensive and efficient tool for modeling complex, time-dependent inelastic behavior in 3D frame structures under impact loading.



***Keywords*:** Strain-rate hardening, plastic node method, beam-column element, dynamic analysis, large displacement, finite element method, hardening rules


## 7.1 Introduction

A variety of situations exist, in which a structure may be subjected to large dynamic loads, which can cause permanent deformation or damage to the structure. Therefore, structural dynamics and impact mechanics have an important role in engineering design. Earlier investigations on structural impacts have been well described by Jones [10]. The development of theoretical methods for impact mechanics has been aided by an idealization of


Yong Bai. Plasticity and Ductile Fracture in Structural Engineering, (173–206) © 2026 Scrivener Publishing LLC

real complex material behavior as a rigid perfectly plastic material behavior. These methods are classified as rigid-plastic analysis methods. Theoretical predictions based on rigid-plastic analyses may give some important information about the impact of plastic behavior in a simple form. The results are often in good agreement with the corresponding experimental results. However, it is difficult to make more realistic modeling of the plastic deformations because they are interspersed with elastic deformation. Plastic flow causes a change in shape and size, and the plastic regions may disappear and re-appear. The structure may invoke strain hardening as well as strain-rate hardening when it is yielded due to time-dependent loading.

General solutions for arbitrary types of structures subjected to arbitrary impacts can be obtained by numerical methods such as finite element methods. Considerable progress has been made in both the theoretical aspects as well as in the development of general-purpose computer programs for dynamic plastic analysis. Unfortunately, there is insufficient theoretical knowledge on the effect of strain rate on material properties and on consistent constitutive modeling of plasticity. Benchmark tests using a number of well-known computer programs require substantial computer speed and capacity and show that only a few programs can give reliable solutions [19]. In addition, such programs are not particularly well-suited and convenient to use for the analysis of complex structures. Therefore, there is a demand for numerical analysis procedures, which can be used to simulate the impact behavior of frame structures with large displacements and strain hardening as well as strain-rate hardening.

This chapter presents a simple and efficient procedure for large displacement plastic analysis of beam-column elements. The elastic stiffness matrix is established by combining a linear stiffness matrix [17], a geometrical stiffness matrix [1], and a deformation stiffness matrix [16]. Furthermore, the effect of plastic deformation is taken into account in an efficient and accurate way by the plastic node method [8, 20, 21]. In the plastic node method, the distributed plastic deformation of the element is concentrated on the nodes using the plastic hinge mechanism. The elastic–plastic stiffness matrices of the elements are derived without requiring numerical integration.

The objective of this chapter is to present a theoretical formulation for the modeling of strain-rate hardening effects and show how these effects can be implemented into three-dimensional finite beam-column elements. The finite beam-column element is ideally suited for the impact analysis of frames with large displacements, strain hardening, and strain-rate hardening. The accuracy and efficiency of the element are examined by comparing the present results with those obtained from experiments by others, rigid-plastic analyses, and existing finite element analysis results (see Part II Chapters 13 to 15). For the fundamental theory of finite element analysis, the readers may refer to Przemieniecki [17], Zienkiewicz [25], Bathe [5], among many other books. To understand plasticity used in the section on the plastic node method, some basic books such as Save and Massonnet [18], Yagawa and Miyazaki [23], Chen and Han [7], and Chakrabarty [6] may be helpful. To aid the understanding of the plastic node method, a basic theory of plasticity is presented for finite element analysis of solids, based on Yagawa and Miyazaki [23].

Part of the formulation presented in this chapter appeared in Bai and Pedersen [3] and Fujikubo *et al* [8]. The new extension is to account for the effect of strain-rate hardening for dynamic analysis.

## 7.2 Elastic Beam-Column with Large Displacements

The element has three translational displacements $u_x$, $u_y$, and $u_z$ and three rotational displacements $\theta_x, \theta_y$, and $\theta_z$; see Figure 7.1.

These displacements are interpolated by using a polynomial interpolation of functions, which are associated with the Timoshenko beam theory. A generalized strain vector is subsequently established in the form

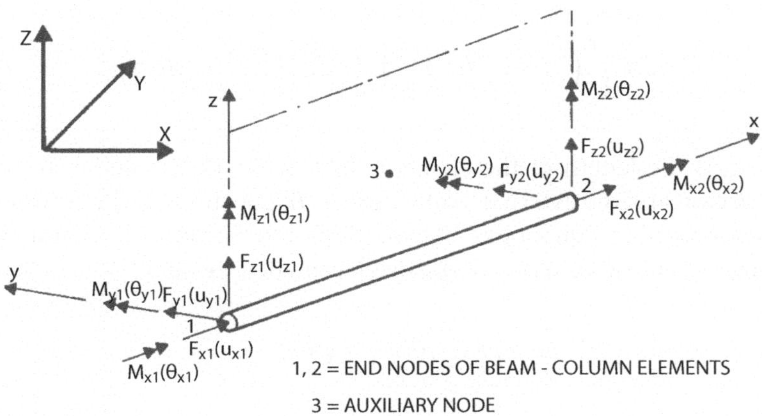


**Figure 7.1** Three-dimensional beam elements with nodal forces.

$$\{\varepsilon\} = \begin{Bmatrix} e_x \\ e_y \\ e_z \\ \kappa_x \\ \kappa_y \\ \kappa_z \end{Bmatrix} = \begin{Bmatrix} u'_x + \frac{1}{2}\left[(u'_x)^2 + (u'_y)^2 + (u'_z)^2\right] \\ u'_y - \theta_z \\ u'_z + \theta_y \\ \theta'_x \\ \theta'_y \\ \theta'_z \end{Bmatrix} \tag{7.1}$$

where $(\ )' \equiv d/ds$ and s denote the axial coordinate of the element.

A generalized elastic stress vector $\{\sigma\}$ is expressed as

$$\begin{aligned} &\{d\sigma\} = [D_E]\{d\varepsilon\} \\ &\{\sigma\} = \{F_x\ F_y\ F_z\ M_x\ M_y\ M_z\}^T \\ &[D_E] = [EA_x\ GA_y\ GA_z\ GI_x\ GI_y\ GI_z] \end{aligned} \tag{7.2}$$

where E is Yong's modulus, G the shear modulus, $A_x$ denotes the area of the cross-section $A_y$ and $A_z$ denotes the effective shear areas; $I_y$ and $I_z$ are moments of inertia, and $I_x$ denotes the torsional moment of inertia.

Applying a virtual work principle, we obtain

$$\{\delta du^e\}^T(\{f\} + \{df\}) = \int_L \{\delta d\varepsilon\}(\{\sigma\} + \{d\sigma\})ds \tag{7.3}$$

where L is the length of the element, $\{u^e\}$ is the elastic nodal displacement vector, and the external load vector is $\{f\}$. Substituting the strains and stresses defined in Equations (7.1) and (7.2) into Equation (7.3) and omitting the second-order terms of the displacements, we get [2]

$$[k_E]\{du^e\} = \{dx\} \tag{7.4}$$

where

$$\left[k_E\right]=\left[k_L\right]+\left[k_G\right]+\left[k_D\right] \tag{7.5}$$

and

$$\{dx\}=\{f\}+\{df\}-\left(\left[k_L\right]+\left[k_G\right]\right)\{u^e\} \tag{7.6}$$

The matrix $\left[k_L\right]$ is a standard linear stiffness matrix [17], $\left[k_G\right]$ is a geometrical stiffness matrix [1], and $\left[k_D\right]$ is a deformation stiffness matrix [16].

## 7.3 The Plastic Node Method

### 7.3.1 History of the Plastic Node Method

The plastic node method was named by Ueda *et al.* [26]. It is a generalization of the plastic hinge method. Ueda and Yao [21] published the plastic node method in an international journal, and Fujikubo [27] published his Ph.D. thesis on this simplified plastic analysis method.

Fujikubo [8] further extended the theory of the plastic node method to account for the effect of strain hardening. In the following sections, the existing theory is further extended to account for the effects of strain-rate hardening.

### 7.3.2 Consistency Condition and Hardening Rates for Beam Cross-Sections

For a beam-column element with strain hardening and strain-rate hardening, the yield condition of its cross-section is expressed as

$$f=Y\left(\{\sigma-.\alpha\}\right)-\sigma_0\left(\bar{\varepsilon}^p,\dot{\bar{\varepsilon}}^p\right)=0 \tag{7.7}$$

where Y is the yield (full plastic) function, $\{\alpha\}$ represents the translation of the yield surface due to kinematic hardening, and $\sigma_0$ is a parameter expressing the size of the yield surface. The vector $\{\alpha\}$ has the same dimension as the generalized stress and is expressed as

$$\{\alpha\} = \{\alpha_{fx}\ \alpha_{fy}\ \alpha_{fz}\ \alpha_{mx}\ \alpha_{my}\ \alpha_{mz}\}^T \tag{7.8}$$

Due to isotropic hardening, the yield surface is expanding as the plastic deformations increase. This expansion of the yield surface is expressed by the stress parameter $\sigma_0$, which is a function of the generalized equivalent plastic strain $\bar{\varepsilon}^P$ and of the plastic strain-rate $\dot{\bar{\varepsilon}}^p$. The equivalent strain $\bar{\varepsilon}^P$ is evaluated as a summation of its increments, which are defined as

$$\sigma_0 d\bar{\varepsilon}^p = \{\sigma - \alpha\}^T \{d\varepsilon^p\} \tag{7.9}$$

where the increments of the generalized plastic strain are taken to be

$$\{d\varepsilon^p\} = \{de_x^p \quad de_y^p \quad de_z^p \quad d\kappa_x^p \quad d\kappa_y^p \quad d\kappa_z^p\}^T \tag{7.10}$$

The equivalent plastic strain-rate $\dot{\bar{\varepsilon}}^p$ is defined as

$$\dot{\bar{\varepsilon}}^p = \frac{d\bar{\varepsilon}^p}{dt} \tag{7.11}$$

where $dt$ is an increment of time $t$.

The increments of the parameter $\sigma_0$ due to isotropic strain hardening and that due to strain-rate hardening are de-coupled to the simplest form

$$d\sigma_0 = dg_1\left(\bar{\varepsilon}^p\right) + dg_2\left(\dot{\bar{\varepsilon}}^p\right) \tag{7.12}$$

where $dg_1\left(\bar{\varepsilon}^P\right)$ expresses the increment of the parameter $\sigma_0$ for a beam cross-section due to isotropic strain hardening and $dg_2\left(\dot{\bar{\varepsilon}}^P\right)$ denotes the increment of the parameter $\sigma_0$ due to strain-rate hardening. Similar equations were used by Yoshimura *et al.* [24] and Mosquera *et al.* [14, 15].

The consistency condition for a yielded cross-section satisfying the yield condition Eq. (7.12) is expressed as

$$df = \left\{\frac{\partial f}{\partial \sigma}\right\}^T \{d\sigma\} - \left\{\frac{\partial f}{\partial \alpha}\right\}^T \{d\alpha\} - \frac{dg_1}{d\bar{\varepsilon}^p} d\bar{\varepsilon}^p - \frac{dg_2}{d\dot{\bar{\varepsilon}}^p} d\dot{\bar{\varepsilon}}^p = 0 \tag{7.13}$$

Here, we introduce a kinematic hardening rate $H'_{sk}$ and an isotropic hardening rate $H'_{si}$ for the full plastic cross-sections, which are defined by

$$H'_{sk} = \{\partial f / \partial \sigma\}^{\mathrm{T}} \{\mathrm{d}\alpha\} / \mathrm{d}\bar{\varepsilon}^P \tag{7.14}$$

and

$$H'_{si} = dg_1 \big/ d\bar{\varepsilon}^p \tag{7.15}$$

Similarly, a strain-rate hardening rate $H'_{sr}$ for the full plastic cross-section is defined as

$$H'_{sr} = dg_2 \big/ d\dot{\bar{\varepsilon}}^p \tag{7.16}$$

With these definitions, the consistency condition Eq. (7.13) may be rewritten as

$$df = \left\{\frac{\partial f}{\partial \sigma}\right\}^T \{d\sigma\} - \left(H'_{sk} + H'_{si}\right) d\bar{\varepsilon}^p - H'_{sr} d\dot{\bar{\varepsilon}}^p = 0 \tag{7.17}$$

The subscript "s" in Eqs. (7.14) thru (7.17) indicates generalized values related to the full beam cross-section. To avoid confusion, the kinematic and isotropic material-hardening rate obtained from uniaxial tests is denoted by $H'_k$ and $H'_i$, respectively.

Introducing a linear interpolation for $d\bar{\varepsilon}^p$, we obtain

$$d\bar{\varepsilon}^p = \dot{\bar{\varepsilon}}^p_{(t+dt)} \theta dt + \dot{\bar{\varepsilon}}^p_{(t)} (1-\theta) dt \tag{7.18}$$

where $\theta$ is a parameter, which will be taken as $1/2$ in the numerical examples.

From Eq. (7.18), we find

$$\dot{\bar{\varepsilon}}^p_{(t+dt)} = \left[d\bar{\varepsilon}^p - (1-\theta)\dot{\bar{\varepsilon}}^p_{(t)}dt\right] / (\theta dt) \tag{7.19}$$

Then, the increment of the equivalent plastic strain-rate $d\dot{\bar{\varepsilon}}^p$ may be estimated as

$$d\dot{\bar{\varepsilon}}^p = \dot{\bar{\varepsilon}}^p_{(t+dt)} - \dot{\bar{\varepsilon}}^p_{(t)} = \left[d\bar{\varepsilon}^p - \dot{\bar{\varepsilon}}^p_{(t)}dt\right] / (\theta dt) \tag{7.20}$$

For simplicity, in the following equations, the subscript "t" will be omitted.

Considering Eq. (7.20), the consistency condition Eq. (7.17) is rewritten in the form

$$\mathrm{d}f = \{\partial f/\partial\sigma\}^{\mathrm{T}}\{\mathrm{d}\sigma\} - (H'_{sk} + H'_{si} + H'_{sr}/(\theta dt))\mathrm{d}\bar{\varepsilon}^P + \left[H'_{sr}/\theta\right]\dot{\bar{\varepsilon}}^p = 0 \tag{7.21}$$

In the following, the hardening rates $H'_{si}, H'_{sk}$ and $H'_{sr}$ in Eq. (7.21) will be discussed.

The isotropic hardening rate for the cross-section is evaluated as follows. Following Ueda and Fujikubo [20] and Fujikubo *et al.* [8], the increments of the generalized stress due to isotropic hardening are estimated to be

$$\{d\sigma\} = \left[H'_{si}\right]\{d\varepsilon^p\} \tag{7.22}$$

and the matrix $\left[H'_{si}\right]$ is obtained by first deriving a relationship between the stress increments and the plastic strain increments for points and after integrating those stresses over the cross-section. If we use the Von Mises yield criteria and neglect the interaction between shear stress and axial stresses, Eq. (7.23) is obtained:

$$\left[H'_{si}\right] = \left[H'_i A_x \;\; H'_i A_y/3 \;\; H'_i A_z/3 \;\; H'_i I_x/3 \;\; H_i I_y \;\; H'_i I_z\right] \tag{7.23}$$

Considering f in Eq. (7.7) as a plastic potential and applying the flow theory of plasticity, we may obtain

$$\{d\varepsilon_P\}=cd\bar{\varepsilon}_P\left\{\frac{\partial f}{\partial\sigma}\right\} \tag{7.24}$$

where

$$c=\sigma_0\Big/\left(\{\sigma-\alpha\}^T\{\partial f/\partial\sigma\}\right) \tag{7.25}$$

The increment of the parameter $\sigma_0$ due to isotropic strain hardening is defined as

$$dg_1=\{\partial f/\partial\sigma\}^T\{d\sigma\} \tag{7.26}$$

Substituting Eqs. (7.24) and (7.22) into Eq. (7.26), the isotropic cross-sectional strain-hardening rate defined in Eq. (7.15) is given as

$$H'_{si}=c\left\{\frac{\partial f}{\partial\sigma}\right\}^T\left[H'_{si}\right]\left\{\frac{\partial f}{\partial\sigma}\right\} \tag{7.27}$$

The kinematic hardening rate for the cross-section will be derived using a similar approach. The yield surface translation increment $\{d\alpha\}$ can be obtained by using Ziegler's rule and the Mises yield criteria [8]:

$$\{d\alpha\}=\left[H'_{sk}\right]\{d\varepsilon^p\} \tag{7.28}$$

where $\left[H'_{sk}\right]$ used in the present chapter is taken as

$$\left[H'_{sk}\right]=\left[H'_k A_x H'_k A_y/3H'_k A_z/3H'_k I_x/3H'_k I_y H'_k I_z\right] \tag{7.29}$$

Substituting Eqs. (7.24) and (7.28) into Eq. (7.14), we obtained

$$H'_{sk} = c\left\{\frac{\partial f}{\partial \sigma}\right\}^T \left[H'_{sk}\right]\left\{\frac{\partial f}{\partial \sigma}\right\} \tag{7.30}$$

Finally, we shall determine the strain-rate hardening rate for the cross-sections. The increment of the parameter $\sigma_0$, due to strain-rate hardening, is estimated by the use of a constitutive equation that expresses the relationship between $g_2$ and the equivalent plastic strain. For instance, the Cowper-Symonds constitutive equation is expressed as (Jones, 1989)

$$\sigma_{0x} = \sigma_y\left[1+\left(\dot{\varepsilon}_x^p/D\right)^{1/q}\right] \tag{7.31}$$

where $\sigma_y$ are the yield stress and $\bar{\varepsilon}_x^p$ denotes the plastic strain-rate for a point. Values of D and q that are often used are the following:

For mild steel D=40.4 $sec^{-1}$ q = 5

and for aluminum alloy D=6500 $sec^{-1}$ q = 4

Eq. (7.31) was obtained for uniaxial stress tests. It is assumed that this equation is still a valid approximation when it is applied to multiaxially loaded beam cross-sections. Eq. (7.31) becomes

$$g_2\left(\dot{\bar{\varepsilon}}^p\right) = N_Y\left[1+\left(\dot{\bar{\varepsilon}}^p/D\right)^{1/q}\right] \tag{7.32}$$

where $N_Y = A_x\sigma_y$

Using Eq. (7.32), the strain-rate hardening rate defined in Eq. (7.16) may be given as

$$H'_{sr} = N_Y\left(\dot{\bar{\varepsilon}}^p/D\right)^{\left(\frac{1}{g}-1\right)} / q \tag{7.33}$$

### 7.3.3 Plastic Displacement and Strain at Nodes

The plastic deformations of the element are concentrated at the node in a mechanism similar to the plastic hinge. Referring to Eq. (7.7), the yield condition at the node is expressed as

$$F_i = Y_i\left(\left\{\sigma_i - \alpha_i\right\}\right) - \sigma_{0i}\left(\bar{\varepsilon}_i^{\,p}, \dot{\bar{\varepsilon}}_i^{\,p}\right) = 0 \tag{7.34}$$

where the subscript "i" denotes values at the node no. i. From Eq. (7.21), the consistency condition for node i is expressed as

$$dF_i\left\{\varphi_i\right\}^T\left\{dx\right\} - \left[H'_{sk} + H'_{si} + H'_{sr}/(\theta dt)\right]_i d\bar{\varepsilon}_i^{\,p} + a_i = 0 \tag{7.35}$$

where

$$\left\{\varphi_i\right\} = \left\{\partial F_i/\partial x\right\} \tag{7.36}$$

$$a_i = \left[H'_{sr}/\theta\right]_i \dot{\bar{\varepsilon}}_i^{\,p} \tag{7.37}$$

where $\{x\}$ is the nodal force vector.

Applying the plastic flow theory, the increments of plastic nodal displacement of the element due to plasticity at node i are estimated as [21]

$$\left\{du_i^p\right\} = d\lambda_i\left\{\varphi_i\right\} \tag{7.38}$$

where $d\lambda_i$ is a measure of the magnitude of plastic deformation.

In the following paragraphs, we shall establish a relationship between $d\bar{\varepsilon}_i^{\,p}$ and $d\lambda_i$ using a plastic work procedure [8, 20]. The increment of the plastic work done at the plastic node "i" is expressed as

$$dw_i^p = \left\{x\right\}^T\left\{du_i^p\right\} = \left\{x\right\}^T\left\{\varphi_i\right\}d\lambda_i \tag{7.39}$$

The increment of the plastic work done in the actual plastic region around node i is evaluated as

$$dw_i^{p*} = \int_{L_I^P}\left\{\sigma\right\}^T\left\{d\varepsilon^p\right\}ds = \int_{L_i^P} c\left\{\sigma\right\}^T\left\{\frac{\partial f}{\partial\sigma}\right\}d\bar{\varepsilon}^{\,p}ds \tag{7.40}$$

From Eq. (7.21), the increment of the equivalent plastic strain at a coordinate s can be expressed as a function of the value at node i in the form

$$d\bar{\varepsilon}^{p} = g(s)d\bar{\varepsilon}_i^{p} \tag{7.41}$$

where

$$g(s) = \frac{\left[H'_{sk} + H'_{si} + H'_{sr}/(\theta dt)\right]_i}{H'_{sk} + H'_{si} + H'_{sr}/(\theta dt)} \frac{\{\partial f/\partial \sigma\}^T \{d\sigma\} + (H'_{sr}/\theta)\dot{\bar{\varepsilon}}^{p}}{\{\partial f_i/\partial \sigma_i\}^T \{d\sigma_i\} + (H'_{sr}/\theta)_i \dot{\bar{\varepsilon}}_i^{p}}$$

Substituting Eq. (7.41) into Eqs. (7.40) and (7.42) is obtained:

$$dw_i^{p*} = d\bar{\varepsilon}_i^{p} \int_{L_i^p} c\{\sigma\}^T \left\{\frac{\partial f}{\partial \sigma}\right\} g(s)ds \tag{7.42}$$

Equating the plastic work increments $dw_i^p$ in Eq. (7.39) and $dw_i^{p*}$ in Eq. (7.42), Eq. (7.43) is obtained:

$$d\bar{\varepsilon}_i^{p} = h_i d\lambda_i \tag{7.43}$$

where

$$h_i = \{x\}^T \{\varphi_i\} / \int_{L_i^p} c\{\sigma\}^T \left\{\frac{\partial f}{\partial \sigma}\right\} g(s)ds \tag{7.44}$$

A simpler alternative approach for determining the strain-hardening rate at a plastic node is to establish relationships between the plastic nodal displacements and the generalized plastic strain vector at the node in the form of

$$\left\{d\bar{\varepsilon}_i^{p}\right\} = \left\{du_i^p\right\} / L_{di} = \{\varphi_i\} d\lambda_i / L_{di} \tag{7.45}$$

where $L_{di}$ denotes an equivalent length of the plastic region.

The increment of the equivalent plastic strain at the node can be evaluated by substituting Eq. (7.45) into Eq. (7.9) and obtain Eq. (7.43):

$$h_i = \{\sigma_i - \alpha_i\}^T \{\varphi_i\} / (L_{di}\sigma_{0i}) \tag{7.46}$$

Integration along the axial axis of the element becomes unnecessary when Eq. (7.46) is applied to calculate $d\varepsilon_i^p$ instead of Eq. (7.44). This results in an extremely simple numerical procedure. Unfortunately, the actual regions where the plastic flow occurs causes a change in shape, size, and may disappear/re-appear. Evidently, the equivalent length of the plastic region for each stress component should be different and considered to be a function of time. However, for simplicity, we would like to find a constant value that will provide adequate approximations. Then, length $L_{di}$ can simply be approximated as

$$L_{di} = \alpha_D H \tag{7.47}$$

or

$$L_{di} = \alpha_L L \tag{7.48}$$

where $\alpha_D$ and $\alpha_L$ are coefficients, and H is the diameter for a circular cross-section or width (or height) for a rectangular cross-section, etc. This approach will be used in the case where a structural member is modeled by only one element. Substitution of Equation (7.43) into Equation (7.35) gives

$$dF_i = \{\varphi_i\}^T \{dx\} - H'_{ni} d\lambda_i + a_i = 0 \tag{7.49}$$

where

$$H_{ni} = \left[ H'_{sk} + H'_{si} + H'_{sr} / (\theta dt) \right]_i h_i \tag{7.50}$$

### 7.3.4 Elastic-Plastic Stiffness Equation for Elements

When both nodes 1 and 2 are plastic, the following matrix equation may be established from Eq. (7.49):

$$[\Phi]^T\{dx\}-[H']\{d\lambda\}+\{A\}=0 \tag{7.51}$$

where

$$[\Phi]=[\{\varphi_1\}\{\varphi_2\}]$$

$$[H']=[H'_{n1}\ H'_{n2}]\ (2\times 2\text{ diagonal matrix})$$

$$\{d\lambda\}=\{d\lambda_1\ d\lambda_2\}^T$$

and

$$\{A\}=\{a_1\ a_2\}^T$$

From Eq. (7.38), the increments of the plastic nodal displacement $\{du^p\}$ are given as

$$\{du^p\}=[\Phi]\{d\lambda\} \tag{7.52}$$

The increments of the total nodal displacement $\{du\}$ are expressed by the summation of the elastic and plastic components as

$$\{du\}=\{du^e\}+\{du^p\} \tag{7.53}$$

Substitution of Eqs. (7.52) and (7.53) into Eq. (7.4) gives

$$[k_E](\{du\}-[\Phi]\{d\lambda\})=\{dx\} \tag{7.54}$$

Solving Eqs. (7.51) and (7.54) with respect to $\{d\lambda\}$, we obtain

$$\{d\lambda\}=([H']+[\Phi]^T[k_E][\Phi])^{-1}([\Phi]^T[k_E]\{du\}+\{A\}) \tag{7.55}$$

Substituting $\{d\lambda\}$ into Eq. (7.54) gives the elastic–plastic stiffness equation:

$$[k_p]\{du\} = \{dx\} + \{dx\ \} \tag{7.56}$$

where

$$[k_P] = [k_E] - [k_E][\Phi]\left([H'] + [\Phi]^T [k_E][\Phi]\right)^{-1} [\Phi]^T [k_E] \tag{7.57}$$

$$\{dx'\} = [k_E][\Phi]\left[[H'] + [\Phi]^T [k_E][\Phi]\right]^{-1} \{A\} \tag{7.58}$$

If the sign of $\{d\lambda_1\}$ or $\{d\lambda_2\}$ is found to be negative, unloading occurs at the plastic node and the node should then be treated as elastic. It is noted that the effects of large displacements and strain hardening, as well as strain-rate hardening, have been taken into account in the derived elastic–plastic stiffness equation.

## 7.4 Transformation Matrix

In this section, a new transformation matrix $[T_t]$ is described, which transfers element displacements measured in the global coordinate system XYZ to element displacements measured in the local coordinate xyz, at time *t*. The transformation matrix is evaluated as

$$[T_t] = [\Delta T][T_t - dt] \tag{7.59}$$

where $[T_t - da]$ is a matrix, which transfers the element displacements to the local coordinate system at time *t-dt*. [$\Delta T$] is a matrix that transfers the element displacements measured in the local coordinate system at the time *t-dt* to the local coordinate system at time *t*.

The transformation matrix [ΔT] is shown by Figure 7.2 and is composed of submatrices [*tt*] which transform the displacement vectors. The submatrix [*tt*] is evaluated as

$$[tt] = [t_a] + [t_b] \tag{7.60}$$

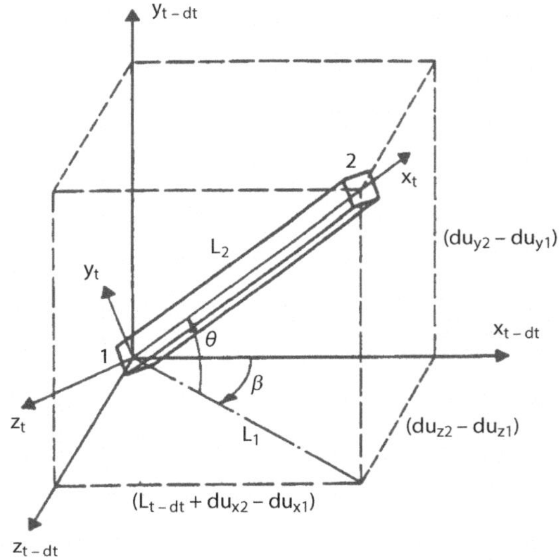


**Figure 7.2** Transformation matrix.

where

$$
\begin{aligned}
\left[t_a\right] &= \begin{bmatrix} 1 & 0 & 0 \\ 0 & \cos\alpha & \sin\alpha \\ 0 & -\sin\alpha & \cos\alpha \end{bmatrix} \\
\left[t_b\right] &= \begin{bmatrix} \cos\beta\cos\theta & \sin\theta & \sin\beta\cos\theta \\ -\cos\beta\sin\theta & \cos\theta & -\sin\beta\sin\theta \\ -\sin\beta & 0 & \cos\beta \end{bmatrix}
\end{aligned}
\tag{7.61}
$$

By considering the increments of nodal displacement from time *t-dt* to time *t*, measured in the local coordinate system at time *t-dt*, we obtain

$$
\begin{aligned}
sin\ \beta &= \left(du_{z2} - du_{z1}\right)/L_1 \\
cos\ \beta &= \left(L_{t-dt} + du_{x2} - du_{x1}\right)/L_1 \\
sin\ \theta &= \left(du_{y2} - du_{y1}\right)/L_2 \\
cos\ \theta &= L_1/L_2
\end{aligned}
\tag{7.62}
$$

where

$$L_1 = \left[ \left( L_{t-dt} + du_{x2} - du_{x1} \right)^2 + \left( du_{z2} - du_{z1} \right)^2 \right]^{\frac{1}{2}}$$
$$L_2 = \left[ L_1^{\,2} + \left( du_{y2} - du_{y1} \right)^2 \right]^{\frac{1}{2}} \tag{7.63}$$

$L_{t-dt}$ is the distance between nodes 1 and 2 at time *t-dt*. Furthermore, the angle $\alpha$ is calculated as

$$\alpha = \frac{1}{2}\left[ t_{b11}\left( d\theta_{x1} + d\theta_{x2} \right) + t_{b12}\left( d\theta_{y1} + d\theta_{y2} \right) + t_{b13}\left( d\theta_{z1} + d\theta_{z2} \right) \right] \tag{7.64}$$

## 7.5 Stress-Based Plasticity Constitutive Equations

### 7.5.1 General

This appendix is written based on a Japanese book authored by Yagawa and Miyazaki [23]. When the formulation presented in this chapter was made, the author had been inspired by this book and Yamada [22]. The objective of this appendix is to describe the basics of plasticity that may be useful to help understand the mathematical formulation presented in the main body of this chapter.

In the uniaxial tensile test, when the stress is small, the material behavior is elastic. The proportional constant *E* is Young's modulus. If the load is released, the stress will become 0, and the material will return to its original condition. On the other hand, when the stress exceeds a limit, permanent deformation may occur. The permanent deformation is called plastic deformation.

Figure 7.3 shows a typical stress–strain diagram of metallic materials. The material is in elastic behavior range until the yield point *A*, and the stress $\sigma$ and strain $\varepsilon$ are in proportion.

This proportion relationship is called Hook's law. After going over point *A*, the gradient of the stress-strain curve decreases, and the gradient $H'_0$ is called tangent modulus. If unloading occurs at point *B*, the stress will decrease along with $B \rightarrow C$, which is parallel to *OA*. The residual strain is called plastic strain, $\varepsilon^P$. On the other hand, the recovered strain

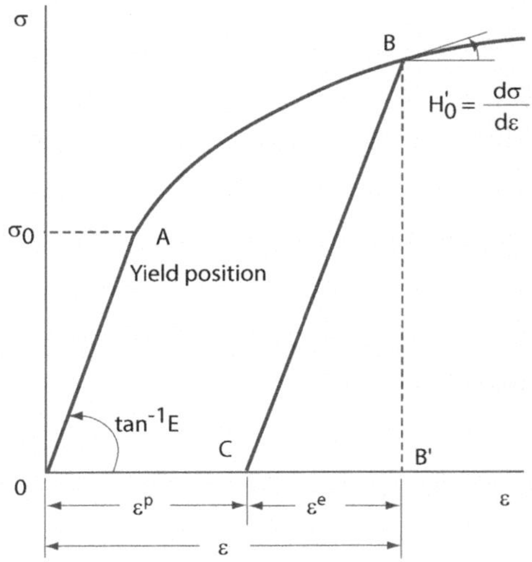


**Figure 7.3** Stress and strain diagram.

corresponding with *CB'* is called elastic strain, $\varepsilon^e$. The total strain is the sum of the elastic strain and plastic strain.

$$\varepsilon = \varepsilon^e + \varepsilon^p \tag{7.65}$$

Figure 7.4 shows the relationship between stress and plastic strain. The gradient *H'* in this stress–plastic strain curve is called the strain-hardening rate. Referring to Figures 7.3 and 7.4, the following relationship may be obtained.

$$d\varepsilon = \frac{d\sigma}{H_0'} = d\varepsilon^e + d\varepsilon^p = \frac{d\sigma}{E} + \frac{d\sigma}{H'} \tag{7.66}$$

where $H_0'$ and *H'* can be expressed as follows:

$$H' = \frac{EH_0'}{E - H_0'}, \qquad H_0' = \frac{EH'}{E + H'} \tag{7.67}$$

As shown in Figure 7.5, when the stress is beyond the yield point of the material, plastic strain occurs; if the load is released after point *B*, and

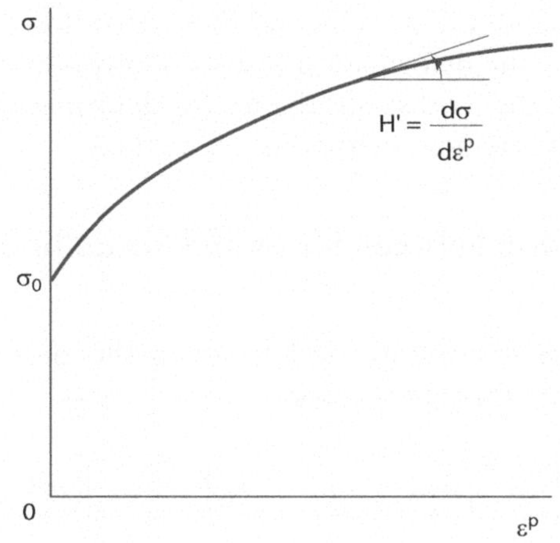


**Figure 7.4** Curve of strain hardening.

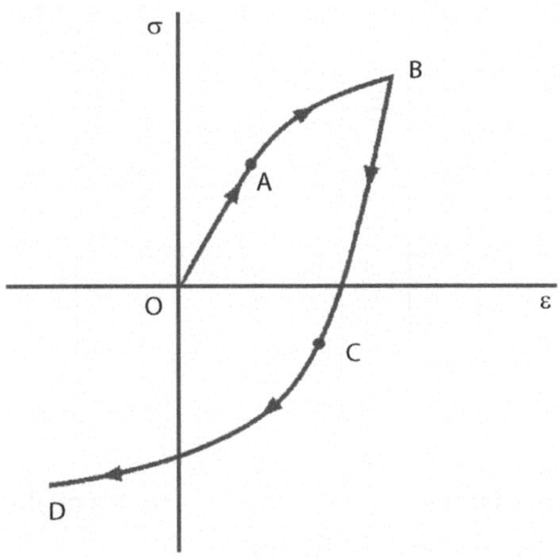


**Figure 7.5** Bauschinger's effect.

a compressive load is applied, the relationship between stress and strain will follow curve BCD and the material yields at compressive stress (point C) that is lower than its initial yield stress. This phenomenon is called Bauschinger's effect.

Although there is a one-to-one correspondence between stress and strain in the elastic region as described by Hook's law, this correlation does not exist in the plastic region. This means that if the strain is above a certain level, it is dependent on the deformation history along with the stress. For an elastic–plastic solid, the incremental theory (or flow theory) is widely used to account for the deformation history. However, for the simplification of calculation, the total strain theory (or deformation theory) is also used when the finite element method is not applied.

### 7.5.2 Relationship between Stress and Strain in the Elastic Region

The relationship between stress and strain in the elastic region can be repressed in a matrix form as follows:

$$\left.\begin{array}{ll} \{\sigma\} = \left[D^e\right]\{\varepsilon\} \quad or \quad \{\varepsilon\} = \left[C^e\right]\{\sigma\} \\ \left[D^e\right] = \left[C^e\right]^{-1} \end{array}\right\} \tag{7.68}$$

where

$$\{\sigma\} = \begin{Bmatrix} \sigma_x \\ \sigma_y \\ \sigma_z \\ \tau_{xy} \\ \tau_{yz} \\ \tau_{zx} \end{Bmatrix}, \quad \{\varepsilon\} = \begin{Bmatrix} \varepsilon_x \\ \varepsilon_y \\ \varepsilon_z \\ \gamma_{xy} \\ \gamma_{yz} \\ \gamma_{zx} \end{Bmatrix} \tag{7.69}$$

For an isotropic material, $\left[D^e\right], \left[C^e\right]$ are formulated with Young's modulus, E, and Poisson's ratio in the form

$$\left[D^e\right]=\frac{E\left(1-\nu\right)}{\left(1+\nu\right)\left(1-2\nu\right)}\begin{bmatrix} 1 & \frac{\nu}{1-\nu} & \frac{\nu}{1-\nu} & 0 & 0 & 0 \\ & 1 & \frac{\nu}{1-\nu} & 0 & 0 & 0 \\ & & 1 & 0 & 0 & 0 \\ & & & \frac{1-2\nu}{2\left(1-\nu\right)} & 0 & 0 \\ & & & & \frac{1-2\nu}{2\left(1-\nu\right)} & 0 \\ & & & & & \frac{1-2\nu}{2\left(1-\nu\right)} \end{bmatrix} \tag{7.70}$$

$$\left[C^e\right]=\frac{1}{E}\begin{bmatrix} 1 & -\nu & -\nu & 0 & 0 & 0 \\ & 1 & -\nu & 0 & 0 & 0 \\ & & 1 & 0 & 0 & 0 \\ & & & 2\left(1+\nu\right) & 0 & 0 \\ & & & & 2\left(1+\nu\right) & 0 \\ & & & & & 2\left(1+\nu\right) \end{bmatrix} \tag{7.71}$$

In the elastic region, the relationship between stress increment and strain increment may be written based on Equation (7.68) as follows:

$$\left\{\Delta\sigma\right\}=\left[D^e\right]\left\{\Delta\varepsilon\right\} \quad or \quad \left\{\Delta\varepsilon\right\}=\left[C^e\right]\left\{\Delta\sigma\right\} \tag{7.72}$$

where $\Delta$ is an increment.

### 7.5.3 Yield Criterion

The stress condition for the initiation of plastic deformation is called yield criterion and is generally written as a yield function $f$

$$f\left(J_1,J_2,J_3\right)=0 \tag{7.73}$$

where $J_1, J_2, J_3$ are the invariants and expressed as

$$\left.\begin{aligned} J_1 &= \sigma_x + \sigma_y + \sigma_z \\ J_2 &= -\left(\sigma_x\sigma_y + \sigma_y\sigma_z + \sigma_z\sigma_x\right) + \tau_{xy}^2 + \tau_{yz}^2 + \tau_{zx}^2 \\ J_3 &= \sigma_x\sigma_y\sigma_z - \sigma_x\tau_{yz}^2 - \sigma_y\tau_{zx}^2 - \sigma_z\tau_{xy}^2 + 2\tau_{yz}\tau_{zx}\tau_{xy} \end{aligned}\right\} \tag{7.74}$$

The geometrical surface for the yield criterion in a stress space is called the yield surface. Because the first approximation of the yield function has no relation with the hydrostatic pressure for the metallic materials, the yield criterion can be expressed as

$$f\left(J_2', J_3'\right) = 0, \qquad J_1' = 0 \tag{7.75}$$

where $J_1', J_2', J_3'$ are called the invariants of deviatoric stress as shown below:

$$\left.\begin{aligned} &\sigma_x' = \sigma_x - \sigma_m, \qquad \sigma_y' = \sigma_y - \sigma_m, \qquad \sigma_z' = \sigma_z - \sigma_m, \\ &\sigma_m = \frac{\sigma_x + \sigma_y + \sigma_z}{3}, \\ &\tau_{xy}' = \tau_{xy}, \qquad \tau_{yz}' = \tau_{yz}, \qquad \tau_{zx}' = \tau_{zx} \end{aligned}\right\} \tag{7.76}$$

The most widely used yield criterion for metallic materials is Mises's yield criterion, in which the function $f$ in Equation (7.75) is only expressed as a function of the second invariant of deviatoric stress, $J_2'$.

$$f = \sqrt{3J_2'} - \sigma_0 \tag{7.77}$$

where $\sigma_0$ is the yield stress for the uniaxial loading. Here, $J_2'$ is

$$
\begin{aligned}
J_2' &= -\left(\sigma_x'\sigma_y' + \sigma_y'\sigma_z' + \sigma_z'\sigma_x'\right) + \tau_{xy}'^2 + \tau_{yz}'^2 + \tau_{zx}'^2 \\
&= \frac{1}{2}\left[\sigma_x'^2 + \sigma_y'^2 + \sigma_z'^2 + 2\left(\tau_{xy}'^2 + \tau_{yz}'^2 + \tau_{zx}'^2\right)\right]^{\frac{1}{2}} \\
&= \frac{1}{6}\left[\left(\sigma_x - \sigma_y\right)^2 + \left(\sigma_y - \sigma_z\right)^2 + \left(\sigma_z - \sigma_x\right)^2 + 6\left(\tau_{xy}^2 + \tau_{yz}^2 + \tau_{zx}^2\right)\right]
\end{aligned}
\tag{7.78}
$$

If the equivalent stress $\bar{\sigma}$ is defined as

$$
\begin{aligned}
\bar{\sigma} = \sqrt{3J_2'} &= \sqrt{\frac{3}{2}}\left[\sigma_x'^2 + \sigma_y'^2 + \sigma_z'^2 + 2\left(\tau_{xy}'^2 + \tau_{yz}'^2 + \tau_{zx}'^2\right)\right]^{\frac{1}{2}} \\
&= \frac{1}{\sqrt{2}}\left[\left(\sigma_x - \sigma_y\right)^2 + \left(\sigma_y - \sigma_z\right)^2 + \left(\sigma_z - \sigma_x\right)^2 + 6\left(\tau_{xy}^2 + \tau_{yz}^2 + \tau_{zx}^2\right)\right]^{\frac{1}{2}}
\end{aligned}
\tag{7.79}
$$

The yield criterion becomes

$$
\bar{\sigma} = \sigma_0 \tag{7.80}
$$

The multiaxial stress condition can then correspond to the uniaxial stress condition.

When the stress is larger than the yield criterion of the material, the hardening and plastic deformation occurs, and the yield function $f = 0$ must be satisfied. However, if $f < 0$, unloading occurs, and the material is in the elastic region.

### 7.5.4 Plastic Strain Increment

When plastic deformation occurs, the shape of the yield surface may change following the hardening rule. Here, the isotropic hardening rule and the kinematic hardening rule are described below.

#### *7.5.4.1 Isotropic Hardening Rule*

As shown in Figure 7.6, in the hardening process, the size of the yield surface may increase but no change to the position and shape of the yield surface.

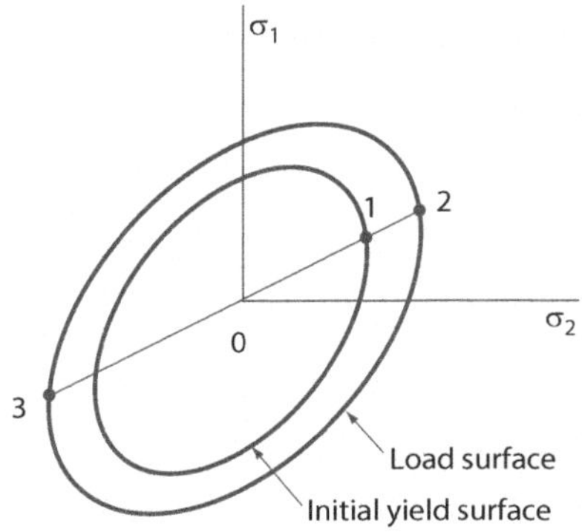


**Figure 7.6** Isotropic hardening rule.

Figure 7.7 shows the relationship between uniaxial stress and strain. After loading along with the curve *OYA*, and unloading to point *B*, and then continuing in the reverse direction to point *C*, $\overline{AB} = \overline{BC}$, $\overline{BC} > \overline{OY}$. If the strain hardening is considered, the yield function in Equation (7.77) becomes

$$f = \sqrt{3J_2'} - \sigma_0 = \bar{\sigma} - \sigma_0\left(\bar{\varepsilon}^p\right) \tag{7.81}$$

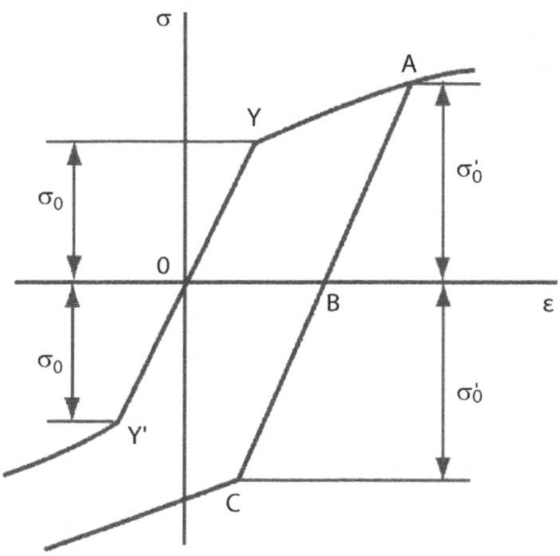


**Figure 7.7** Uniaxial stress and strain relation based on isotropic hardening rule.

where $\bar{\varepsilon}^{p}$ is the equivalent plastic strain and may be expressed as

$$\bar{\varepsilon}^{p} = \int d\bar{\varepsilon}^{p} \tag{7.82}$$

$d\bar{\varepsilon}^{p}$ in the above equation may be estimated as below:

$$\begin{aligned} d\bar{\varepsilon}^{p} = \Delta\bar{\varepsilon}^{p} &= \sqrt{\frac{2}{3}}\left[\Delta\varepsilon_x^{p2} + \Delta\varepsilon_y^{p2} + \Delta\varepsilon_z^{p2} + \frac{1}{2}\left(\Delta\gamma_{xy}^{p2} + \Delta\gamma_{yz}^{p2} + \Delta\gamma_{zx}^{p2}\right)\right]^{\frac{1}{2}} \\ &= \frac{\sqrt{2}}{3}\left[\left(\Delta\varepsilon_x^{p} - \Delta\varepsilon_y^{p}\right)^2 + \left(\Delta\varepsilon_y^{p} - \Delta\varepsilon_z^{p}\right)^2 + \left(\Delta\varepsilon_z^{p} - \Delta\varepsilon_x^{p}\right)^2 \right. \\ &\left. + \frac{3}{2}\left(\Delta\gamma_{xy}^{p2} + \Delta\gamma_{yz}^{p2} + \Delta\gamma_{zx}^{p2}\right)\right]^{1/2} \end{aligned} \tag{7.83}$$

If the plastic strain increment in uniaxial loading in the $x$-direction is defined as $\Delta\varepsilon_x^{p}$, and the condition for incompressibility of plastic strain

$$\Delta\varepsilon_x^{p} + \Delta\varepsilon_y^{p} + \Delta\varepsilon_z^{p} = 0 \tag{7.84}$$

is used, the following equation may be obtained:

$$\Delta\varepsilon_y^{p} = \Delta\varepsilon_z^{p} = -\frac{\Delta\varepsilon_x^{p}}{2}, \qquad \Delta\gamma_{xy}^{p} = \Delta\gamma_{yz}^{p} = \Delta\gamma_{zx}^{p} = 0 \tag{7.85}$$

Substituting Equation (7.84) into Equation (7.85), we obtain

$$\Delta\bar{\varepsilon}^{p} = \Delta\varepsilon_x^{p} \tag{7.86}$$

This means that the increment of equivalent plastic strain is a conversion of the plastic strain increment in the multiaxial stress condition to that of the uniaxial stress condition.

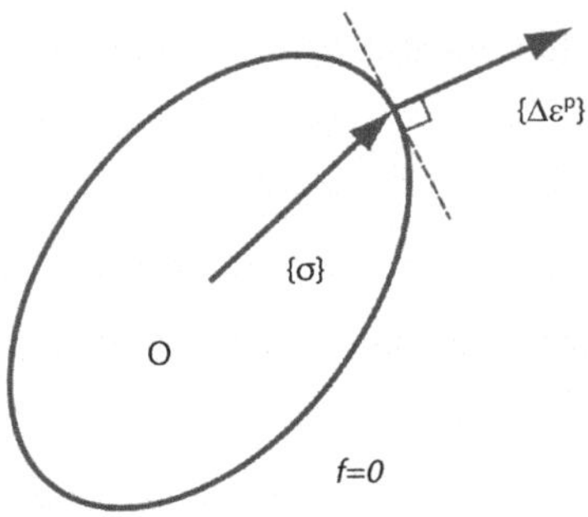


**Figure 7.8** Flow rule.

The plastic strain increment may be obtained from the flow rule. If the plastic potential is defined as yield function $f$, the plastic strain increment is expressed as

$$\left\{\Delta\varepsilon^{p}\right\} = \Delta\lambda\left\{\frac{\partial f}{\partial\sigma}\right\} = \Delta\lambda\frac{3}{2\bar{\sigma}}\left\{\sigma'\right\} \tag{7.87}$$

where, yield function $f$ is expressed in Equation (7.81). $\Delta\lambda$ (:>0) is an undetermined scalar constant, and $\left\{\sigma'\right\}$ is a vector of deviatoric stress as below:

$$\left\{\sigma'\right\} = \begin{Bmatrix} \sigma'_x \\ \sigma'_y \\ \sigma'_z \\ 2\tau_{xy} \\ 2\tau_{yz} \\ 2\tau_{zx} \end{Bmatrix} \tag{7.88}$$

Equation (7.88) means that the plastic strain increment is in the perpendicular direction of the yield surface, $f = 0$, as shown in Figure 7.8.

### 7.5.4.2 *Kinematic Hardening Rule*

In the kinematic hardening rule, although the size of the yield surface does not change due to the hardening process, the position of its center moves as shown in Figure 7.9. The relationship between stress and strain for a uniaxial stress case is shown in Figure 7.10. From the relation of $\overline{YY'} = \overline{AC}, \overline{BC} < \overline{OY}$, the Bauschinger's effect may be qualitatively expressed in the figure.

The yield function for kinematic hardening rule is defined as

$$f = f\left(\{\sigma\} - \{\alpha_0\}\right) \tag{7.89}$$

where $\{\alpha_0\}$ is the center of yield surface and may be expressed as

$$\{\alpha_0\} = \begin{Bmatrix} \alpha_x \\ \alpha_y \\ \alpha_z \\ \alpha_{xy} \\ \alpha_{yz} \\ \alpha_{zx} \end{Bmatrix} \tag{7.90}$$

There are two ways to determine $\{\alpha_0\}$: Prager kinematic hardening rule and Ziegler kinematic hardening rule:

$$\{\Delta\alpha_0\} = C\{\Delta\varepsilon^p\} \qquad \text{:Prager} \tag{7.91}$$

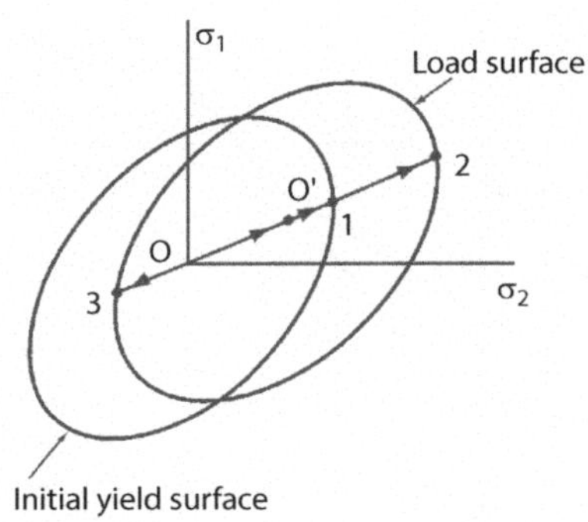


**Figure 7.9** Kinematic hardening rule.

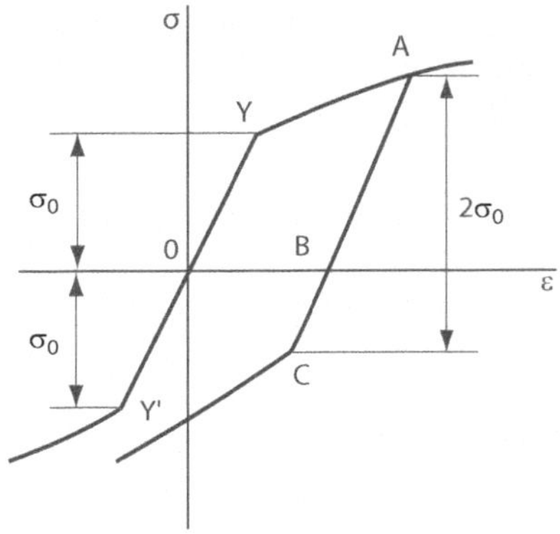


**Figure 7.10** Uniaxial stress and strain relation based on kinematic hardening rule.

$$\{\Delta\alpha_0\} = \Delta\mu\left(\{\sigma\} - \{\alpha_0\}\right) \qquad \text{:Ziegler} \tag{7.92}$$

As shown in Figure 7.11, Prager kinematic hardening rule moves in the direction perpendicular to the yield surface; Ziegler kinematic hardening rule moves along the direction from the center of the yield surface $\{\alpha_0\}$ to the stress point $\{\sigma\}$

From Equation (7.89), the yield condition becomes

$$f\left(\{\sigma_\alpha\}\right) = \bar{\sigma}_\alpha - \sigma_0 = 0 \tag{7.93}$$

where

$$\{\sigma_\alpha\} = \{\sigma\} - \{\alpha_0\} \tag{7.94}$$

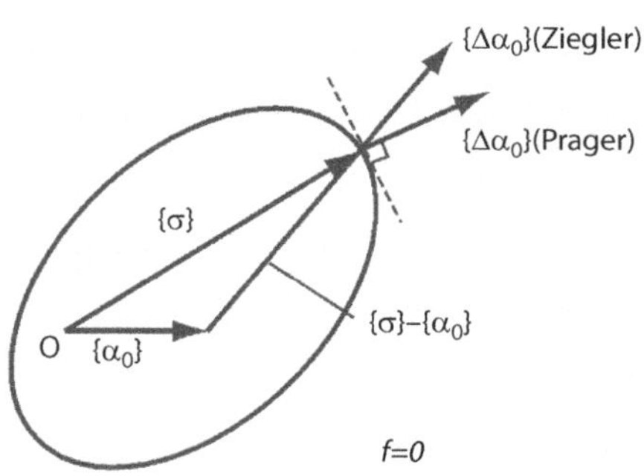


**Figure 7.11** Move direction of center for yield surface.

$\bar{\sigma}_\alpha$ is the equivalent stress which considers the movement of yield surface and is expressed as below:

$$\bar{\sigma}\alpha = \frac{1}{\sqrt{2}}\Bigg[\left\{(\sigma_x - \alpha_x) - (\sigma_y - \alpha_y)\right\}^2 + \left\{(\sigma_y - \alpha_y) - (\sigma_z - \alpha_z)\right\}^2$$
$$+\left\{(\sigma_z - \alpha_z) - (\sigma_x - \alpha_x)\right\}^2$$
$$+6\left\{(\tau_{xy} - \alpha_{xy})^2 + (\tau_{yz} - \alpha_{yz})^2 + (\tau_{zx} - \alpha_{zx})^2\right\}\Bigg]^{1/2} \tag{7.95}$$

By setting yield function $f$ as the plastic potential, plastic strain increment may be expressed as

$$\{\Delta\varepsilon^p\} = \Delta\lambda\left\{\frac{\partial f}{\partial\sigma}\right\} = \Delta\lambda\frac{3}{2\bar{\sigma}_\alpha}\{\sigma'_\alpha\} \tag{7.96}$$

where

$$\{\sigma'_\alpha\} = \{\sigma'\} - \{\alpha'_0\} \tag{7.97}$$

### 7.5.5 Stress Increment–Strain Increment Relation in the Plastic Region

Total strain increment is the sum of elastic strain increment and plastic strain increment,

$$\{\Delta\varepsilon\} = \{\Delta\varepsilon^e\} + \{\Delta\varepsilon^p\} \tag{7.98}$$

On the other hand, the relationship between stress increment and elastic strain increment may be expressed as below:

$$\{\Delta\sigma\} = [D^e]\{\Delta\varepsilon^e\} \tag{7.99}$$

Substituting this equation into Equation (7.98), we obtain

$$\{\Delta\sigma\} = [D^e](\{\Delta\varepsilon\} - \{\Delta\varepsilon^p\}) \tag{7.100}$$

If the associated flow rule according to the yield function and plastic potential is used, the plastic strain increment $\{\Delta\varepsilon^p\}$ can be expressed as

$$\{\Delta\varepsilon^p\} = \Delta\lambda\left\{\frac{\partial f}{\partial\sigma}\right\} \tag{7.101}$$

In general, the yield function *f is* a function of stress and plastic strain and may be written as

$$f = f(\{\sigma\},\{\varepsilon^p\}) \tag{7.102}$$

When plastic deformation occurs, the following equation may be obtained:

$$\Delta f = \left\{\frac{\partial f}{\partial\sigma}\right\}^T \{\Delta\sigma\} + \left\{\frac{\partial f}{\partial\varepsilon^p}\right\}^T \{\Delta\varepsilon^p\} = 0 \tag{7.103}$$

Substituting Equation (7.101) into Equations (7.100) and (7.103), we obtain

$$\{\Delta\sigma\} = [D^e]\left(\{\Delta\varepsilon\} - \Delta\lambda\left\{\frac{\partial f}{\partial\sigma}\right\}\right) \tag{7.104}$$

$$\left\{\frac{\partial f}{\partial\sigma}\right\}^T \{\Delta\sigma\} + \left\{\frac{\partial f}{\partial\varepsilon^p}\right\}^T \left\{\frac{\partial f}{\partial\sigma}\right\}\Delta\lambda = 0 \tag{7.105}$$

Eliminating $\{\Delta\sigma\}$ from Equations (7.104) and (7.105), $\Delta\lambda$ is obtained from the following equation:

$$\Delta\lambda = \frac{\left\{\frac{\partial f}{\partial \sigma}\right\}^T \left[D^e\right]\{\Delta\varepsilon\}}{-\left\{\frac{\partial f}{\partial \varepsilon^p}\right\}^T \left\{\frac{\partial f}{\partial \sigma}\right\} + \left\{\frac{\partial f}{\partial \sigma}\right\}^T \left[D^e\right]\left\{\frac{\partial f}{\partial \sigma}\right\}} \tag{7.106}$$

Substituting Equation (7.106) into Equation (7.104), we obtain

$$\begin{aligned}\{\Delta\sigma\} &= \left(\left[D^e\right] - \frac{\left[D^e\right]\left\{\frac{\partial f}{\partial \sigma}\right\}\left\{\frac{\partial f}{\partial \sigma}\right\}^T \left[D^e\right]}{-\left\{\frac{\partial f}{\partial \varepsilon^p}\right\}^T \left\{\frac{\partial f}{\partial \sigma}\right\} + \left\{\frac{\partial f}{\partial \sigma}\right\}^T \left[D^e\right]\left\{\frac{\partial f}{\partial \sigma}\right\}}\right)\{\Delta\varepsilon\} \\ &= \left(\left[D^e\right] + \left[D^p\right]\right)\{\Delta\varepsilon\}\end{aligned} \tag{7.107}$$

Here, $\left[D^p\right]$ is expressed as

$$\left[D^p\right] = -\frac{\left[D^e\right]\left\{\frac{\partial f}{\partial \sigma}\right\}\left\{\frac{\partial f}{\partial \sigma}\right\}^T \left[D^e\right]}{-\left\{\frac{\partial f}{\partial \varepsilon^p}\right\}^T \left\{\frac{\partial f}{\partial \sigma}\right\} + \left\{\frac{\partial f}{\partial \sigma}\right\}^T \left[D^e\right]\left\{\frac{\partial f}{\partial \sigma}\right\}} \tag{7.108}$$

and this must be considered when the material is in the plastic condition.

Substituting Equation (7.106) into Equation (7.101), the plastic strain increment is expressed in the following equation.

$$\left\{\Delta\varepsilon^p\right\} = \frac{\left\{\frac{\partial f}{\partial \sigma}\right\}\left\{\frac{\partial f}{\partial \sigma}\right\}^T \left[D^e\right]\{\Delta\varepsilon\}}{-\left\{\frac{\partial f}{\partial \varepsilon^p}\right\}^T \left\{\frac{\partial f}{\partial \sigma}\right\} + \left\{\frac{\partial f}{\partial \sigma}\right\}^T \left[D^e\right]\left\{\frac{\partial f}{\partial \sigma}\right\}} \tag{7.109}$$

In the process of plastic deformation, $\Delta\lambda$ in Equation (7.101) must have a positive value. Therefore, by checking the sign of $\Delta\lambda$ in Equation (7.106), the unloading condition can be detected.

## 7.6 Deformation Matrix

The deformation matrix $\left[k_D\right]$ is symmetric, and nonzero terms are given below:

$$k_D(1,2) = k_D(7,8) = -k_D(1,8) = -k_D(2,7) = -\rho_z^2(EA/L)(\theta_{z1} + \theta_{z2})/10 \tag{7.110}$$

$$k_D(1,3) = k_D(7,9) = -k_D(1,9) = -k_D(3,7) = \rho_y^2(EA/L)(\theta_{y1} + \theta_{y2})/10 \tag{7.111}$$

$$k_D(1,5) = -k_D(5,7) = -\alpha_y + \rho_y^2 EA(-4\theta_{z1} + \theta_{z2})/30 \tag{7.112}$$

$$k_D(1,6) = -k_D(6,7) = -\alpha_z + \rho_z^2 EA(-4\theta_{z1} + \theta_{z2})/30 \tag{7.113}$$

$$k_D(1,11) = -k_D(7,11) = \alpha_y + \rho_y^2 EA(\theta_{y1} - 4\theta_{y2})/30 \tag{7.114}$$

$$k_D(1,12) = -k_D(7,12) = -\alpha_z + \rho_z^2 EA(\theta_{z1} - 4\theta_{z2})/30 \tag{7.115}$$

where

$$\eta_y = EI_y/(GA_z L^2) \quad \eta_z = EI_z/(GA_y L^2) \tag{7.116}$$

$$\rho_y = 1/(1 + 12\eta_y) \quad \rho_z = 1/(1 + 12\eta_z) \tag{7.117}$$

$$\alpha_y = 2\rho_y^2 EA\eta_y(1 + 6\eta_y)(\theta_{y1} - \theta_{y2}) \tag{7.118}$$

$$\alpha_z = 2\rho_z^2 EA\eta_z \left(1+6\eta_z\right)\left(\theta_{z1}-\theta_{z2}\right) \tag{7.119}$$

# Biblography


1. Archer, J.S., Consistent Matrix Formulations for Structural Analysis Using Finite Element Techniques. *AIAA J.*, 3, 1910–1918, 1965.
2. Bai Y. and Pedersen, P.T., Earthquake Response of Offshore Structures. *Proc. of the 10th Int. Conference on Offshore Mechanics and Arctic Engineering, OMAE '91*, Stavanger, 1991.
3. Bai, Y. and Pedersen, P.T., Collision Response of Offshore Structures and Bridges. *International Symposium on Marine Structures, ISMS '91*, Shanghai, 1991.
4. Bai, Y., SANDY - A Structural Analysis Program for Static and Dynamic Response of Nonlinear System. User's Manual, Version 2, Department of Ocean Engineering, The Technical University of Denmark, March 1991.
5. Bathe, K.J., *Finite Element Methods*, Springer, 1986.
6. Chakrabarty, J., *Theory of Plasticity*, McGraw-Hill Book Company, 1987.
7. Chen, W.F. and Han, D.J., *Plasticity for Structural Engineers*, Springer, 1987.
8. Fujikubo, M., Bai, Y., Ueda, Y., Dynamic Elastic-Plastic Analysis of Offshore Framed Structures by Plastic Node Method Considering Strain-Hardening Effects. *Int. J. Offshore Polar Engng. Conf.*, 1, 3, 220–227, 1991.
9. Hill, R., *The Mathematical Theory of Plasticity*, Oxford Univ. Publisher, 1950.
10. Jones, N., *Structural Impacts*, Cambridge University Press, 1989.
11. Liu, J.H. and Jones, N., Experimental Investigation of Clamped Beams Struck Transversely by a Mass. *Int. J. Impact Eng.*, 6, 4, 303–335, 1987.
12. Liu, J.H. and Jones, N., Dynamic Response of a Rigid Plastic Clamped Beam Struck by a Mass at Any Point on the Span. *Int. J. Solids Struct.*, 24, 3, 251–270, 1988.
13. Messmer, S. and Sayir, M., Dynamic Elastic-Plastic Behavior of a Frame. *Eng. Comput.*, 5, 231–240, 1988.
14. Mosquera, J.M., Symonds, P.S., Kolsky, H., On Elastic-Plastic Rigid-Plastic Dynamic Response with Strain Rate Sensitivity. *Int. J. Mech. Sci.*, 27, 741–749, 1985.
15. Mosquera, J.M., Symonds, P.S., Kolsky, H., Impact Tests on Frames and Elastic-Plastic Solutions. *J. Eng. Mech. ASCE*, 111, 11, 1380–1401, 1985.
16. Nedergaard, H. and Pedersen, P.T., Analysis Procedure for Space Frame with Material and Geometrical Nonlinearities. *Europe-US Symposium - Finite Element Methods for Nonlinear Problems*, edited by Bergan, Bathe and Wunderlich, Springer, pp. 211–230, 1986.
17. Przemieniecki, J.S., *Theory of Matrix Structural Analysis*, McGraw-Hill Inc., 1968.

18. Save, M.A. and Massonnet, C.E., *Plastic Analysis and Design of Plates, Shells and Disks*, North-Holland Publishing Company, 1972.
19. Symonds, P.S. and Yu, T.X., Counter-Intuitive Behavior in a Problem of Elastic-Plastic Beam Dynamics. *J. Appl. Mech.*, 52, 517–522, 1985.
20. Ueda, Y. and Fujikubo, M., Plastic Collocation Method Considering Strain Hardening Effects. *J. Soc. Nav. Archit. Jpn.*, 160, 306–317, 1986, (in Japanese).
21. Ueda, Y. and Yao, T., The Plastic Node Method: A New Method of Plastic Analysis. *Comput. Methods Appl. Mech. Eng.*, 34, 1089–1104, 1982.
22. Yamada, Y., Yoshimura, N., Sakurai, T., Plastic Stress-Strain Matrix and its Application for the Solution of Elastic-Plastic Problems by the Finite Element Method. *Int. J. Mech. Sci.*, 10, 343–354, 1968.
23. Yagawa, G. and Miyazaki, N., *Heat Induced Stress, Creep and Heat Transfer Analysis Based on the Finite Element Analysis*, Science Publisher (in Japanese), 1985.
24. Yoshimura, S., Chen, K.L., Atluri, S.N., A Study of Two Alternate Tangent Modulus Formulations and Attendant Implicit Algorithms for Creep as well as High-Strain-Rate Plasticity. *Int. J. Plast.*, 3, 391–413, 1987.
25. Zienkiewicz, O.C., *The Finite Element Method*, McGraw-Hill Book Company, 1977.
26. Ueda, Y. Randomly transitional phenomena in the system governed by Duffing's equation. *J. Stat. Phys.*, 20, 181–196, 1979.
27. Fujikubo, M., Akifuji, K., Tamehiro, M., Study on Deployment Method of Mooring Chains of Semi-submersible Offshore Units, *J. Soc. Nav. Archit. Jpn.*, 1987, 162, 523–534, 1987.

8

# Metal Forming Processing Plasticity

***Abstract***
This chapter provides an in-depth overview of metal forming processes with emphasis on their practical applications, mechanics, and material behavior. It begins by introducing manufacturing classification, highlighting bulk and sheet metal forming methods such as rolling, forging, extrusion, deep drawing, and powder forming. The discussion extends to forming mechanics, describing the role of tensile, compressive, and shear forces in shaping metals. Sheet metal formability is explored through strain distribution, deformation modes, and failure mechanisms, including necking, wrinkling, and tearing. The chapter introduces strain diagrams and proportional strain paths to evaluate deformation states, using true strain measures for accuracy. Material anisotropy in rolled sheets is analyzed through the Lankford coefficient and its angular variation within the sheet plane, with implications for forming quality and earing behavior. Finally, a case study on the carcass layer of flexible pipes showcases the integration of forming theory into industrial applications. The chapter underscores the importance of understanding strain paths, yield behavior, and anisotropic effects for optimizing forming performance and product integrity in metal processing industries.

***Keywords*:** Metal forming, sheet formability, bulk deformation, anisotropy, strain analysis, deep drawing, carcass production

## 8.1 Introduction

Metal forming is a key manufacturing method that holds great industrial value due to its many benefits, including low cost, enhanced mechanical performance, operational versatility, high output, and efficient material use. Most everyday products are made up of parts that begin as raw materials and are shaped using various production methods. For example, a simple pen consists of multiple assembled pieces, while a car is built from over 15,000 parts created using different fabrication techniques [1]. In general, manufacturing refers to the transformation of raw materials into finished goods through processes like casting, forming, machining, joining, and surface treatment. Today's manufacturing landscape also includes

Yong Bai. Plasticity and Ductile Fracture in Structural Engineering, (207–260) © 2026 Scrivener Publishing LLC

cutting-edge methods such as micromachining, nanotechnology-based fabrication, and ultra-precise manufacturing. To keep up with industrial demands, engineers must carefully choose suitable materials and tools to reduce both costs and energy use [2]. Environmental impacts, including emission reduction and waste management, are also crucial factors in method selection. Modern strategies—such as lean and agile manufacturing, adaptive process control, and group technology—help improve product quality while lowering expenses. Furthermore, the integration of computer technology, robotics, and simulation tools allows engineers to fine-tune production parameters ahead of full-scale operations, leading to more efficient and high-quality manufacturing [3].

## 8.2 Classification

Manufacturing processes play a vital role in transforming raw materials into finished components through a wide variety of methods, each tailored to meet specific functional and structural requirements. These processes are typically grouped into categories such as casting, forming, machining, joining, coating, dividing, and property modification. Among these, forming processes stand out for their extensive industrial application and ability to produce components with high structural integrity, minimal material waste, and enhanced mechanical properties. Forming is indispensable in numerous sectors, including automotive, aerospace, energy, defense, and consumer goods, where complex geometries and reliable performance are required.

Forming operations are central to the production of both everyday consumer items and high-precision engineering components. A single automobile, for instance, integrates thousands of formed parts such as wheel rims, chassis members, engine valves, and structural panels. In daily life, countless household items—ranging from cookware and containers to bicycles, tools, and fans—are created using forming techniques. This widespread application stems from the process's capacity to shape raw materials through plastic deformation by applying mechanical forces like tension, compression, bending, or shear. Unlike machining, which removes material, forming reshapes it while conserving volume, resulting in cost efficiency and structural benefits.

In manufacturing, forming is broadly divided into various types based on factors such as the shape and size of the raw material, the dominant force applied, and the temperature conditions during deformation. Commonly, forming operations are distinguished into bulk forming and sheet metal forming, depending on the material's surface area-to-volume ratio. In addition, modern manufacturing practices have introduced powder forming,

Table 8.1 Metal forming types.

| Formed by compressive stress | Tensile and compressive stresses | Forming under tensile stress | Bending and shearing stresses |
|---|---|---|---|
| Open die forging<br>Closed die forging<br>Rolling<br>Coining<br>Extrusion | Deep drawing<br>Spinning<br>Stripping<br>Wrinkle bulging | Stretch forming<br>Stretching<br>Expanding | Bending<br>Shearing<br>Punching<br>Blanking |

which enables the production of near-net-shape components with tailored properties and minimal waste, though it presents its own challenges, particularly in managing porosity.

The method of force application further classifies forming operations into distinct categories, which are systematically presented in technical classifications such as those found in Table 8.1. This categorization helps engineers and designers select appropriate forming strategies based on material behavior and design requirements. Furthermore, forming processes are also defined by temperature: cold, warm, and hot forming, with hot forming typically carried out above the material's recrystallization temperature (ranging from 0.5 Tm to 0.8 Tm, where Tm is the melting point). This classification by temperature affects not only the energy consumption and tool wear but also the microstructural evolution and mechanical properties of the formed component.

In modern manufacturing systems, the integration of forming techniques must also consider factors such as energy efficiency, process automation, and environmental impact. Advanced simulation tools, intelligent process control, and hybrid forming methods are increasingly being adopted to meet the stringent demands of performance, quality, and sustainability. Thus, forming remains a critical element of the manufacturing landscape, enabling the efficient production of components across a vast spectrum of applications.

## 8.3 Forming Processes

### 8.3.1 Bulk-Forming Process

Bulk-forming processes are fundamental manufacturing techniques in which substantial plastic deformation is applied to a workpiece to achieve a specific shape or cross-sectional geometry. These methods are predominantly based on compressive forces, although tensile and shear stresses may also play a role depending on the specific process. Unlike sheet-forming operations, bulk-forming processes are used on materials with low surface-area-to-volume ratios, making them highly effective for shaping thick sections and large-volume components. They are widely used in industries such as automotive, aerospace, defense, construction, and heavy machinery for the production of components requiring strength, reliability, and geometric precision.

One of the most widely used bulk-forming techniques is rolling, a process that involves compressive deformation wherein a metal workpiece is passed between two rotating cylindrical rolls. The primary goal of rolling is to reduce the thickness of the material and to alter its cross-section, as depicted schematically in Figure 8.1. Rolling is used not only to produce semi-finished products—such as bars, sheets, and plates—but also to manufacture finished structural elements like angles, channels, I-beams, and custom sections. Depending on the temperature of the material during processing, rolling can be categorized into hot rolling and cold rolling. Hot rolling is carried out above the material's recrystallization temperature, enabling significant shape change with lower force requirements and improved workability. On the other hand, cold rolling is performed at or near room temperature, offering improved surface finish, tighter

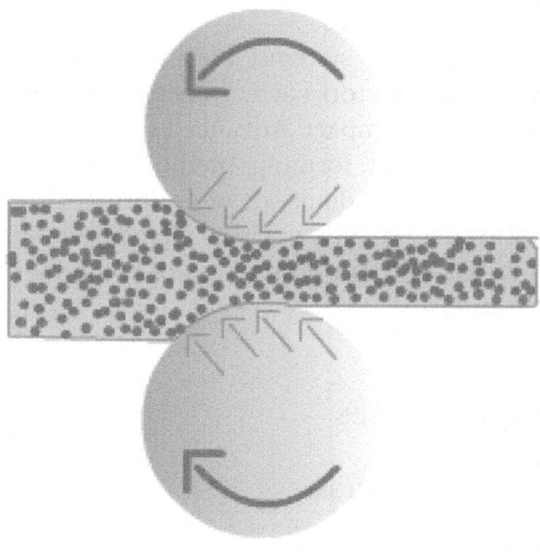

**Figure 8.1** Rolling process.

dimensional tolerances, and enhanced mechanical strength through strain hardening.

Another essential bulk-forming operation is forging, a process wherein a workpiece, typically referred to as a billet, is plastically deformed into a final shape through the application of compressive and sometimes tensile forces. This deformation is achieved using a die-punch setup, as illustrated in Figure 8.2. Forging can be classified into open-die and closed-die forging, depending on the level of confinement applied to the material during the process.

In open-die forging, the material is compressed between flat or slightly contoured dies that do not completely enclose the workpiece. This process is often used for the initial rough shaping of large metal blocks, converting them into intermediate forms suitable for subsequent processing. Operations like drawing out, upsetting, and spreading are common in this stage.

In contrast, closed-die forging (also known as impression-die forging) confines the workpiece entirely within a die cavity that mirrors the final part shape. The billet is squeezed under high pressure until it completely fills the cavity, thereby achieving a near-net-shape product. This method is ideal for manufacturing high-precision components with intricate geometries, such as valve bodies, pump parts, small gears, connecting rods, and hand tools like spanners. Closed-die forging is often used in mass production due to its ability to yield consistent, repeatable results with minimal material waste.

A specialized form of forging known as coining is used when surface detail is crucial. In this process, very high compressive stresses are applied to the surface of the material using an embossing punch, transferring detailed patterns or impressions onto the workpiece. Common applications of coining include the fabrication of currency coins, commemorative

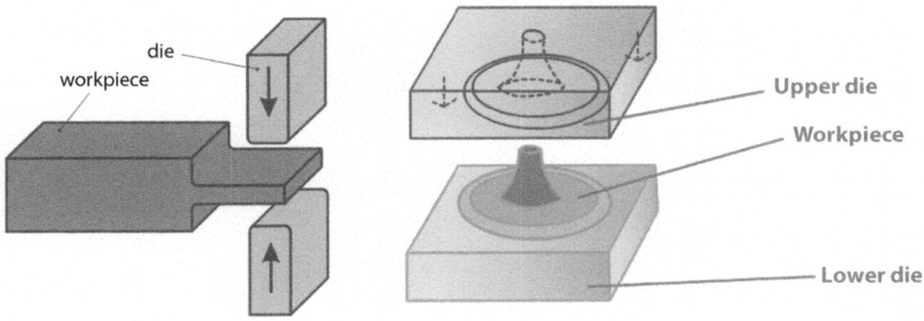


**Figure 8.2** Forging process.

medallions, badges, and ornamental parts, where high fidelity to the mold pattern and excellent surface quality are essential.

Another highly significant bulk-forming method is extrusion, which involves pushing a metal billet through a die to produce long products with uniform or variable cross-sections. As shown in Figure 8.3, forward (or direct) extrusion entails pushing the billet in the same direction as the material flow through the die. In contrast, Figure 8.4 depicts backwards (or indirect) extrusion, where the die moves toward the stationary billet, and the material flows in the opposite direction of the applied punch force. Both extrusion types can be conducted under hot or cold conditions, depending on material properties and product requirements. The extrusion process is widely used to produce shafts, tubes, rods, cups, cans, and preforms for gear blanks. Due to its ability to manufacture components with complex cross-sections and excellent surface finishes, extrusion remains indispensable in modern industry.

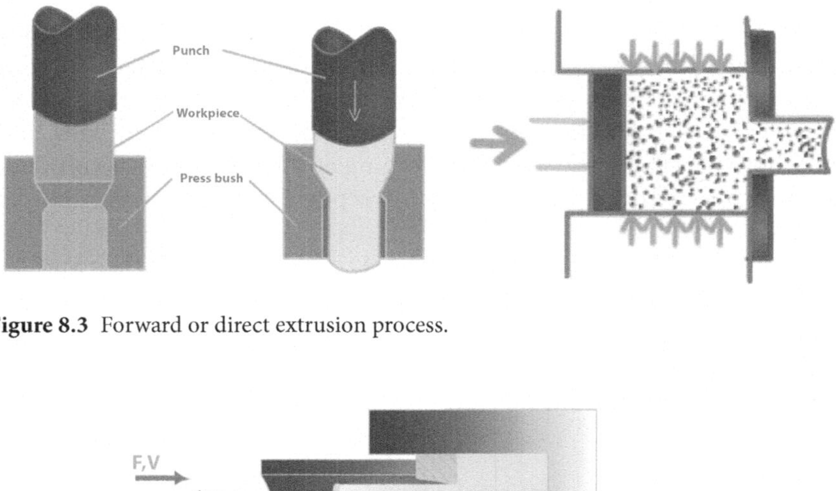


**Figure 8.3** Forward or direct extrusion process.

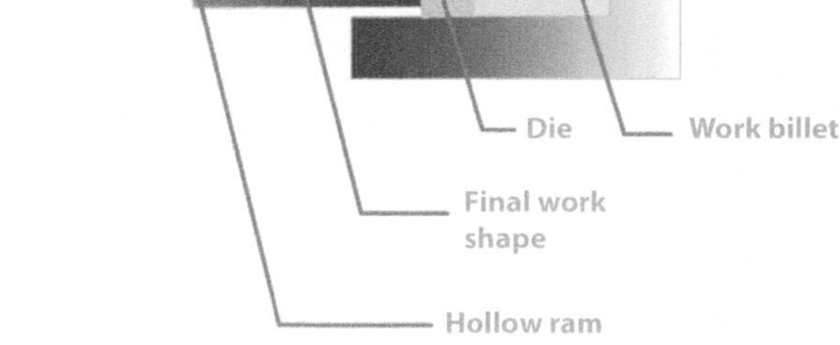


**Figure 8.4** Backwards or indirect extrusion.

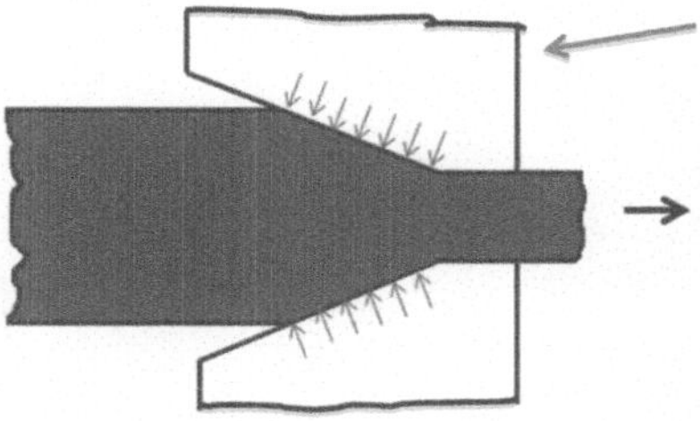


**Figure 8.5** Wire drawing.

Finally, the wire drawing process is employed when converting thicker rods into long, thin wires. This operation relies on tensile forces to pull the material through a series of progressively smaller dies, gradually reducing its diameter and increasing its length. A typical schematic is presented in Figure 8.5. This method is extensively used for the production of electrical wires, cables, tension wires, and musical instrument strings, where consistent diameter and high tensile strength are critical. A related process, tube drawing, follows similar principles and is used to produce seamless metal tubes that are essential in applications demanding high strength and leak-proof performance, such as hydraulic systems, heat exchangers, and high-pressure pipelines.

In summary, bulk-forming processes are central to modern manufacturing due to their ability to shape large volumes of material efficiently, improve mechanical performance through controlled deformation, and produce a wide array of structural and functional components. Their adaptability, scalability, and cost-effectiveness make them vital across virtually all sectors of engineering production.

### 8.3.2 Sheet Metal Operations

Sheet metal forming encompasses a range of processes in which thin, flat metal sheets, strips, or plates are plastically deformed into desired shapes through the application of mechanical forces. Unlike bulk forming, where the volume of the material is large relative to the surface area, sheet metal forming involves high surface-area-to-volume ratio workpieces, and deformation typically occurs without significant changes in material thickness. These processes are widely used in industries such as automotive, aerospace, household appliances, consumer electronics, and packaging, where lightweight, durable, and geometrically complex components are required.

One of the most important and widely used sheet metal operations is deep drawing, a process that transforms a flat sheet or blank into a hollow,

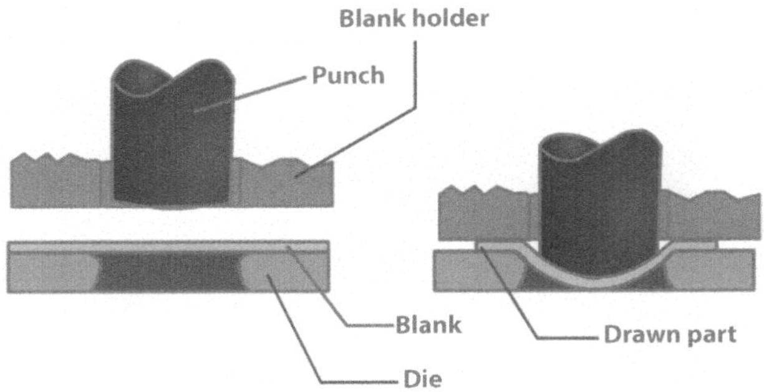


**Figure 8.6** Deep drawing.

cup-like, or cylindrical component without appreciable change in thickness. As illustrated in Figure 8.6, the sheet metal is clamped and then drawn into a die cavity using a punch.

During this process, tensile and compressive forces act simultaneously on the material. The edges of the blank experience compressive hoop stress, while the central region undergoes tensile stress as it stretches over the punch. Deep drawing is capable of producing intricate geometries and is commonly used for manufacturing automotive fuel tanks, kitchen sinks, cooking pots, beverage cans, and pressure vessels.

In cases where the desired final shape is too complex or deep to be achieved in a single stage, the process may be carried out in multiple stages, including redrawing and multi-draw operations, which gradually form the blank into the final component through successive passes. The success of deep drawing depends on various factors, including material ductility, blank holder force, die design, and lubrication.

An advanced variation of this method is hydromechanical deep drawing, which combines the action of a mechanical punch with the hydrostatic pressure of a pressurized fluid acting on the opposite side of the sheet. This method provides improved formability, enables greater depth-to-diameter ratios, and helps minimize wrinkling or tearing during drawing. It is especially useful for thin-walled, high-precision components where dimensional accuracy is critical.

Other notable sheet metal processes include flanging, which involves the formation of flanges or collars by bending the edge of the sheet material. This is particularly useful for creating mounting features or reinforcing the edges of components.

Spinning, another significant process, involves shaping a sheet metal blank into a symmetrical, hollow part by rotating it against a mandrel using a roller, as shown schematically in Figure 8.7. The process combines tensile and compressive stresses as the material is gradually formed around the mandrel. Spinning is frequently used to produce funnel-shaped objects, satellite dishes, lampshades, and pressure vessels, especially when small to medium production volumes are involved. The process offers excellent material utilization, low tooling cost, and the ability to fabricate complex shapes with fine surface finishes.

Embossing is another specialized sheet metal technique used to create raised or recessed patterns on the surface of the sheet. This is achieved by pressing the sheet between a die and an embossing punch, as depicted in Figure 8.8. Embossing is commonly applied for decorative designs, brand markings, serial numbers, and texture patterns in both industrial and commercial products.

Bending, a widely utilized operation in sheet metal fabrication, refers to deforming the sheet along a straight axis to obtain an angular profile.

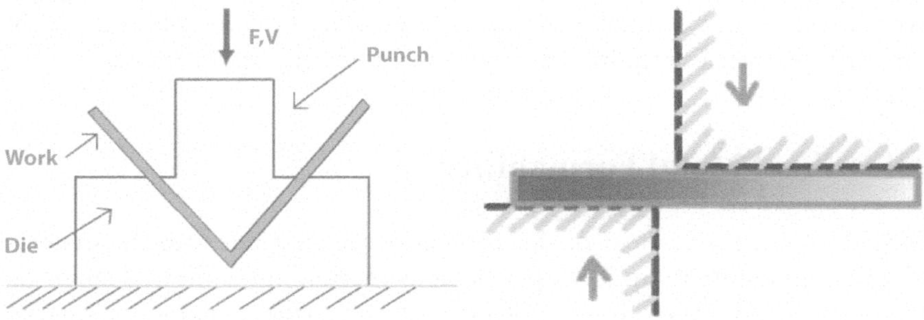


**Figure 8.7** Bending drawing.

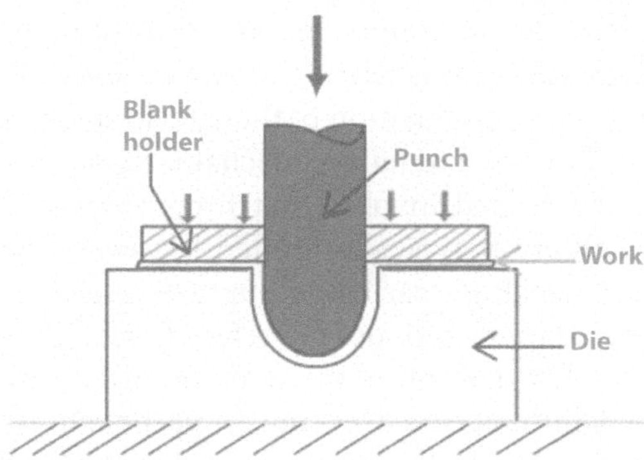


**Figure 8.8** Shearing drawing.

Several bending techniques are available depending on the complexity and application, including rotary bending, swivel bending, roll bending, and die bending. Rotary and roll bending utilize curved dies or rollers to apply the necessary force, while die bending can be performed using flat or shaped dies, enabling sharp or contoured bends. For applications requiring high surface detail and minimal spring-back, die coining is often used, where a sheet is pressed between closed dies with sufficient force to permanently set the bend angle.

These operations form the backbone of precision sheet metal fabrication, enabling the creation of a wide variety of components with controlled geometry, excellent repeatability, and minimal material waste. The selection of the appropriate method depends on factors such as sheet thickness, material properties, production volume, desired tolerances, and cost efficiency.

In summary, sheet metal forming operations offer exceptional versatility for manufacturing lightweight, strong, and esthetically pleasing products. By combining tensile, compressive, and shear forces in various configurations, these processes enable engineers to produce both simple and highly complex geometries required in modern engineering applications.

## 8.4 Sheet Metal Formability

Formability in the context of sheet metal processing refers to the material's inherent ability to undergo plastic deformation into a desired shape or geometry without the onset of defects such as tearing, necking, or surface irregularities. This property is a critical performance metric in the evaluation of sheet metal workability and directly influences the reliability, cost-effectiveness, and quality of the formed components. However, the concept of formability is not uniform across all forming processes; it varies significantly depending on whether the process involves deep drawing, stretching, bending, or a combination of these techniques, as each imposes distinct strain distributions and failure mechanisms on the material.

In forming processes that involve stretching, such as stretch forming or stamping, formability is typically limited by several key phenomena, including localized necking, tearing or rupture, surface wrinkling, surface roughness degradation, and loss of surface finish. These defects act as limiting conditions that determine the maximum deformation the sheet can sustain before failure occurs. Necking, in particular, is a critical mode of instability not only in stretching operations but also in any forming technique where significant in-plane tensile strains are present. Therefore, understanding and predicting necking behavior is essential for ensuring successful and defect-free forming operations.

Unlike traditional mechanical properties such as yield strength or hardness, formability is not a single, easily quantifiable value. Instead, it is a complex and multifactorial material property that depends on a wide range of intrinsic and extrinsic factors. These include the material's strain hardening behavior, anisotropy, strain rate sensitivity, uniform elongation, and homogeneity of deformation. Due to this complexity, various analytical and empirical models have been developed to quantify formability in practical forming scenarios.

One significant advancement in this area came from the work of Cockcroft and Latham [4], who initially proposed energy-based criteria to describe fracture and deformation behavior. Building upon this foundation, Marciniak [5] introduced a more refined approach that involves the use of formability indices, which are mathematical functions tailored to specific forming conditions. These indices offer a numerical framework to evaluate how different materials respond under specific process parameters and can be used in both experimental studies and computational simulations.

The formability index can be described as a function of several material-specific parameters, as given by the equation:

$$F_{\mathrm{kl}} = F(n, m, r, e_{\mathrm{u}}, f) \tag{8.1}$$

where $m$, $n$, $e_u$, $f$, and $r$ are the strain rate sensitivity, the hardening coefficient, the ultimate strain, the non-homogeneity coefficient, and the anisotropy coefficient, respectively.

This equation highlights that the formability index is not fixed but depends intricately on the material behavior, and it changes dynamically with different process setups. The variation or sensitivity of the formability index to each parameter can be captured through a differential expression:

$$dF = \frac{\partial F}{\partial n}dn + \frac{\partial F}{\partial m}dm + \frac{\partial F}{\partial r}dr + \frac{\partial F}{\partial \varepsilon_u}d\varepsilon_u + \frac{\partial F}{\partial f}df \tag{8.2}$$

Each partial derivative in this expression quantifies how sensitive the formability index is to a specific mechanical or structural property of the material. These sensitivities may be derived through analytical modeling or experimental calibration, depending on the complexity of the forming operation and the desired level of precision [6].

Historically, the understanding and analysis of formability emerged alongside the evolution of industrial metal forming techniques during the late 19th century. As sheet metal applications expanded rapidly, particularly in transportation and construction, researchers began to investigate the material behaviors governing successful forming. Early pioneers such as Bessemer, Parkes, Adamson, Considère [7], and Erichsen [8] laid the theoretical and experimental groundwork that has since evolved into modern formability science. Their work contributed to the development of standard testing protocols and evaluation criteria still in use today.

Over the decades, researchers have proposed numerous methodologies to assess the formability of sheet materials. These methods can be broadly classified into four main categories, each providing unique insights into how sheet metal behaves under forming loads:

1. Simulative tests — These are process-specific tests that replicate actual forming operations to assess the material's behavior under realistic conditions.
2. Mechanical property-based techniques — These use standard mechanical tests, such as tensile testing, to estimate formability parameters.
3. Limiting dome height (LDH) methods — These tests involve forming a hemispherical dome to determine the maximum height (and hence strain) before failure.
4. Forming limit diagrams (FLD) — Widely used in industrial settings, FLDs map the critical strain combinations at which necking or tearing occurs, providing a visual tool for process optimization.

Each of these methods plays a crucial role in material selection, tooling design, and process planning, ensuring that the chosen material and forming parameters lead to defect-free parts.

As depicted in Figure 8.9, a sheet element can be analyzed in three stages to understand the strain evolution during forming. Initially (a), the undeformed sheet is marked with reference grids, typically in circular or square patterns. After deformation (b), these grids stretch into ellipses, characterized by major and minor diameters $d_1$ and $d_2$, which reflect the principal strain directions. Finally (c), the traction vectors (T) indicate the distributed forces acting per unit width, representing the mechanical loads driving the deformation process.

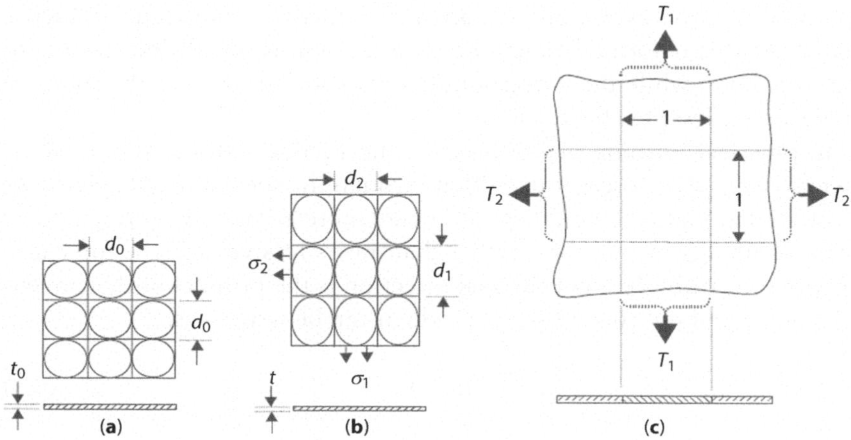


**Figure 8.9** A sheet element is illustrated in three stages: (a) the initial, undeformed state, where both circular and square grids are marked; (b) the deformed state, where the circular grids are stretched into ellipses with major and minor diameters $d_1$ and $d_2$, respectively; and (c) the applied tractions, $T$, representing the forces acting per unit width of the sheet.

Together, the evaluation of sheet metal formability serves as a cornerstone for advancing forming technology, minimizing defects, and ensuring that material behavior aligns with design intent across a wide range of engineering applications.

### 8.4.1 Uniform Strain Distribution in Sheet Metal Forming

The theoretical framework is now applied to analyze a segment of sheet metal undergoing uniform and proportional deformation, a scenario commonly encountered in many practical forming operations. This process is schematically illustrated in Figure 8.9. In the initial undeformed configuration, the sheet has a uniform thickness $t_0$ and is marked with a reference grid to observe strain distribution. This grid can either be composed of circular marks with a diameter $d_0$ or a square mesh with spacing also equal to $d_0$, as shown in Figure 8.9(a). These reference markings are essential in experimentally tracking material deformation during forming processes.

As deformation proceeds uniformly, the originally circular markers transform into elliptical shapes due to differential elongation along the principal strain directions. These ellipses are characterized by their major and minor diameters, denoted by $d_1$ and $d_2$, respectively. Alternatively, if a square grid is used and it aligns with the material's principal directions of strain, the squares evolve into rectangular elements, as illustrated in Figure 8.9(b). At this deformed stage, the sheet thickness is reduced from $t_0$ to a new value $t$, while the corresponding principal stresses in the material plane are represented by $\sigma_1$ and $\sigma_2$.

In practical forming operations, the mechanical deformation of sheet metal is caused by forces transmitted through the sheet along its plane. As illustrated in Figure 8.9(c), these forces are more accurately represented by stress resultants or tractions, rather than raw stress values. The force acting per unit width in any principal direction is the product of the current thickness $t$ and the current stress $\sigma$, which can be written as:

$$T = \sigma t \tag{8.3}$$

In this expression, $T$ is referred to as the tension, traction, or stress resultant, depending on the context. Because it integrates both the instantaneous stress level and the current material thickness, this quantity offers a more accurate representation of the true load being carried by the sheet at any moment during deformation. As such, this formulation is adopted consistently in deformation analyses that follow.

It is important to note that while the term "tension" is widely used, the value of $T$ is not strictly limited to tensile loading. If $T$ becomes negative in a given direction, it implies a compressive stress state. This convention is generally acceptable in sheet metal forming, as it is assumed that at least one of the two principal directions experiences tensile stress during any effective forming operation. This aligns with the physical limitation that sheet metal cannot be plastically formed by compressive loading at the edges—tensile stress is essential to initiate and sustain plastic flow.

Under this established convention, Principal Direction 1 corresponds to the direction experiencing the highest positive principal stress, and the associated major tension $T_1 = \sigma_1 \cdot t$ is always considered positive. In stretch forming, which is dominated by biaxial tensile deformation, the minor tension $T_2 = \sigma_2 \cdot t$ also remains tensile. However, in other forming processes like deep drawing or flanging, the minor direction may be subjected to compressive stress. When $T_2$ becomes negative, local thickness increase may occur due to lateral squeezing of the material. Nevertheless, if the compressive force becomes excessive, it can lead to surface instabilities such as wrinkling, which is a major defect in formed parts.

In real materials, the true stress $\sigma$ continues to increase with plastic strain due to strain hardening, although the rate of increase typically diminishes as deformation progresses. On the other hand, the traction or tension value $T$, which depends on both stress and thickness, often reaches a peak value during the forming process. This is because despite rising stress levels, the continuous reduction in thickness $t$ reduces the overall load-bearing capacity. The maximum value of $T$ acts as a limiting condition and is a key parameter in evaluating sheet formability, particularly in load-carrying analysis and failure prediction during complex forming sequences.

### 8.4.2 Strain Diagram in Sheet Metal Forming

A crucial first step in the analysis of any sheet metal forming process is the determination of the strain distribution across the formed part. Understanding how strain evolves during deformation allows engineers to identify critical regions, predict potential failure modes, and optimize the forming process. One of the most common and effective techniques to determine strain distribution is the use of a pre-marked reference grid, such as circular or square patterns applied to the sheet surface prior to forming. The deformation of these markers provides measurable changes that can be analyzed to determine local strain states, as previously illustrated in Figure 8.10.

For example, in the deep drawing process, which is commonly used to form cylindrical cups, symmetry about the central axis simplifies the analysis. As shown in Figure 8.10(a), only a single radial section of the cup needs to be studied to characterize the overall strain behavior. This representative section is marked in Figure 8.10(b), where strain measurements are taken along the line extending from the center of the cup toward the outer wall. When the measured strain values at different points along this line are plotted in principal strain space, they generate a curve, or strain locus, which represents the local deformation state throughout the process. This is demonstrated in Figure 8.10(c).

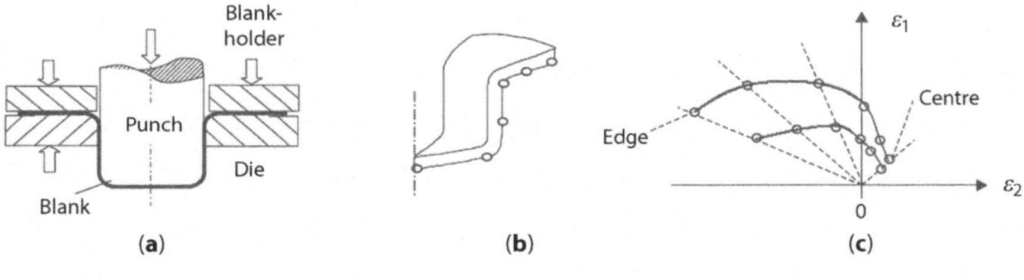


**Figure 8.10** Formation of a cylindrical cup using the deep drawing process. (b) A section of the cup indicating where strain measurements are taken. (c) Graphs showing strain distributions at two different stages of the drawing process.

As the forming operation proceeds and deformation progresses, the strain locus evolves. It typically expands outward, reflecting the accumulation of plastic strain in the material. However, this expansion is often non-uniform, as certain areas of the sheet may undergo further deformation while others reach a state of strain saturation or even process-induced limitations such as necking or wrinkling. Each forming process yields a unique pattern of strain distribution—a kind of fingerprint, often referred to as the strain signature. This signature is valuable in identifying and diagnosing defects, evaluating formability, and validating numerical simulations. The upcoming sections delve deeper into the interpretation of strain loci and how they relate to material performance and process limits.

The data points on the strain diagram, such as those depicted in Figure 8.10(c), are derived by evaluating the transformation of the circular grid patterns described earlier in Figure 8.9. As deformation transforms a circular grid into an ellipse, the major and minor axes of the resulting ellipse, denoted as $d_1$ and $d_2$, can be used to calculate the in-plane principal strains at that point. Additionally, by measuring the thickness of the sheet at the deformed location, the through-thickness strain can also be determined.

The principal strains—those in the directions of maximum and minimum normal deformation—are calculated using the logarithmic (true strain) relationships derived from the final dimensions. These are given by

$$\varepsilon_1 = \ln\frac{d_1}{d_0}, \varepsilon_2 = \ln\frac{d_2}{d_0}, \varepsilon_3 = \ln\frac{t}{t_0} \tag{8.4}$$

It is often assumed that the strain path followed during deformation is linear, implying that the ratio of the principal strains remains constant throughout the process. This assumption simplifies analysis and modeling, and the strain path ratio $\beta$ is then defined as

$$\beta = \frac{\varepsilon_2}{\varepsilon_1} \tag{8.5}$$

$$\beta = \frac{\ln\left(d_2 / d_0\right)}{\ln\left(d_1 / d_0\right)} \tag{8.6}$$

While this linear strain path assumption is valid in many practical forming operations, it is important to recognize that deviations can and do occur, especially in complex, multi-stage processes or in regions undergoing strain reversal. In such cases, the actual strain trajectory becomes nonlinear, and more advanced analytical or numerical techniques must be used to properly describe material behavior.

In situations where thickness strain is of interest, it can be computed either directly by comparing the current thickness $t$ with the original $t_0$, or indirectly *via* the assumption of constant volume deformation (i.e., plastic incompressibility). According to this assumption, the total volume remains unchanged during deformation, which yields the relationship:

$$\varepsilon_3 = \ln\frac{t}{t_0} = -(1+\beta)\varepsilon_1 = -(1+\beta)\ln\frac{d_1}{d_0} \tag{8.7}$$

From this, the current thickness after deformation can be expressed as

$$t = t_0 \exp(\varepsilon_3) = t_0 \exp\left[-(1+\beta)\varepsilon_1\right] \tag{8.8}$$

Alternatively, this same conclusion can be derived from volume conservation based on area transformation, such that

$$t = t_0 \frac{d_0^2}{d_1 d_2} \tag{8.9}$$

Using the relationships above, both the principal strains and the strain ratio can be calculated. The deformation process is most effectively characterized using the principal strains, that is,

$$\varepsilon_1 = \ln\left(\frac{d_1}{d_0}\right); \varepsilon_2 = \ln\left(\frac{d_2}{d_0}\right) = \beta\varepsilon_1; \varepsilon_3 = \ln\left(\frac{t}{t_0}\right) = -(1+\beta)\varepsilon_1 \tag{8.10}$$

Using these relationships, both principal strains and the strain ratio $\beta$ can be determined from direct geometric measurements. This analytical framework allows for a detailed and accurate description of the material's strain history, which is essential for predicting outcomes such as necking, failure, or thinning during forming.

In this context, the strain diagram provides an effective visual tool to represent the evolution of deformation. Each point plotted on this diagram corresponds to a final state of major and minor principal strains, along with the assumed linear strain path that connects the undeformed state to the current one. This is further illustrated in Figure 8.11, which presents

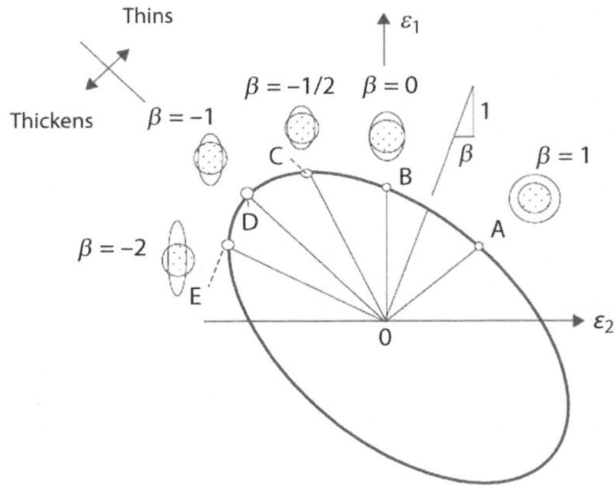


**Figure 8.11** The strain diagram illustrates various deformation modes, each associated with a specific strain ratio.

a generalized strain diagram not specific to any single forming operation. Instead, it serves as a reference framework for understanding various modes of deformation, each defined by a unique strain ratio $\beta$.

In Figure 8.11, the ellipse shown represents a contour of constant effective strain ε. All points lying on this contour have undergone the same equivalent plastic strain, resulting in a uniform flow stress $\sigma_f$ across the material. This concept is valuable in comparing different strain paths and assessing material strength, hardening behavior, and failure risk under various loading scenarios.

In summary, the strain diagram serves as a powerful analytical and diagnostic tool in sheet metal forming, enabling engineers to track deformation progress, predict forming limits, and design processes that ensure optimal material utilization while avoiding defects.

### 8.4.3 The Deformation Modes

In sheet metal forming analysis, deformation modes are most effectively visualized using a strain diagram, such as the one shown in Figure 8.11, where principal strain space is used to map the various paths that material elements may follow during plastic deformation. By convention, the first principal direction—often labeled Direction 1—is defined as the orientation along which the material experiences the largest (most positive) principal stress, $\sigma_1$, and hence the greatest major strain, $\varepsilon_1$. Under this assumption, all meaningful forming paths appear to the left of the diagonal in the strain space where the strain ratio $\beta = \varepsilon 2/\varepsilon 1 = 1$. According to this convention, the major principal stress $\sigma_1$ is either tensile or zero—that is, $\sigma_1 \geq 0$. The stress–strain relationship in forming processes can be described using the appropriate flow criteria, and for principal stress states, it takes the general form:

$$\alpha = \frac{2\beta + 1}{2 + \beta}, and, \beta = \frac{2\alpha - 1}{2 - \alpha} \tag{8.11}$$

In the extreme case where $\sigma_1 = 0$, yield $\alpha = -\infty$ and $\beta = -2$. Therefore, in practical sheet forming operations, all straining paths will fall within the range from point 0A to 0E in Figure 8.11, corresponding to a strain ratio range of $-2 \leq \beta \leq 1$.

The path labeled 0A represents equal biaxial stretching. This type of deformation occurs at the center of a sheet when it is stretched over a hemispherical punch, as shown in Figure 8.12. In this case, the in-plane strains are the same in all directions, causing a circular grid mark to expand uniformly while maintaining its shape. Since the strain ratio $\beta = 1$, the thickness strain is given by $\varepsilon_3 = -2\varepsilon_1$, meaning the sheet thins faster with increasing $\varepsilon_1$ than in any other deformation mode. The equivalent strain is given by

$$\bar{\varepsilon} = \sqrt{\frac{2}{9}\left\{\left(\varepsilon_1 - \varepsilon_1\right)^2 + \left(\varepsilon_2 - \varepsilon_3\right)^2 + \left(\varepsilon_3 - \varepsilon_1\right)^2\right\}} \tag{8.12}$$

Therefore, the effective strain is $\bar{\varepsilon} = 2\varepsilon_1$, indicating that the material work-hardens quickly as the major strain increases.

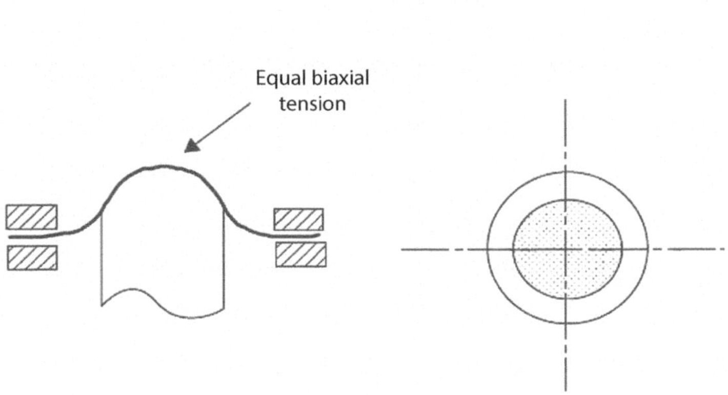


**Figure 8.12** Uniform biaxial stretching at the apex of a formed dome.

In this deformation path, the sheet undergoes extension in one principal direction while the transverse strain remains nearly zero, known as the plane strain condition. The corresponding strain ratio $\beta \approx 0$, and this mode is commonly observed along the vertical sidewalls of elongated parts, such as trough-shaped components (Figure 8.13). Here, grid circles transform into ellipses with one axis unchanged. Under plane strain, the material is especially susceptible to splitting, making this a critical failure mode in forming operations.

This mode mirrors the conditions in a standard uniaxial tensile test, where the sheet is pulled along one direction, and the other principal stress component $\sigma_2 = 0$. The minor strain is negative (contraction) while the major strain is positive (elongation). This condition often occurs at free edges—for example, during hole extrusion or at the perimeter of a stretching sheet (Figure 8.14). The strain path lies somewhere between the plane strain and drawing paths.

At point D in the diagram, the major and minor strains are equal in magnitude but opposite in direction, causing the material to stretch in one direction while contracting in the other with no net change in thickness. This process is known as drawing or pure shear deformation and is

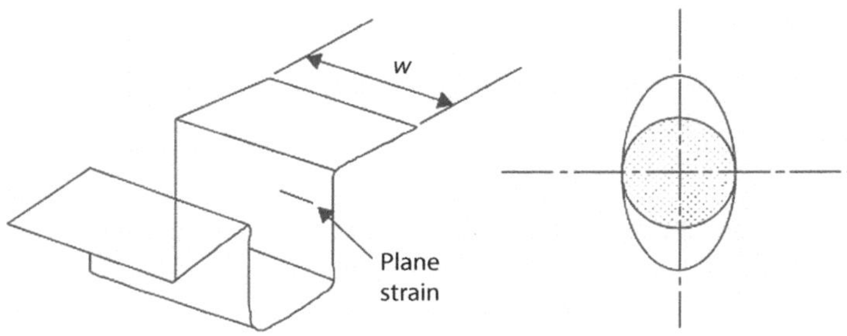


**Figure 8.13** Plane strain deformation occurring along the sidewall of an elongated component.

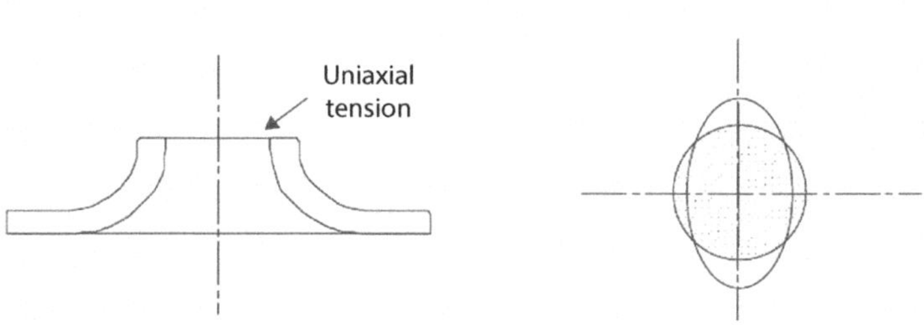


**Figure 8.14** Uniaxial stretching at the edge of an extruded hole.

commonly seen in the flange regions of deep-drawn parts, such as cups (Figure 8.15). According to Equation (8.10), the thickness strain in this process is zero. From Equation (8.12), the effective strain is given by $\bar{\varepsilon} = (2/\sqrt{3})\varepsilon_1 = 1.1555\varepsilon_1$, indicating that work hardening progresses gradually. In this deformation mode, the risk of splitting is low, and in many practical forming operations, high levels of strain can be achieved safely.

The deformation process shown at point *E* represents an extreme condition that occurs when the major stress $\sigma_1$ is zero such as at the edge of a deep-drawn cup, as illustrated in Figure 8.16. In this case, the minor stress $\sigma_2$ is compressive, equal to $\sigma_2 = -\sigma_f$. The effective strain and stress are given by $\bar{\varepsilon} = -\varepsilon_2$ and $\bar{\sigma} = -\sigma_2$, respectively. This type of deformation causes the sheet to thicken, and it is often associated with a high risk of wrinkling. Plotting strains in diagrams like Figure 8.11 is highly valuable for evaluating sheet forming processes. The location of a point on the diagram provides insight into how the sheet's thickness is changing. If the point lies

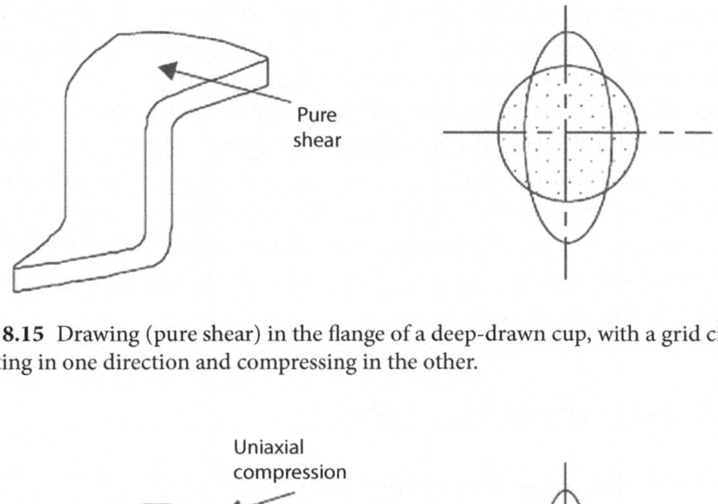


**Figure 8.15** Drawing (pure shear) in the flange of a deep-drawn cup, with a grid circle elongating in one direction and compressing in the other.

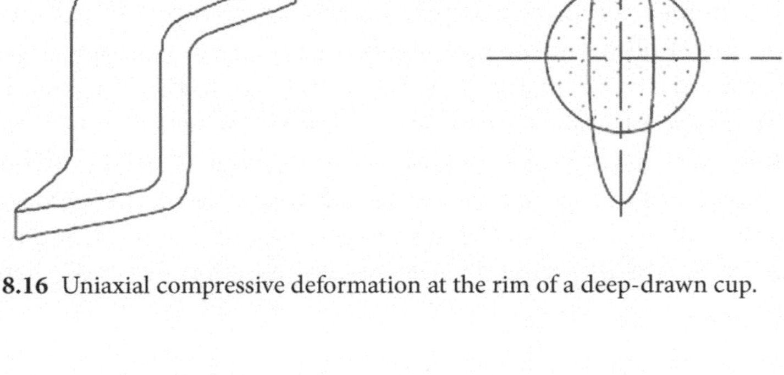


**Figure 8.16** Uniaxial compressive deformation at the rim of a deep-drawn cup.

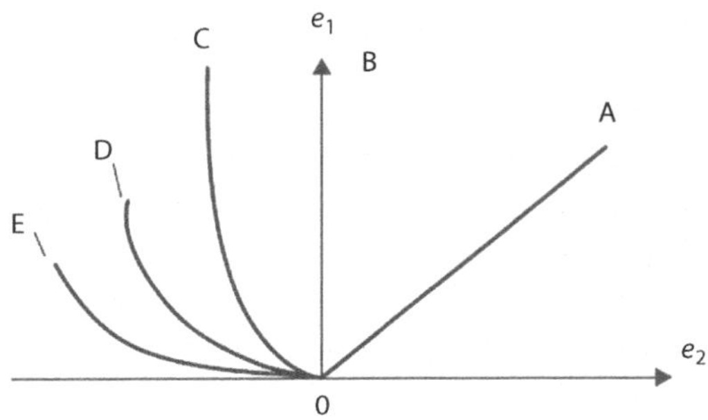


**Figure 8.17** Various proportional strain paths from Figure 8.10 represented in an engineering strain diagram.

to the right of the drawing line (i.e., when $\beta>-1$), the sheet is undergoing thinning. Conversely, if the point is below the drawing line (i.e., $\beta<-1$), the sheet material is thickening.

In the sheet metal industry, the data shown in Figure 8.11 is often represented using engineering strain. Figure 8.17 displays strain paths with constant true strain ratios plotted in terms of engineering strain. It becomes clear that many of these proportional deformation paths do not appear as straight lines, highlighting the limitations of engineering strain as a reliable measure of deformation. Therefore, true strain will be used throughout this work for most analyses. However, since engineering strain diagrams are still commonly used in practice, it is important to understand and be familiar with both representations. Unless stated otherwise, true strain diagrams will be the standard in this text.

## 8.5 Anisotropy of Sheet Metal

### 8.5.1 Uniaxial Anisotropy Coefficients

Because of their crystallographic structure and the nature of the rolling process, sheet metals typically display notable anisotropy in their mechanical properties. Rolling introduces a specific type of anisotropy characterized by symmetry in the material's behavior relative to three perpendicular planes. This type of material response is known as orthotropy. The lines where these symmetry planes intersect are referred to as the orthotropy axes. As illustrated in Figure 8.18, for rolled sheet metals, these axes correspond to the rolling direction (RD), transverse direction (TD), and normal direction (ND). The directional variation in the material's plastic behavior

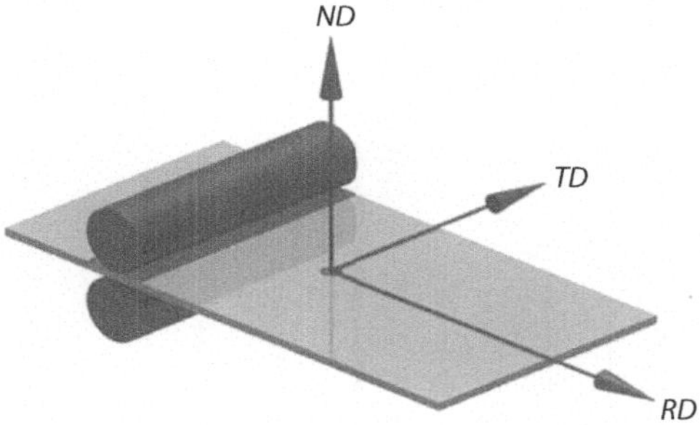


**Figure 8.18** Orthotropy axes of the rolled sheet metals: LD—longitudinal direction; TD—transversal direction; ND—normal direction.

is measured using a value known as the Lankford parameter or anisotropy coefficient [9]. This coefficient is determined through uniaxial tensile tests conducted on sheet metal strips. The anisotropy coefficient, denoted as *r*, is defined by the following relationship:

$$r = \frac{\varepsilon_{22}}{\varepsilon_{33}} \tag{8.13}$$

where $\varepsilon_{22}$ and $\varepsilon_{33}$ represent the strains in the width and thickness directions, respectively.

For isotropic materials, the anisotropy coefficient *r* equals 1, meaning the material exhibits equal strain in both width and thickness directions. When $r>1$, the strain in the width direction is greater, indicating higher resistance to thinning. Conversely, if $r<1$, the material deforms more in the thickness direction, meaning thinning is less resisted. Using the notation presented in Figure 8.19, Equation (8.13) can be rewritten as

$$r = \frac{\ln \frac{w}{w_0}}{\ln \frac{t}{t_0}} \tag{8.14}$$

where $w_0$ and $w$ represent the initial and final widths, while $t_0$ and $t$ denote the initial and final thicknesses of the specimen, respectively.

Since the specimen's thickness is much smaller than its width typically by at least an order of magnitude, the measurement errors for thickness

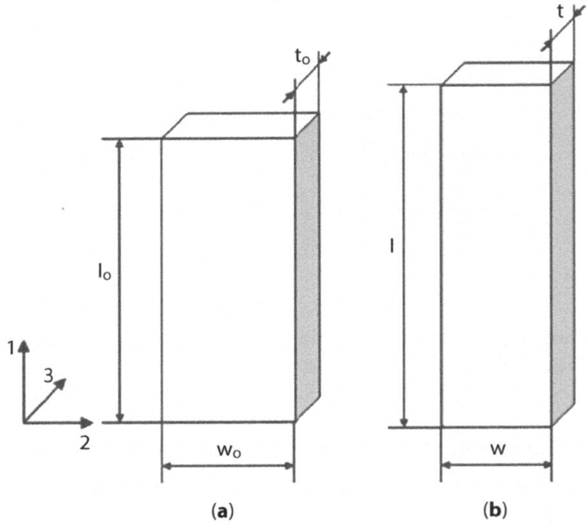


Figure 8.19 Geometry of the specimen: (a) before and (b) after deformation.

and width strains can differ significantly. To reduce this discrepancy, the previous relationships are replaced with an equivalent expression involving quantities of the same order of magnitude, specifically, the length and width of the specimen. This adjustment is made under the assumption of constant volume during deformation:

$$\varepsilon_{11} + \varepsilon_{22} + \varepsilon_{33} = 0 \tag{8.15}$$

This leads to the following reformulated version of Equation (8.13):

$$r = -\frac{\varepsilon_{22}}{\varepsilon_{11} + \varepsilon_{22}} \tag{8.16}$$

Accordingly, Equation (8.14) takes the form

$$r = \frac{-\ln\dfrac{w}{w_0}}{\ln\dfrac{l}{l_0} + \ln\dfrac{w}{w_0}} \tag{8.17}$$

In this equation, $l_0$ and $l$ represent the initial and final gauge lengths, with $l_0$ defined according to standard specifications [10]. Equation (8.17) can be rearranged in the following form:

$$r = \frac{\ln \frac{w}{w_0}}{\ln \frac{l_0 . w_0}{l . w}} \tag{8.18}$$

This expression is commonly used in practical applications to calculate the anisotropy coefficient. Experimental results indicate that the anisotropy coefficient $r$ varies depending on the in-plane orientation. When a tensile specimen is cut such that its longitudinal axis forms an angle $\theta$ with the rolling direction, the resulting coefficient is denoted as $r_\theta$, as shown in Figure 8.20. The subscript $\theta$ specifies the angle between the specimen's axis and the rolling direction.

Another key factor in evaluating the effectiveness of a yield criterion is its ability to predict how the uniaxial yield stress and the plastic anisotropy coefficient vary within the plane of the sheet metal. To examine this capability, we will derive the relationships that describe how these parameters change with respect to angle $\theta$, which is measured from the rolling direction, as illustrated in Figure 8.21.

Let $Y_\theta$ represent the uniaxial yield stress in the direction that forms an angle $\theta$ with the rolling direction. Under uniaxial loading conditions,

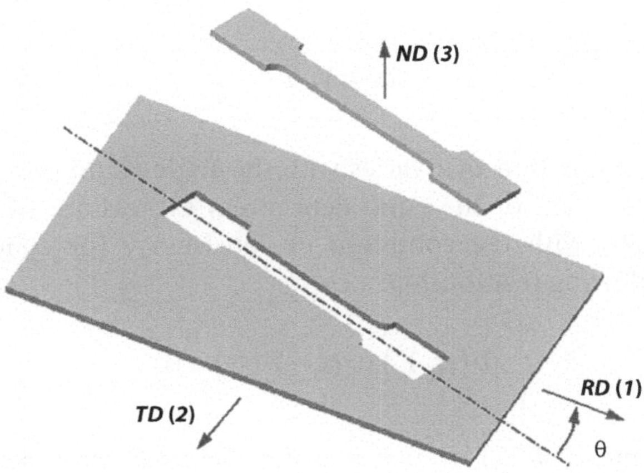


**Figure 8.20** Tensile specimen pre-elevated at the angle θ (measured from the rolling direction).

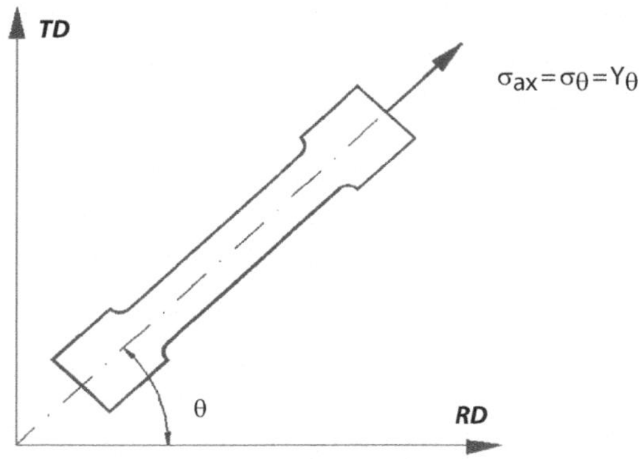


**Figure 8.21** Uniaxial yield stress corresponding to a direction in the plane of the sheet metal.

the components of the stress tensor for this inclined orientation can be expressed as follows:

$$\begin{aligned} \sigma_{11} &= Y_\theta \cos^2\theta \\ \sigma_{22} &= Y_\theta \sin^2\theta \\ \sigma_{21} &= \sigma_{12} = Y_\theta \sin\theta\cos\theta \end{aligned} \tag{8.19}$$

Substituting Equation (8.19) into the expression for equivalent stress, and considering the homogeneity of the function, we arrive at the following result:

$$\sigma = Y_\theta . F_\theta \tag{8.20}$$

Here, $F_\theta$ is a function that varies with the angle $\theta$ and is defined based on the specific form of the equivalent stress expression. By combining Equation (8.20) with the condition of consistency (or coherence), we obtain the following relationship:

$$\Phi(\sigma, Y) := \sigma - Y(h) = 0 \tag{8.21}$$

In this context, $\Phi(\sigma, Y)$ represents the yield function corresponding to the chosen yield criterion, where $Y$ is the yield stress and $h$ is a scalar parameter that characterizes the accumulated plastic strain in the material. From this, we derive the following expression:

$$Y_\theta = \frac{Y(h)}{F_\theta} \tag{8.22}$$

Equation (8.22) describes the uniaxial yield stress for a specific in-plane direction defined by the angle $\theta$. If the yield stress in the rolling direction is chosen as the reference value (i.e., $Y(h) = Y_0$), then the relationship simplifies to the following form:

$$Y_\theta = \frac{Y_0}{F_\theta} \tag{8.23}$$

In this case, the yield stress in any given in-plane direction depends solely on the yield stress along the rolling direction and the function $F_\theta$, which is linked to the specific yield criterion used in the model. The method for determining $F_\theta$ will be detailed in the following subchapters for each type of yield criterion. Similarly, we can derive an expression that describes how the plastic anisotropy coefficient varies within the plane of the sheet metal. Consider a tensile specimen oriented at an angle $\theta$ relative to the rolling direction, as shown in Figure 8.21. According to Equation (8.13), the instantaneous plastic anisotropy coefficient $r_\theta$ is defined as the ratio of the plastic strain rate in the width direction (which lies at an angle $\theta$+90° to the rolling direction), denoted $\dot{\varepsilon}_{\theta+90}$, to the plastic strain rate in the thickness direction, $\dot{\varepsilon}_{33}$.

$$r_\theta = \frac{\dot{\varepsilon}_{\theta+90}}{\dot{\varepsilon}_{33}} \tag{8.24}$$

By considering the incompressibility condition referred to Equation (8.15) along with the expressions for the strain rate components in the principal directions, we can proceed to derive the following relationship:

$$\dot{\varepsilon}_\theta = \dot{\varepsilon}_{11}\cos^2\theta + \dot{\varepsilon}_{22}\sin^2\theta + \dot{\varepsilon}_{12}\sin\theta\cos\theta \tag{8.25}$$

$$\dot{\varepsilon}_{33} = -\left(\dot{\varepsilon}_{11} - \dot{\varepsilon}_{22}\right) \tag{8.26}$$

This leads to the expression that defines the coefficient of plastic anisotropy corresponding to the direction $\theta$:

$$r_\theta = \frac{\dot{\varepsilon}_{11}\cos^2\theta + \dot{\varepsilon}_{22}\sin^2\theta + \dot{\varepsilon}_{12}\sin\theta\cos\theta}{\dot{\varepsilon}_{11} + \dot{\varepsilon}_{22}} - 1 \tag{8.27}$$

If the associated flow rule is applied, Equation (8.27) can be reformulated using the stress components as follows:

$$r_\theta = \frac{\sigma_{11}\dfrac{\partial\sigma}{\partial\sigma_{11}} + \sigma_{22}\dfrac{\partial\sigma}{\partial\sigma_{22}} + \sigma_{12}\dfrac{\partial\sigma}{\partial\sigma_{12}}}{\sigma_{11}\dfrac{\partial\sigma}{\partial\sigma_{11}} + \sigma_{22}\dfrac{\partial\sigma}{\partial\sigma_{22}}} - 1 \tag{8.28}$$

By combining Equation (8.28) with Euler's identity, we arrive at the following result:

$$r_\theta = \frac{\sigma}{Y_\theta\left(\dfrac{\partial\sigma}{\partial\sigma_{11}} + \dfrac{\partial\sigma}{\partial\sigma_{22}}\right)} - 1 \tag{8.29}$$

Finally, by substituting Equation (8.20) into the previous expression, we obtain the following relationship:

$$r_\theta = \frac{F_\theta}{\dfrac{\partial\sigma}{\partial\sigma_{11}} + \dfrac{\partial\sigma}{\partial\sigma_{22}}} - 1 \tag{8.30}$$

This formula expresses the coefficient of plastic anisotropy as a function of the specimen's orientation angle. To apply it, one must know the expression for the equivalent stress and the corresponding function $F_\theta$, both of which depend on the specific yield criterion used in the plasticity model. The directional variation of the anisotropy coefficient within the sheet plane will be explored in the upcoming subchapters for different yield criteria. The average of the $r$-values measured in different in-plane directions is known as the coefficient of normal anisotropy ($r_n$). When the anisotropy coefficient is determined from specimens cut at three standard angles 0°,

45°, and 90° to the rolling direction, the average $r_n$ is calculated using the following formula:

$$r_n = \frac{r_0 + 2r_{45} + r_{90}}{4} \tag{8.31}$$

The degree to which normal anisotropy varies with respect to the angle from the rolling direction is quantified by the following parameter:

$$\Delta r = \frac{r_0 - 2r_{45} + r_{90}}{2} \tag{8.32}$$

which is identified as planar anisotropy.

This quantity is directly linked to the earring amplitude seen in deep-drawn cups. Specifically, when the anisotropy coefficient remains constant in all in-plane directions, the earing effect does not occur. To visualize planar anisotropy more intuitively, a polar coordinate plot is often used. Figure 8.22 presents such a diagram for an aluminum alloy (AA5182-0). These polar diagrams are favored because they clearly reveal the sheet metal's tendency to form ears. The symmetry of the curve with respect

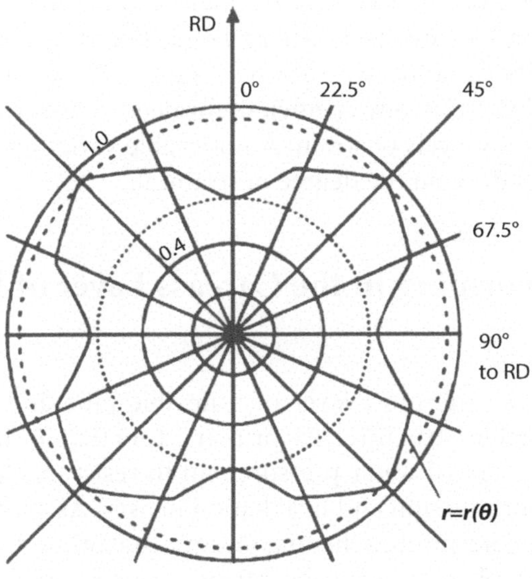


**Figure 8.22** Directional variation of the anisotropy coefficient within the sheet metal plane [11].

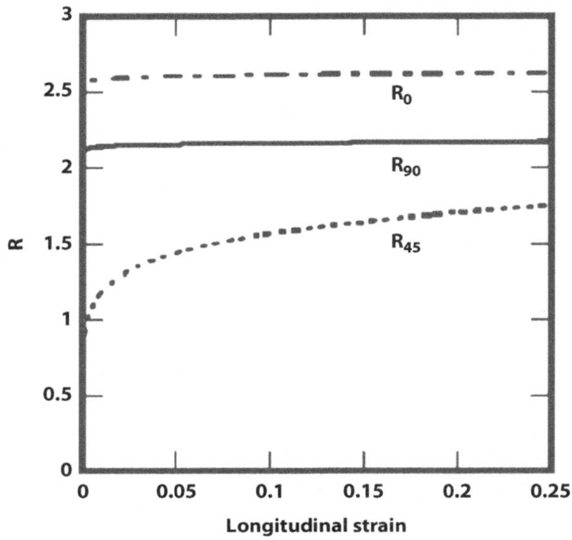


**Figure 8.23** Anisotropy coefficient variation with increasing deformation [12].

to the coordinate axes indicates the orthotropic nature of the material. By convention, $r$-values are typically measured at 20% elongation for consistency in comparison. Modern tensile testing equipment allows for real-time measurement of the necessary data and can calculate the anisotropy coefficient during the test itself. This enables the evaluation of $r$ at 20% elongation and its variation with strain. Figure 8.23 illustrates the planar distribution of the anisotropy coefficient for deep-drawing quality (DDQ) mild steel at various levels of strain. A noticeable variation in the $r_{45}$ value is observed depending on the degree of straining.

## 8.6 Metal Forming in the Carcass Layer of Flexible Pipes

The carcass is the innermost layer of a flexible pipe and consists of an interlocking, flexible structure. Its primary function is to prevent collapse caused by external water pressure and to resist mechanical damage during both the production and installation phases. As such, the carcass is a key element in determining the maximum installation depth of the flexible pipe and also plays a significant role in ensuring its overall flexibility [13]. Carcass production is a continuous two-stage process. The first stage

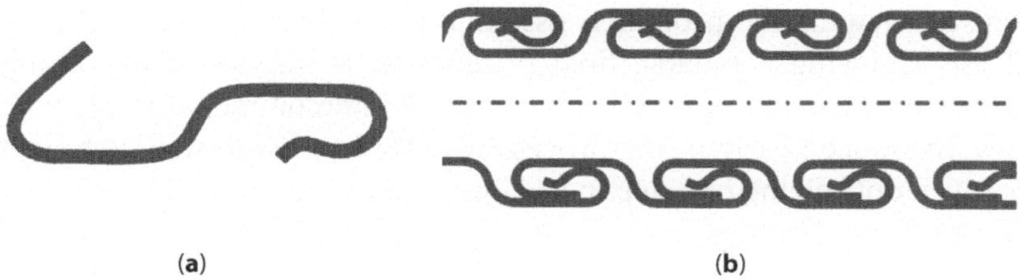


**Figure 8.24** (a) Cross-section of the roll-formed profile used for interlocking carcass production, and (b) side view cross-section of the interlocked carcass. The direction of production is from left to right.

involves roll forming stainless steel strips into an S-shaped profile, as shown in Figure 8.24(a). In the second stage, this profile is wound helically onto a mandrel, where it interlocks with itself, as illustrated in Figure 8.24(b). Both stages are carried out on a single-machine setup, where the roll forming line rotates around the mandrel to produce carcass lengths that can span several kilometers. Mandrel diameters typically range from ø50 mm to ø400 mm. This rotating setup restricts the number of rolls forming tools that can be used, which in turn influences the risk of profile buckling and twisting. Additionally, this setup results in higher process loads, as greater bending is needed at each forming stage [14].

Figure 8.25 demonstrates the entire carcass production sequence. While analyzing the metal-forming stage alone might yield valuable insights, it is significantly affected by several preceding operations. Therefore, adopting

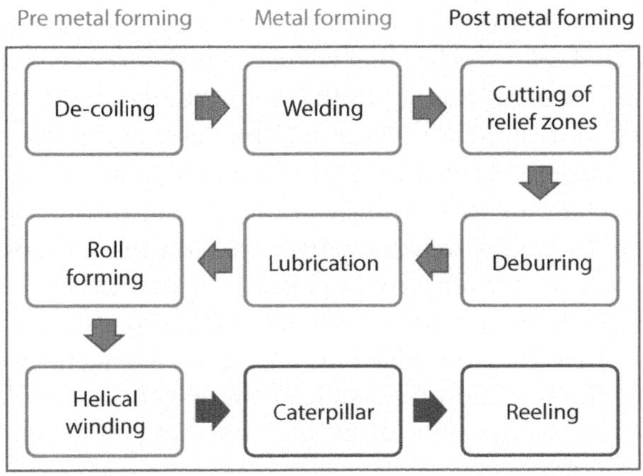


**Figure 8.25** Step-by-step flow diagram of the carcass material pathway during production.

a holistic view of the process is essential for a thorough understanding and for achieving a reliable final product. It is important to highlight that carcass production involves a series of interconnected steps, where issues introduced early in the process may lead to failures at later stages. Moreover, optimizing one step in isolation can unintentionally cause problems in subsequent stages.

### 8.6.1 Pre-Metal Forming

Pre-metal forming describes a number of essential coil and strip handling steps that are performed before the actual metal-forming process begins. Since the entire strip is utilized in the carcass process, proper handling of the strip edges is critical. The carcass materials arrive in coil form, and the first step is de-coiling, where the material is spooled onto machine reels known as pancakes. Coils are attached to a de-coiler, and the strip is fed into the carcass machine. Defective coil sections from the steel manufacturer are identified and cut out during this stage. Between the de-coiler and the pancakes, strips are joined *via* automatic TIG welding without filler material, using shielding gas on both sides of the weld. Proper strip alignment and preparation of the ends are essential to ensure continuous carcass production. Duplex stainless steels are typically used for carcass production due to their strength, corrosion resistance, and cost-effectiveness. However, welding these materials can be challenging because the mechanical properties in the weld zone are generally inferior to those of the base material. Weld fractures are more common in thinner strips, potentially due to insufficient heat input or excessively rapid cooling caused by the low heat capacity of the fixture in contact with the strip. To address the reduced weld strength of duplex steels, relief zones are sometimes cut alongside the weld, removing material that is wound onto the mandrel before the weld to relieve stress during helical winding. The design and placement of the relief-cut are crucial for function and process stability. While the strip is being coiled, deburring of the edges is carried out through edge-rolling. If the burrs are too large, they may interfere with the interlocking mechanism and result in uneven profile closure. In one experimental case, a strip without deburring caused unwanted carcass rotation in the production direction. Therefore, the edge burr significantly affects torque requirements during winding and may also lead to the generation of edge swarf. A properly deburred edge minimizes these risks and contributes to a stable production process.

## 8.6.2 Metal Forming

Metal forming in the carcass manufacturing process involves several critical stages, starting with lubrication. To extend tool life and prevent pickup formation on the tools, lubrication of the strip is carried out before it enters the roll-forming line. Various lubricants can be used, including mineral oils, biodegradable vegetable oils, and emulsions. Inadequate lubrication leads to galling, where stainless steel cold-welds to the tool surface, causing profile scoring. The first forming stage is roll forming, where the flat strip is continuously shaped by a series of contoured rolls. As shown in finite element analyses (FEA), this method applies longitudinal stretching, longitudinal and transverse bending, and shear to shape the strip into the desired open s-shaped profile. This initial profile, illustrated in Figure 8.24(a), is later interlocked during the helical winding phase to form the closed profile (Figure 8.24(b)). To guide the strip properly into the first roll-forming station, a strip guide ensures correct edge alignment for accurate profile geometry. Machine sizes vary, offering between 4 and 12 roll forming stations, with 4- and 6-step configurations being the most common. Fewer forming steps compared to conventional roll forming lead to higher plastic strain, potentially causing profile skewness. However, in carcass production, this skewness is later corrected during helical winding due to stress relaxation. After roll forming, the strip can be transferred to the winding stage *via* two methods: loop feed and direct feed. The loop feed approach uses a dancer arm to maintain a buffer of pre-formed profile, allowing the winding machine to draw profile freely, which enhances process flexibility and eases interlocking. This separation of forming and winding stages avoids axial prestress. In contrast, the direct feed method connects the roll forming line directly to the winding process, creating a continuous, rigid transfer path. This allows controlled prestressing of the strip but reduces degrees of freedom, and the profile must be fed precisely to avoid pulling or tool wear. The final and most complex forming stage is helical winding, where the profile is bent and interlocked around a mandrel using winding tools (WTs). These tools, arranged around the mandrel, guide the strip along a helical path, defining both the pitch and wall thickness of the carcass. As the profile enters WT1, interlocking with the previous turn begins, with main bending and locking occurring between WT1 and WT2 through three-point bending. The upright position of the profile tongue during the first rotation introduces high strain before it interlocks, making the edge quality of the strip critical—any notches could initiate fracture. The interlocking must allow a sufficient gap between the profile legs for stable engagement, as shown in Figure 8.26. A phenomenon

**Figure 8.26** Interlock detail view: a sufficient gap between the two profile legs is essential to ensure a secure interlock — the gap shown in the image is appropriate.

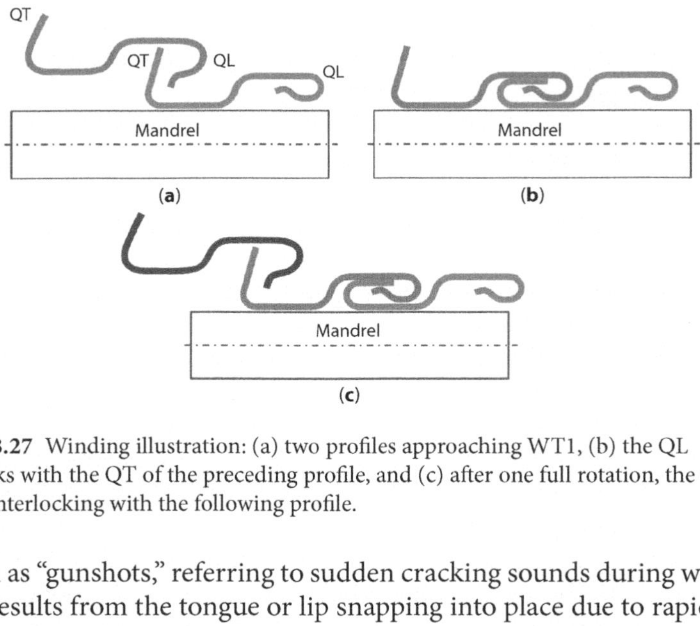


**Figure 8.27** Winding illustration: (a) two profiles approaching WT1, (b) the QL interlocks with the QT of the preceding profile, and (c) after one full rotation, the QL begins interlocking with the following profile.

known as "gunshots," referring to sudden cracking sounds during winding, often results from the tongue or lip snapping into place due to rapid stress release. This is depicted in Figure 8.27, where interlocking progresses from the initial approach of two profiles to full engagement after one mandrel rotation. These sounds are often linked to tongue buckling and indicate potential defects in winding quality.

## 8.6.3 Post-Metal Forming

Post-metal forming operations are essential to ensure proper handling and stability of the finished carcass. One of the most critical components in this

**Figure 8.28** The caterpillar unit prevents the carcass from rotating and facilitates its transfer from the mandrel to the reel.

stage is the caterpillar unit, which prevents the carcass from rotating freely during the helical winding process and enables its controlled removal from the mandrel. The caterpillar consists of rubber-padded belts that grip the carcass at a fixed distance from the mandrel and winding tools, as illustrated in Figure 8.28. This device not only stabilizes the winding operation but also influences the final pitch of the carcass, as its speed directly affects the axial advancement during winding. Although the caterpillar could be considered part of the metal forming process due to its direct impact on winding, it is classified here as a post-forming operation for simplicity, given that it primarily serves a supporting role. Following this, the final stage of the carcass process is reeling. The fully formed carcass is wound onto large reels for storage and transportation.

**Figure 8.29** Carcass wound onto a reel, with the reel flange having an approximate radius of 4 meters.

These reels are capable of accommodating several kilometers of carcass, depending on the profile and carcass size. As shown in Figure 8.29, the reel's flange radius is typically around 4 meters, ensuring adequate capacity and stability for handling large lengths of carcass efficiently.

## 8.7 Metal Forming Effect on Residual Stress of the Carcass Layer in Flexible Pipes

The demand for energy has been increasing rapidly in recent years. The engineers need to explore oil and gas in the deeper locations of seas and oceans. Pipelines in deep water must endure tremendously high external pressures at depths of hundreds of meters, so they must be engineered to withstand extreme circumstances against extreme loading and environmental factors. Pipes with enhanced properties, including toughness, elongation, and strain capacity, may be more suitable for deep-water pipelines. The best replacement for traditional steel pipes is flexible pipelines, which have multiple layers with various structures. Due to their numerous structural configurations, they can tolerate higher internal and external pressure. Bai [15] believed flexible pipes are economical because of their easier transportation and installation due to their higher bending flexibility compared to homogeneous steel pipes. A typical nine-layer unbounded flexible pipe with the names of all layers is demonstrated in Figure 8.30. Under adverse service circumstances, such as significant water depths, the pipe's weight is increased not just by its length but also by additional metal

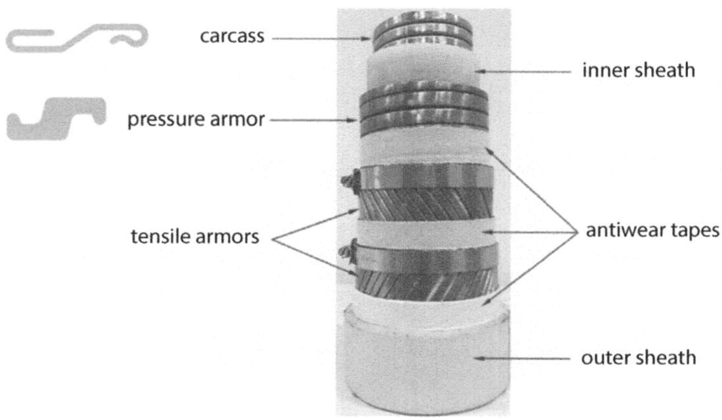


**Figure 8.30** Typical unbounded flexible pipe.

layers included in its cross-section. In fact, using an interlocked carcass is essential to prevent collapse while working with high hydrostatic pressure. Furthermore, the tensile armor layers supply the necessary tensile strength, while the pressure armor layer offers strength against burst pressure. More reinforcements lead to a greater gravitational load, which is reflected at the flexible riser's suspension point. With the increase of water depth, extreme hydrostatic pressure becomes one of the most important issues, so it is important to properly assess the mechanical behavior of unbonded flexible pipes under external pressure. The carcass is the innermost interlocked layer that provides a path for oil and gas flow. However, the main function of the carcass is to resist water pressure loads and accidental loads during installation and operation [16]. Extreme hydrostatic pressure can lead to collapse failure of flexible pipes, so obtaining the critical collapse pressure is important to the anti-collapse design. The carcass is the main feature ensuring flexibility, and it is a vital component of the pipe that has a direct influence on the depth of the pipeline project [17]. Therefore, determining the collapse strength resistance of this layer is a vital task in engineering applications. Several studies have been performed to determine the collapse failure of a carcass, mostly by using numerical simulation or by assuming it is an equivalent homogeneous ring of a particular thickness. The equivalent model's basic idea is to reduce a complex carcass section to a rectangular section with a uniform thickness and then replace it with that section in the elastic buckling theoretical equation to get the critical collapse pressure [18].

Zhang *et al.* [19] calculated the carcass's equivalent thickness using the area equality approach. Martins *et al.* [20] used the approach of equal bending stiffness per unit length to determine the equivalent thickness because they believed that the collapse of the carcass caused bending deformation. Fernando *et al.* [21] studied the pressure layer function, and they conducted finite element studies taking into account both 3D and 2D (axisymmetric) models. Their purpose is to evaluate the conforming effects on the pressure layer profile in the 3D model with the help of the related 2D model. The 3D model only takes into account a small portion of the pressure layer and does not take into account the entire pitch because the goal of their analysis is not to anticipate buckling loads. Two models, a 3D ring model with solid elements and a full 3D model, were subjected to collapse investigations by Neto and Martins [22]. The findings suggest that the carcass structure's helical shape is unimportant. Nogueira and Netto [23] proposed a model made out of beam elements to streamline the study, but because of the limits of beam elements, the critical pressure was overestimated. Neto *et al.* [24] included a correction factor in the comparable calculation, taking into account the gap in the geometry of the carcass cross-section, based on the equivalent bending stiffness. Neto and Martins [25] created a ring model that uses the original carcass structure's equal bending stiffness per unit length. The critical pressure for the perfect linear elastic carcass layer was more accurately estimated by the model. In the aforementioned studies, the material nonlinearity is not taken into consideration because the proposed methods are applicable only in the elastic range. Tang *et al.* [26] overcome these limitations by creating a representative element model of a carcass that incorporates Dirichlet boundary conditions and determining the equivalent thickness using strain energy equivalence. They found that their model is conservative in comparison with other equivalence methods. Besides, Li *et al.* [27] used the strain energy equivalent approach to compute the carcass's equivalent elastic modulus, equivalent yield stress, and equivalent thickness and improved previous results.

The equivalent model overlooks the external structural restrictions and does not take into account the geometric imperfections, nonlinear mechanical behavior during the deformation process of collapse, and residual stresses and the frictional interaction between adjacent surfaces of the carcass. As a result, when compared to the experimental data, the majority of the results produced by the comparable model contain mistakes. The critical pressure is difficult to calculate analytically because of the intricate interlocking cross-section and the internal contacts/friction of the carcass structure.

An expensive and large testing apparatus is needed to experimentally determine the carcass layer's critical hydrostatic pressure. Additionally, it is challenging to keep an exact eye on the carcass structure's buckling behavior. Compared to these limitations of collapse tests, the radial compression experiment is more straightforward and convenient to execute. Additionally, due to the interlocked and helical structure of the carcass, the manufacturing process is complex. During this procedure, different regions of the carcass layer may be under tension, compression, or bending. These frequently added loadings produce a noticeable amount of stress in different nodes of the carcass and remain in the properties as residual stress. The residual stress can affect the collapse strength of the carcass significantly. This subject is investigated in the current paper by employing a radial compression test until full collapse and comparing the result with the established numerical model. Studying the process that makes a carcass layer from a coiled strip to a rolled carcass with pitch is the only way to obtain the most accurate amount of residual stress. A method is proposed in this study to simulate all the production processes.

### 8.7.1 Experiments

A 6-inch carcass, which is covered with HDPE material by the Ningbo-OPR factory, as shown in Figure 8.31, is prepared for a radial compression experiment to evaluate the radial stiffness. The PE layer is used to prevent the possible movement of the carcass during the test and distribute the compression load on the exterior surface of the carcass uniformly. A schematic diagram illustrating the general cross-sectional profile of the carcass is presented in Figure 8.32, and the set of parameters considered is listed in Table 8.2.

As observed, a carcass cross-section of minimal complexity is selected to provide more accurate results from the test. The radial compression test was chosen because it allows for the determination of the endurance of the carcass layer, a parameter that can also be assessed through collapse tests. However, the radial compression test is more straightforward and convenient to execute. The geometric parameters of the specimen are listed in Table 8.3.

**Figure 8.31** The pipe specimen.

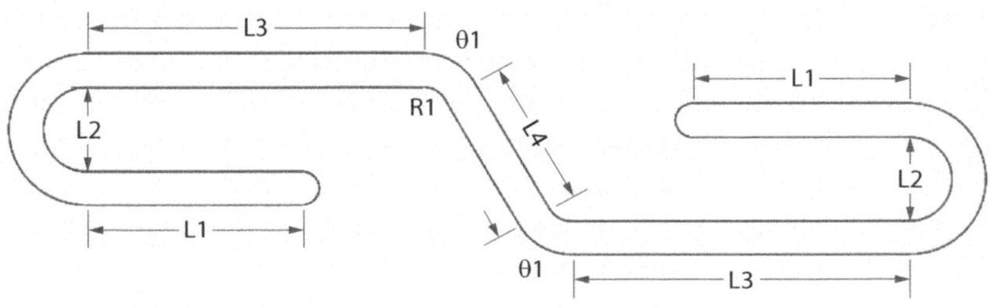


**Figure 8.32** The carcass profiles.

**Table 8.2** Geometrical parameters of 4-inch SSRTP.

| Parameter | Value | Parameter | Value |
|---|---|---|---|
| $L_1$ (mm) | 2.50 | $L_2$ (mm) | 100 |
| $L_3$ (mm) | 3.90 | $L_4$ (mm) | 1.73 |
| $R_1$ (mm) | 0.80 | $\theta_1$ (deg) | 60 |

**Table 8.3** Geometric parameters of the specimen.

| Parameter | Value | Parameter | Value |
|---|---|---|---|
| Inner radius | 75.60 (mm) | Carcass ribbon thickness | 0.40 (mm) |
| Outer radius | 88.00 (mm) | Carcass profile width | 11.33 (mm) |
| Inner PE thickness | 2.40 (mm) | Carcass profile pitch | 6.40 (mm) |
| Outer PE thickness | 10.0 (mm) | Carcass winding angle | 89 (deg) |

A material test for both the carcass and PE layers is performed. An electronic universal machine is used for the uniaxial test. HDPE material, according to ASTM, is cut into dumbbell shapes. As for the carcass material, the initial shape of the strips before the metal-forming process is utilized, since the thickness of the strip is extremely thin. The axial displacement rates for the HDPE and undeformed carcass strip are set to 1.0 mm/min and 0.5 mm/min, respectively. In the experimental setup, two distinct extensometers are employed to measure the strains of high-density polyethylene (HDPE) and steel materials. Due to the inherently ductile nature of HDPE, which undergoes significant plastic deformation, it is imperative to remove the extensometer once the material enters the phase of extensive plastic deformation. Results are commonly represented in terms of engineering stresses and strains, which facilitate the characterization of yield stress, ultimate tensile strength, and fracture point. However, for the development of flow curves necessary in finite element analysis (FEA), it is essential to employ true stresses and strains. By assuming an offset strain at the yielding stress of 0.2%, the engineering stress–strain relationship can be adjusted employing the Ramberg-Osgood equation as follows:

$$\varepsilon_{norm} = \frac{\sigma_{norm}}{E} + 0.002\left(\frac{\sigma_{norm}}{\sigma_y}\right)^n \tag{8.33}$$

where $\varepsilon_{nom}$, $\sigma_{nom}$, $\sigma_y$, and E represent the nominal strain, nominal stress, yield stress, and Young's modulus, respectively. n is a material constant determined through calibration. By fitting the Ramberg–Osgood model to the ultimate tensile strength, the constant n is obtained. Consequently, the true strain and true stress can be evaluated as follows:

$$\varepsilon_{true} = \ln\left(1 + \varepsilon_{norm}\right) \tag{8.34}$$

$$\sigma_{true} = \sigma_{norm}\left(1 + \varepsilon_{norm}\right) \tag{8.35}$$

The calibrated engineering and true stress–strain curves are depicted in Figure 8.33. The material properties of the tested specimen are listed in Table 8.4. The Young's modulus of high-density polyethylene (HDPE) is calculated using the cut-line modulus approach, defined by selecting two distinct points at 0.05% and 0.25% of the actual strain, in accordance with the guidelines specified in ISO 527-1. This methodology ensures a standardized and accurate determination of the material's elastic behavior under controlled conditions.

Since the specimen is placed under radial compression, the pipe should be long enough to avoid boundary effects and ensure uniform stress distribution in the experiment. By testing three different lengths, the pipe containing

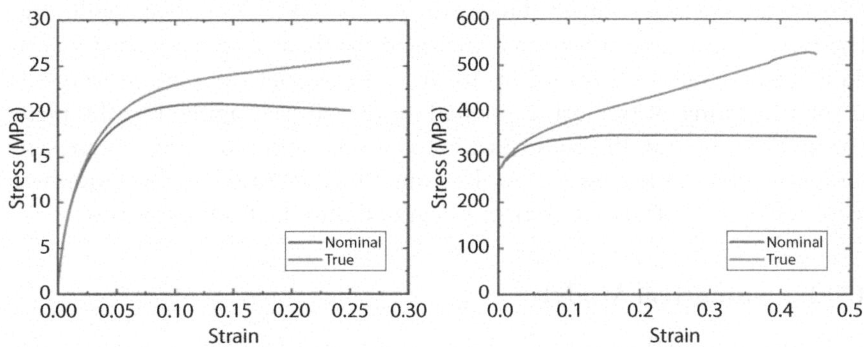


**Figure 8.33** Stress-strain curve of (a) HDPE and (b) carcass steel.

**Table 8.4** Properties of the test specimen.

| Parameter | HDPE | Carcass |
|---|---|---|
| Young's Modulus (MPa) | 1134 | 165000 |
| Poisson's ratio | 0.45 | 0.26 |
| Yield strength (MPa) | \ | 335.92 |
| Limit strength (MPa) | \ | 545.43 |

**Figure 8.34** The experiment setup.

a carcass layer with almost 11 pitches was selected. The experimental configuration, as illustrated in Figure 8.34, involves positioning the test specimen within a universal testing machine between two stainless steel plates, each measuring 30 mm in thickness and 400 × 200 mm$^2$ in area. The lower plate was firmly attached to the testing machine, while the upper plate was connected to an electric actuator responsible for vertical displacement. The actuator system was controlled to ensure precise and reliable test conditions. The radial compression test was employed for the test specimens at a constant loading rate of 2 mm/min. In order to restrict the pipe movements at the beginning of the test, a preloading of 100 N is applied to the plate. Throughout the test, the compression force was captured using a force sensor embedded in the actuator. At the same time, the radial deformation was recorded by a displacement sensor integrated into the testing machine.

### 8.7.2 Numerical Model

A 3D FE model of the interlocked carcass with its related pitch and lay angle covered with an external polymeric layer is established by commercial software, ABAQUS. The carcass model initially is built with the same pitch numbers as the test specimen (with a mean of 10.8 pitches). Then, only two pitches of the carcass are modeled, which is very common in the study of layers with helical structures, and the results are compared with the lengthy model. By multiplying the scale factor based on the number of pitches, which is 5.4 for our model, the results are matched completely well with each other. Hence, in order to decrease the time of calculations, the two-pitched model is chosen for further study. A polymer layer is also established around the carcass layer to behave as a load transmitter. The FE model uses solid elements for the polymeric sheath, and to keep the model simplified and raise the speed of the analysis, the carcass is modeled with shell elements. The carcass profile selected for this study aligns with the specimen that was evaluated in the preceding section. From the fact that the two plates are made of steel with a much higher strength than the pipe model, they are defined as analytical rigid components in the software, and due to the ABAQUS feature, there is no need for entry material properties for these sections. The analysis is performed as a geometrically non-linear problem using the static general step to settle contact problems. The FE model achieved relatively accurate quasi-static analysis results since the ratio between kinetic energy and strain energy for the entire model was low enough during the simulation. The S4R, which is a 4-node doubly curved element with reduced integration and hourglass control, is chosen for meshing the carcass. The polymeric sheath has been meshed with C3D8R (an 8-node linear brick with reduced integration and hourglass control), which ensures the accuracy of the calculation and improves efficiency due to its high precision in various linear and nonlinear complex problems. The detailed mesh is depicted in Figure 8.35.

**Figure 8.35** The detailed mesh.

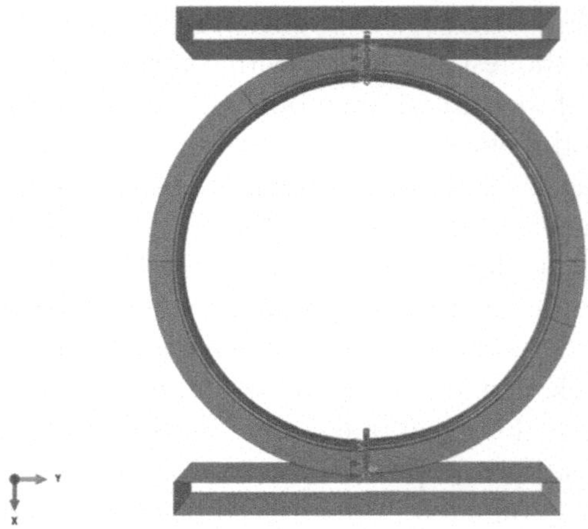

**Figure 8.36** The boundary conditions.

At the initiation of the carcass layer formation, the surfaces at the beginning and termination of the carcass, alongside the PE layer, possess the capacity for unrestricted movement. Consequently, it is imperative to constrain these surfaces comprehensively to prevent any undesired displacement. Therefore, the whole model is rotated by 90 degrees. The boundary conditions are applied on the top and bottom sides of the pipe model, as displayed in Figure 8.36; both sides are restricted to moving in all directions except the x-axis. Besides, the bottom plate is fixed against all degrees of freedom, and the displacement is applied to the reference point of the top plate. Hence, the plate moves onto the outer surface of the polymeric layer, and the loading is transferred to the carcass through its outer surface and the inner surface of the polymeric layer. To prevent any penetration phenomenon between surfaces, all potential contacts were taken into consideration. Normal behavior was specified as hard contact, and after contact, separation is allowed. In the tangential direction, the penalty function technique is applied, and the contact between layers is considered to be surface-to-surface contact. The friction coefficient between the PE and the carcass, and the PE and the plates, is set to 0.1, as given in the experimental results. It should be noted that no ovality or initial imperfection is presented in this simulation, as its essential objective is to acquire the radial displacement.

### 8.7.3 Results

At the end of the test, the pipe cross-section is compared with the accordance time of the FE model; both the experiment and numerical model show an eight-shaped cross-section, as can be seen in Figure 8.37. While the PE is elliptical, the carcass layer is demonstrating an eight-shape due to its interlocked structure. The relationship between the loading and displacement for the test and FE is extracted as demonstrated in Figure 8.38. The total displacement is determined to be constrained to 50 mm because up to this extent of displacement, different sections of the carcass do not have any contact with one another. The slope and tendency of the numerical curve agree quite well with the experiment, but the value of the applied load is higher in the FE. This proves the existence of a noticeable amount of residual stress in the carcass layers, which is inevitable in the manufacturing process. Therefore, the amount of residual stress of the carcass profile is approximately 10% of the main material stress (which is as same as the value that Tang *et al.* [28] extracted from their numerical model in ANSYS).

**Figure 8.37** The collapse mode of (a) test and (b) FE.

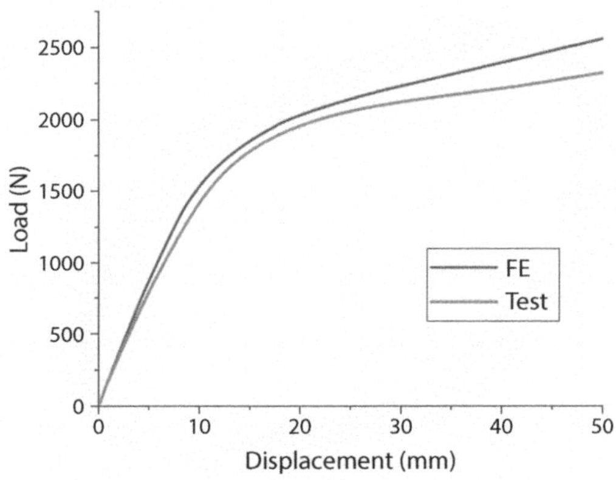


**Figure 8.38** The comparison between test and FE results.

The Von-Mises stress distribution, when the plate descends by 50 mm, has been extracted, and as illustrated in Figure 8.39, this distribution is non-uniform, so the stresses situated at the mid-span of the carcass have been extracted. The outer surface of the layer faces harsher loading conditions; therefore, four shell elements experiencing the most intense stress conditions are chosen on the outer surface of the carcass near the central region of the upper section along the pitch length. Their stress variations are depicted in Figure 8.40. The behavior of these elements is not similar to each other due to their location in the carcass cross-section.

Point 1 is linked to the semi-circular section of the profile, rendering it more robust than Points 2 and 3, which are positioned at the center. Additionally, point 4 is associated with the inclined line of the carcass, making it the most robust element among the four assessed elements. This shows that the carcass layer function relies significantly on its cross-section profile.

The functionality of the carcass is influenced by ovalization; it becomes non-functional once its ovality increases to 3% due to API. To assess the ovality of specimens, one must measure the maximum diameter ($D_{max}$) and

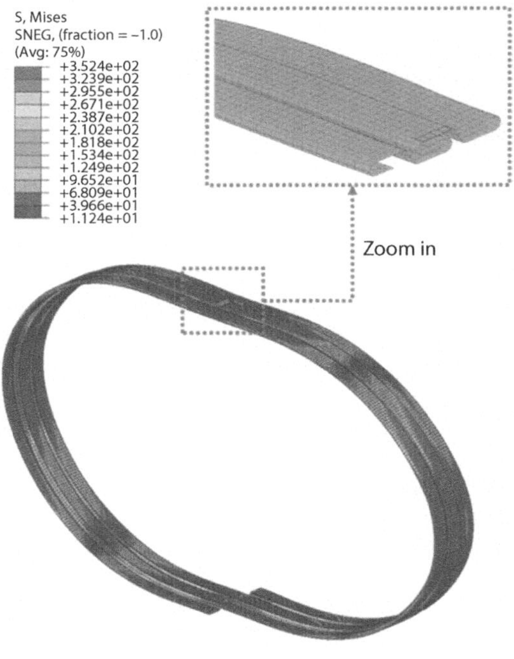


**Figure 8.39** Mises stress distribution of carcass layer.

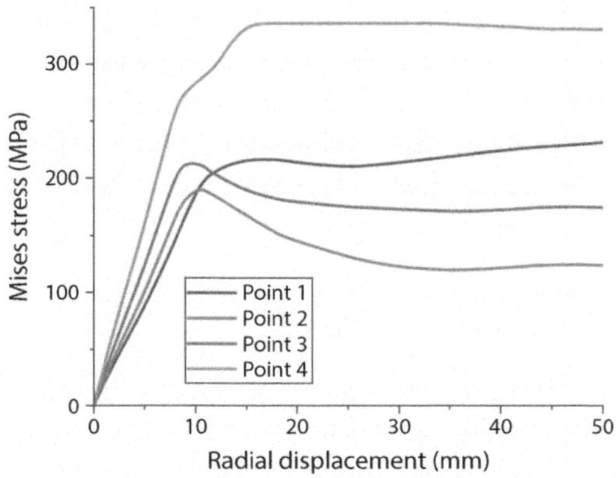


**Figure 8.40** Displacement–stress relationship on the external surface of the carcass.

minimum diameter ($D_{min}$) of their cross-sections at various intervals. The degree of ovality is then determined as follows:

$$\Delta = \frac{D_{\max} - D_{\min}}{D_{\max} + D_{\min}} \tag{8.36}$$

Figure 8.41 shows the relationship of ovality with the displacement of the plate. Furthermore, the form of the carcass cross-section corresponding to

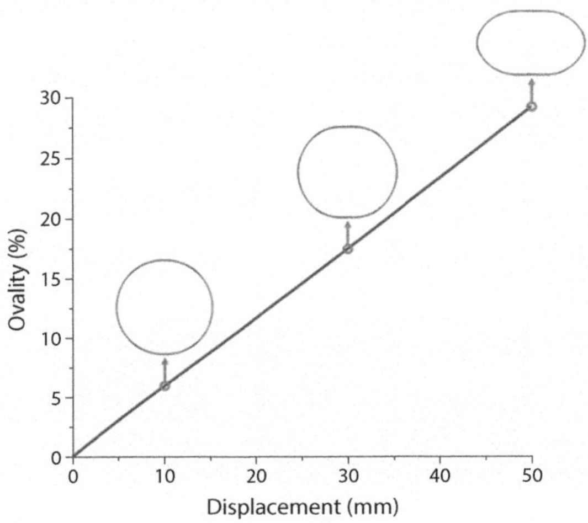


**Figure 8.41** Displacement–ovality relationship of the carcass.

the displacement at three distinct points is derived. The efficiency of the carcass layer is compromised when the ovality results in a plate displacement exceeding 5 mm. Consequently, upon analyzing the ovality value in conjunction with the Mises stress values, it becomes evident that the structural integrity of the carcass layer is compromised before reaching its yield strength.

### 8.7.4 Residual Stress in Carcass

The general production process of a carcass layer is demonstrated in Figure 8.41. The production of the carcass layer for flexible pipes begins with pre-metal forming steps, where stainless steel strips are decoiled and welded together for continuous length, and relief zones are cut near the welds to prevent fracture during winding. The strip edges are then deburred to avoid interference with interlocking (Figure 8.42(a)). Next, in the metal forming stage, lubrication is applied to minimize tool wear and prevent pickup. The strips are then shaped into an open S-profile through a series of roll-forming steps using contoured rolls (Figure 8.42(b)). Depending on the machine configuration, the formed profile is fed into the helical winding process either directly or through a loop buffer. In winding, the profile is wrapped helically around a mandrel using winding tools, which initiate and complete the interlock between successive profiles. This stage is critical, as the profile undergoes bending and deformation to achieve the desired geometry, pitch, and wall thickness (Figure 8.42(c)). Interlocking occurs progressively between winding tools, and the profile is free to adjust slightly, making the winding process partially freeform. After winding, the post-forming stage involves the use of a caterpillar unit, which holds and transports the carcass while maintaining the correct pitch and preventing unwanted rotation. Finally, the completed carcass is reeled onto large

(a) (b) (c)

**Figure 8.42** The manufacturing procedure of the carcass layer.

spools for storage and transport. Each step in this integrated process must be carefully controlled to ensure consistent geometry, reliable interlocking, and mechanical integrity in the final product.

The manufacturing production of the carcass induces a significant amount of stress in the formed sections of the carcass profile. Residual stresses are internal forces that remain locked within a structural component even when it is not subjected to external loads. These stresses are primarily introduced during the manufacturing process, especially due to uneven plastic deformation across the material. Residual stresses can adversely affect structural performance by triggering early yielding, reducing stiffness, and ultimately lowering the load-bearing capacity of the structure. Therefore, there is an important need to estimate the residual stress precisely, since the plastic yielding results from the combined effect of the applied stress and the residual stress, yet figuring out how to accomplish this remains a challenging task. Various experimental techniques are available to measure residual stresses, broadly categorized into destructive and non-destructive methods. Non-destructive approaches include X-ray diffraction, neutron or electron diffraction, ultrasonic testing, and magnetic methods, which assess stress without damaging the specimen. In contrast, destructive techniques involve measuring the deformation that occurs when residual stresses are released by removing material from the structure-sectioning being the most commonly used method in this category. Despite their utility, these experimental methods can be costly and, in many cases, impractical or unsuitable for certain materials, geometries, or applications.

As previously discussed, numerical simulation is the most suitable and practical approach for replicating the carcass production process and evaluating the resulting residual stresses. The simulation process is based on carcass manufacturing, which is conducted as an uninterrupted two-stage process. The initial stage involves the roll forming of stainless-steel strips to an S-shaped carcass cross-section. As shown in Figure 8.43, the simulation begins in ABAQUS with the creation of a series of rollers that precisely match the dimensions of the actual forming machine. A metal strip sheet is then modeled to pass through these rollers, producing an S-shaped profile as the output. The von Mises stress distribution from this step can be observed in Figure 8.44. These residual stresses are extracted from the cross-section and used as input for the following simulation stage. The subsequent stage entails winding the formed carcass profile helically onto a mandrel, in which the profile interlocks with itself, as depicted in Figure 8.45. In order to replicate this process, the S-shaped profile is rotated by applying a rotation velocity amount to make it a ring, enabling

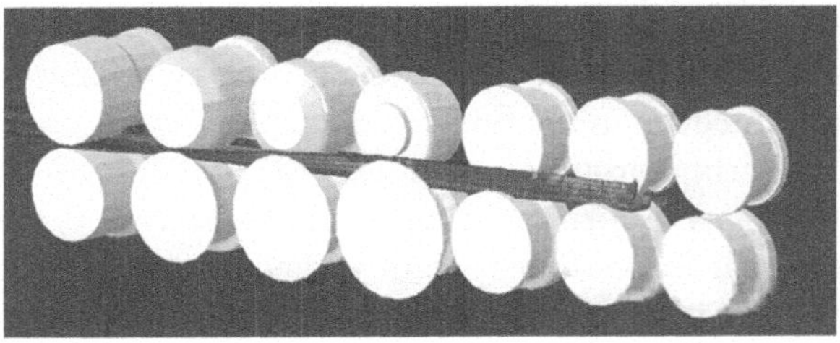

**Figure 8.43** The simulations of rollers for the carcass layer.

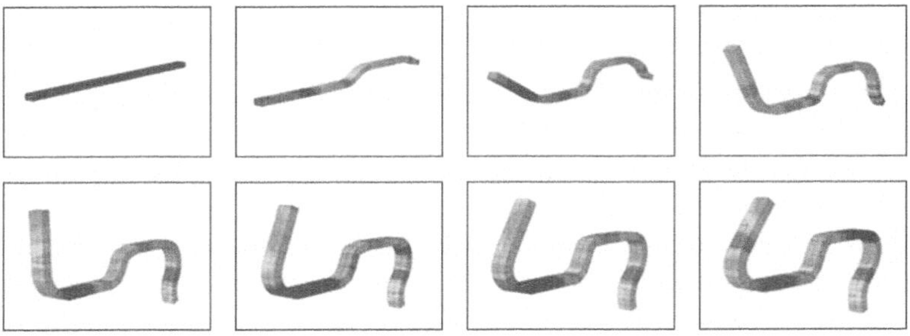

**Figure 8.44** Cross-sectional profile of roll-formed components for the fabrication of interlocking carcass structures.

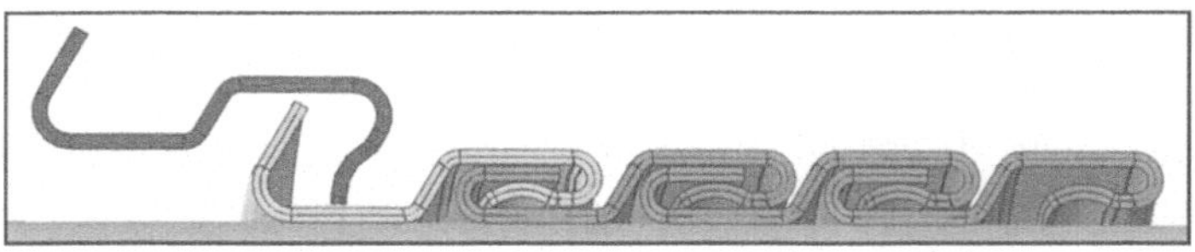

**Figure 8.45** Cross-section of the side of an interlocked carcass.

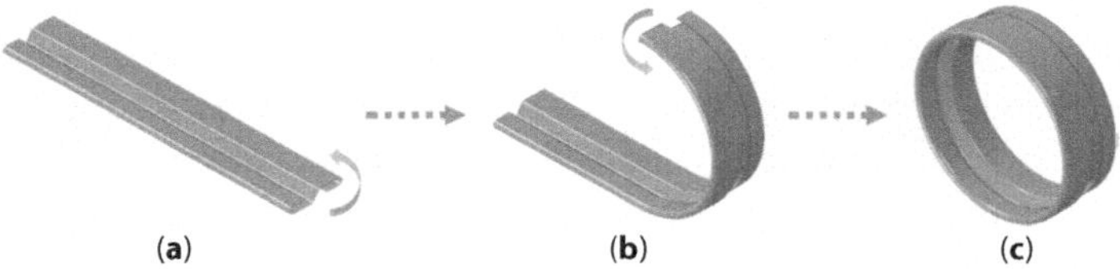


**Figure 8.46** The carcass self-locking process.

the simulation of the helical winding process, as can be seen in Figure 8.46. In this process, the pitch and lay angle of the carcass are disregarded as they exert negligible influence on the amount of residual stress. Finally, the von Mises stress is extracted and saved for the final step. In the last stage, this residual stress is applied to the model that was established before. With the residual stress now incorporated, the simulation results should match properly the experimental data, confirming the model's accuracy. The finite element model is thus validated and ready for further simulation work.

## References


1. Chandramouli, R., Pandey, K.S., Kandavel, T.K., *et al.*, Influence of material flow constraints during cold forming on the deformation and densification behaviour of hypoeutectoid P/M steel ring preforms. *Int. J. Adv. Manuf. Technol.*, 31, 926–932, 2007.
2. Banabic, D., *Sheet Metal Forming Processes*, Springer, Berlin, 2010.
3. Marciniak, Z., Duncan, J.L., Hu, S.J., *Mechanics of sheet Metal Forming*, Edward Arnold, London, 2002.
4. Cockroft, M.G. and Latham, D.J., Ductility and the workability of metals. *J. Inst. Met.*, 96, 33–39, 1968.
5. Marciniak, Z., Duncan, J.L., Hu, S.J., *Mechanics of sheet metal forming*, Butterworth-Heinemann, Oxford, 2002.
6. Marciniak, Z., Assessment of material formability, in: *Advanced Technology of Plasticity. Proceeding of the 2nd ICTP*, Tokyo, pp. 685–694, 1984.
7. Considere, A., Application of iron and steel in constructions. *Ann. Ponts Chaussees*, 9, 574–575 (in French), 1985.
8. Erichsen, A.M., A new test for thin sheets. *Stahl Eisen*, 34, 879–882 (in German), 1914.
9. Lankford, W.I., Snyder, S.C., Bauscher, J.A., New criteria for predicting the press performance of deep-drawing sheets. *Trans. ASM*, 42, 1196–1232, 1950.
10. Wech, P.I., Radtke, L., Bunge, H.J., Comparison of plastic anisotropy parameters. *Sheet Met. Ind.*, 60, 594–597, 1983.
11. Siegert, K., TALAT Lecture 3705 (http://www.eaa.net/eaa/education/TALAT/index.htm).
12. Choi, Y., Walter, M.E., Lee, J.K., Han, C.S., Observations of anisotropy evolution and identification of plastic spin parameters by uniaxial tensile tests. *Int. J. Solids Struct.*, 1, 303–325, Z, 2006.
13. Chen, Reuben, R.L., Owen, D.G., The deformation mechanics of interlock tubes. *Strain*, 28, 3, 99–106, August 1992.

14. Halmos, G.T., *Roll Forming Handbook*, CRC Press - Taylor & Francis Group, LLC, Boca Raton, Florida, 2006.
15. Bai, Y., *Deepwater flexible risers and pipelines*, John Wiley & Sons, Hoboken, NJ 07030, USA, 2020.
16. Saneian, M. and Bai, Y., Assessment of the carcass role in designing deep-water riser connectors, in: *International Conference on Offshore Mechanics and Arctic Engineering*, American Society of Mechanical Engineers, p. V05AT04A033, 2019.
17. Saneian, M. and Bai, Y., Evaluation of the inner liner in unbounded flexible pipelines in the sealing system of end-fitting, in: *International Conference on Offshore Mechanics and Arctic Engineering*, American Society of Mechanical Engineers, p. V003T04A004, 2024.
18. Timoshenko, S.P. and Gere, J.M., *Theory of elastic stability*, Courier Corporation, Mineola, New York, USA, 2012.
19. Zhang, Y., Chen, B., Qiu, L., Hill, T., Case, M., State of the art analytical tools improve optimization of unbonded flexible pipes for deepwater environments, in: *Offshore technology conference*, OTC, pp. OTC–15169, 2003.
20. Martins, C.V.A., Pesce, C.P., Aranha, J.A., Structural behavior of flexible pipe carcass during launching, in: *International Conference on Offshore Mechanics and Arctic Engineering*, pp. 537–546, 2003.
21. Fernando, U.S., Tan, Z., Sheldrake, T., Clements, R., The stress analysis and residual stress evaluation of pressure armour layers in flexible pipes using 3d finite element models, in: *International conference on offshore mechanics and arctic engineering*, pp. 57–65, 2004.
22. Neto, A.G. and Martins, C.V.D.A., A comparative buckling study for the carcass layer of flexible pipes, in: *International Conference on Offshore Mechanics and Arctic Engineering*, pp. 419–428, 2009.
23. Nogueira, V.P.P. and Netto, T.A., A simple alternative method to estimate the collapse pressure of flexible pipes, in: *International Conference on Offshore Mechanics and Arctic Engineering*, pp. 693–699, 2010.
24. Gay Neto, A., Martins, C.V.D.A., Pesce, C.P., Meirelles, C.O.C., Malta, E.R., Barbosa Neto, T.F.F., Godinho, C.A.F., Burst prediction of flexible pipes, in: *International Conference on Offshore Mechanics and Arctic Engineering*, pp. 511–520, 2010.
25. Gay Neto, A. and de Arruda Martins, C., A comparative wet collapse buckling study for the carcass layer of flexible pipes. *J. Offshore Mech. Arct. Eng.*, 134, 031701, 2012.

26. Tang, M., Lu, Q., Yan, J., Yue, Q., Buckling collapse study for the carcass layer of flexible pipes using a strain energy equivalence method. *Ocean Eng.*, 111, 209–217, 2016.
27. Li, X., Jiang, X., Hopman, H., A strain energy-based equivalent layer method for the prediction of critical collapse pressure of flexible risers. *Ocean Eng.*, 164, 248–255, 2018b.
28. Tang, M., Yan, J., Chen, J., Yang, Z., Yue, Q., Residual stress evaluation and buckling analysis of carcass layers in flexible pipes using 3d finite element model, in: *International Conference on Offshore Mechanics and Arctic Engineering*, American Society of Mechanical Engineers, p. V05AT04A046, 2015.

# 9

# Mechanical Behavior of Thin Cylindrical Shells Under Combined Axial Compression and Bending

***Abstract***
This chapter presents a comprehensive experimental and numerical investigation of steel cylindrical shells subjected to combined axial compression and bending, focusing on the buckling behavior of thin-walled tubes. Material tests were conducted to evaluate tensile and compressive properties, revealing significant differences due to cold-drawing effects. Eccentric compression tests were performed on tubes with D/t ratios ranging from 40 to 97, simulating offshore engineering conditions. Two distinct local buckling modes—DENT and COS—were observed, influenced by eccentricity and wall thickness. Finite element models using ABAQUS were validated against experimental results, accurately replicating failure modes and ultimate strengths within a 5% error margin. Initial geometric imperfections were introduced based on elastic buckling analyses, and their impact on post-buckling behavior was examined. A parametric study explored the influence of D/t, slenderness ratio, and eccentricity on load capacity and deformation. Results indicate that increasing any of these parameters reduces ultimate strength and enhances buckling sensitivity. The hoop strain was found to be negligible compared to longitudinal strain, supporting its exclusion from design considerations. This chapter provides valuable insights into structural performance, failure mechanisms, and safe design practices for tubular members under complex loading scenarios.

***Keywords*:** Thin-walled tubes, local buckling, combined loading, finite element analysis, DENT mode, COS mode, eccentric compression

## 9.1 Introduction

Thin wall tubes, such as the bracing members of platform and wind turbines, are widely used in ocean engineering, and it is evident that this trend will accelerate in the coming future. Chen and Han [1] did a summarize of

Yong Bai. Plasticity and Ductile Fracture in Structural Engineering, (261–280) © 2026 Scrivener Publishing LLC

the mechanical behavior of tubular members in offshore structures. Due to the complex conditions during production and installation, bracing members of the ocean platform may be subjected to different loads, among which the combination of bending and axial compression should be studied carefully.

As regulated by API [3], DNV [2], and ABS [17], the general approach in these structural steel codes is to limit the design stress to the 0.2% proof stress and ignore the pronounced strain hardening in the determination of cross-section resistances. In this way, the design state is kept in the elastic range where local buckling will not happen but sometimes make the structure design over-conservative. In practical application, the tubes often experience extended excursions into the inelastic range of the material, giving rise need to consider material nonlinearity. Researches on the inelastic behaviors of cylindrical tubes under bending moment and axial load respectively have been carried out in many papers. Kyriakides and Ju [5] did both experimental and theoretical studies on tubes under the bending moment, including an analytical procedure with iteration method. Bardi and Kyriakides [7] conducted a series of researches on the buckling behavior of circular tubes under axial compression, revealing the buckling modes under pure compression. Yadav and Gerasimidis [6] carried out a comprehensive numerical study on the localization instability phenomenon of thin steel tubes under bending moment. Experimental results and theoretical equations for ultimate strength from these researches contributed to building design codes, which have been proved suitable in actual application.

A few efforts have also been spent on developing an understanding of the interaction between axial loads and bending effects that exist in tubular members. Sohal and Chen [9–11] studied the local buckling behavior of tubular beam-columns by building a simple analytical model by an empirical method. They thought the local buckling occurs when the compressive strain reaches the critical strain, $\varepsilon_{cr}$, which is determined by an empirical equation curve-fitted with the available experimental data. Zhao [12]researched the mechanical behavior of steel tubular members under combined axial compression and bending by both experiments and numerical simulations. Zhao [13] also did similar research to study the buckling behaviors of the aluminum columns under eccentric compression. However, what they concentrated on are tubes with a *D*/*t* ratio smaller than 40. The local buckling behavior of thin cylindrical tubes under combined loads still remains incompletely understood and thus being the first topic of this chapter.

This chapter describes a comprehensive experimental and numerical study of steel cylindrical shells under eccentric compression, with the purpose to investigate the buckling behavior of steel tubes under combined loading. A testing procedure on material tensile and compressive mechanical properties and eccentrically loaded circular beam-columns are proposed here, followed by a numerical study, where the models are firstly validated against the performed experiments. The key design factors of the specimens were set to be practical for ocean engineering applications. The diameter by thickness (*D/t*) ratio is used in this chapter ranged from 40 to 97, which contains thick and thin tubes. The validation of the FE models is performed against the experimental results, and the models are proved to be able to predict the ultimate strengths and buckling behavior of thin-walled tubes. Upon validating the FE models, further analysis is performed with them, and some useful conclusions are obtained.

## 9.2 Experiments

### 9.2.1 Material Tests

Cold-formed carbon steel cylindrical tubes with different *D/t* ratios were tested under combined compression and bending in this study. The test specimens were fabricated by cold drawing from the initial tube stock with an inner diameter of 95.0 mm and a wall thickness of 3.2 mm. The thickness of specimens was changed to a specific value while the inner diameter remained the same, so different D/t ratios can be derived. As reported by Skelton [14], the tensile and compressive properties of metal material can be different during cyclic deformation, which is also known as the Bauschinger effect. In order to correctly describe the material behavior, tensile and compressive properties were both measured carefully before

**Table 9.1** Geometry and mechanical properties of four groups of specimens.

| Specimen number | D (mm) | d (mm) | t (mm) | D/t | E (Mpa) | σ0.2t (Mpa) | nt | σ0.2c (Mpa) | $n_c$ |
|---|---|---|---|---|---|---|---|---|---|
| HA | 97.0 | 95.0 | 1.0 | 97 | 192,658 | 441.3 | 20 | 345.7 | 6 |
| HB | 97.4 | 95.0 | 1.2 | 81.2 | 192,374 | 568.8 | 20 | 367.8 | 6 |
| HC | 98.2 | 95.0 | 1.6 | 61.4 | 187,902 | 531.8 | 30 | 362.9 | 6 |
| HD | 100.0 | 95.0 | 2.5 | 40.0 | 184,460 | 458.5 | 30 | 323.6 | 6 |

the combined loading tests. Tested specimens were labeled with HA, HB, HC, and HD, and geometries of the four cross-sections are summarized in Table 9.1, where *D*, *d*, and *t* are the outer diameter, the inner diameter, and the thickness, respectively.

### 9.2.2 Tensile Property Testing

Tensile coupon tests were adopted to determine the material stress–strain response of each specimen under uniaxial tension. The coupons, which were cut directly from each cylindrical specimen along the longitudinal direction, were dimensioned and machined in accordance with the industry standard for material tensile testing. The tensile fracture occurred when the strain was in the order 0.6 to 1.4%, which shows that the ductility of the material was reduced due to the cold fabrication process. The nominal stress–strain relationship for the four kinds of specimens is shown in Figure 9.1 in solid lines. The measured Young's modulus *E* and tensile yield stress, which is defined to the stress with a 0.2% offset plastic strain $\sigma_{0.2t}$, are shown in Table 9.1.

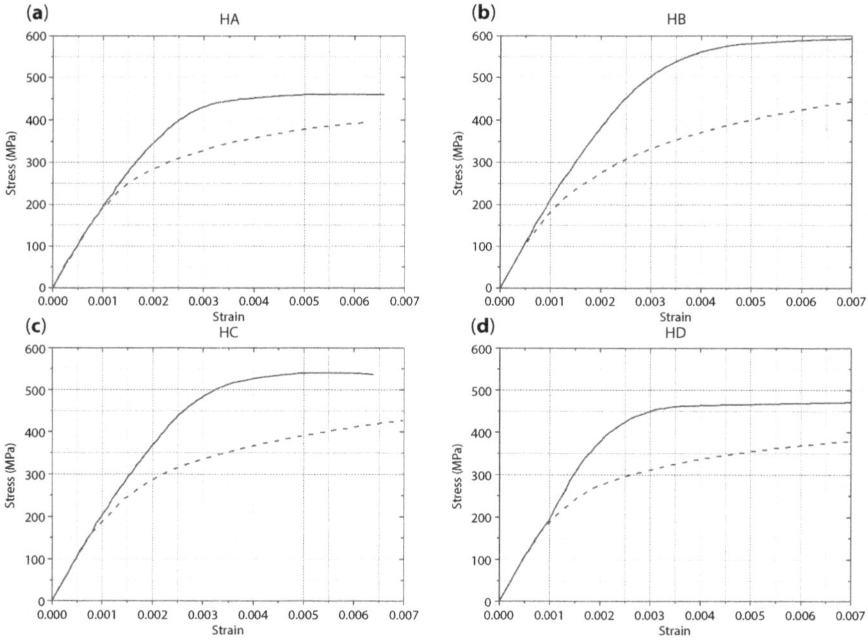


**Figure 9.1** Stress–strain curves for material tests.

### 9.2.3 Compressive Property Testing

The compressive properties were tested by thrusting the stub column specimens cut from four groups of tubes. The length of stub pipe was selected to 300 mm, which made the column stocky enough to avoid the column from buckling. Four pieces of bi-axial strain gauges were put on the central cross-sections of the test specimens. The nominal stress–strain relations are plotted in Figure 9.1 denoted by the dotted lines. Since shell mode buckling might occur in the upper or lower edges, the stress–strain relationship was measured up to the strain level of 0.1%. As shown in the diagram, the compressive and tensile curves coincide with each other in their common elastic range, which illustrates the similarity of Young's modulus. The compressive yield stress $\sigma_{0.2c}$ is summarized in Table 9.1.

### 9.2.4 Material Parameters

In the numerical studies, the material parameters of these four kinds of steel materials have been adopted. The stress–strain relationship beyond the linear region is defined as an Osgood curve in the form

$$\varepsilon = \frac{\sigma}{E} + 0.002\left(\frac{\sigma}{\sigma_y}\right)^n \tag{9.1}$$

where $\sigma_y$ and $n$ are yield stress and strain-hardening parameter respectively. $E$ denotes Young's modulus. All these parameters are recorded in Table 9.1.

It should be noted that FE software requires the material properties to be specified in the format of true stress and strain, which are transformed from the engineering stress and strain from the material tests. The measured engineering stress–strain curves from material tests, which are mentioned above, were converted into true stress–strain curves according to Equations (9.2) and (9.3), where $\sigma_{true}$ and $\varepsilon_{true}$ are true stress and true strain and $\sigma_{nom}$ and $\varepsilon_{nom}$ are nominal stress and nominal strain.

$$\sigma_{true} = \sigma_{nom}(1 + \varepsilon_{nom}) \tag{9.2}$$

$$\varepsilon_{true} = \ln(1 + \varepsilon_{nom}) \tag{9.3}$$

### 9.2.5 Combined Loading Tests

The combined loading tests were conducted by a large-scale model testing machine. Schematic diagrams of the general combined loading test setup and the end fixture are shown in Figure 9.2. The cylindrical plug and sphere support at both ends are used to simulate pin-ended boundary conditions. The nominal initial loading eccentricities were taken to be D/4, D/8, D/16, and D/32 to indicate the impact of eccentricity on buckling behavior. These eccentricities are obtained by changing the position of the plug relative to the loading head. The stain along axial and circumferential directions are both measured by gauges during the experiments. Axial loading can be obtained through the test machine directly, and the end shortening is measured by a displacement meter.

The results from fifteen experiments are reported in Table 9.2. From the experimental results, as eccentricity increases, the ultimate strength decreases. This is because larger eccentricity leads to a bigger additional moment which makes a contribution to the buckling of tubes. The same tendency is observed when the length increases. Tubes with a larger slenderness ratio may collapse due to overall instability, which reduces carrying capacity significantly. Figure 9.3 shows two kinds of local buckling modes observed in tests, DENT and COS modes. The DENT mode is the main buckling mode of the specimens, and the COS mode only appears on specimens with relatively thick wall thickness and large eccentricity. In order to describe the deformation process more clearly, a group of schematic diagrams showing the change in cross-section at different locations along the length of the tube is presented in Figure 9.4. It can be obtained

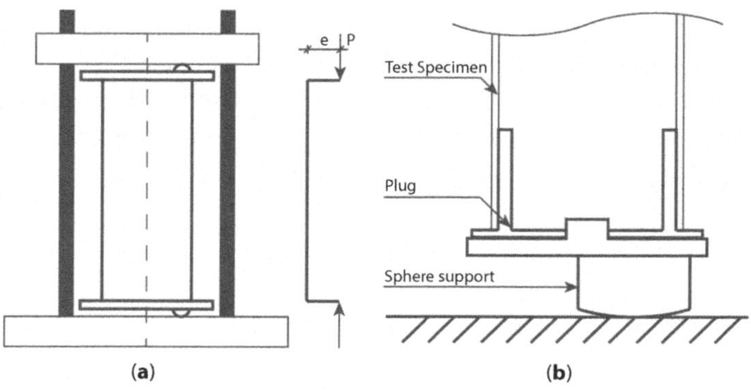


**Figure 9.2** (a) Schematic diagram of the setup for combined loading tests; (b) diagram of end fixtures for eccentric axial thrust tests.

**Table 9.2** Geometry and mechanical properties of four groups of specimens.

| Cross-section | Specimen ID | Load eccentricity e/D | Length L (mm) | Initial deflection $a_n$ (mm) | Ultimate load $N_u$ (KN) | Buckling mode |
|---|---|---|---|---|---|---|
| HA | 1 | 1/4 | 1635 | 0.43 | 56.4 | DENT |
| | 2 | 1/16 | 895 | 0.13 | 95.9 | DENT |
| | 3 | 1/16 | 605 | 0.25 | 105.9 | DENT |
| HB | 1 | 1/8 | 1635 | 1.02 | 78.0 | DENT |
| | 2 | 1/16 | 1635 | 0.61 | 89.3 | DENT |
| | 3 | 1/32 | 1635 | 0.10 | 97.1 | DENT |
| HC | 1 | 1/8 | 1635 | 1.40 | 98.0 | DENT |
| | 2 | 1/16 | 1635 | 0.64 | 116.7 | DENT |
| | 3 | 1/32 | 1635 | 0.44 | 135.0 | DENT |
| HD | 1 | 1/4 | 1635 | 1.44 | 132.0 | COS |
| | 2 | 1/8 | 1635 | 0.87 | 146.7 | DENT |
| | 3 | 1/16 | 1635 | 0.63 | 176.1 | DENT |

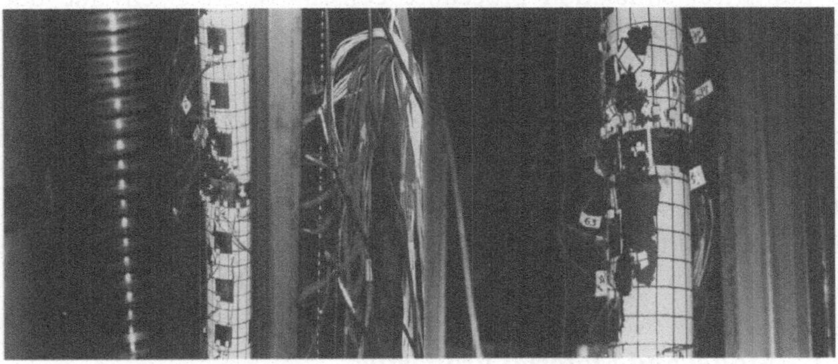

(a) DENT mode (b) COS mode

**Figure 9.3** Buckling modes of tested specimens.

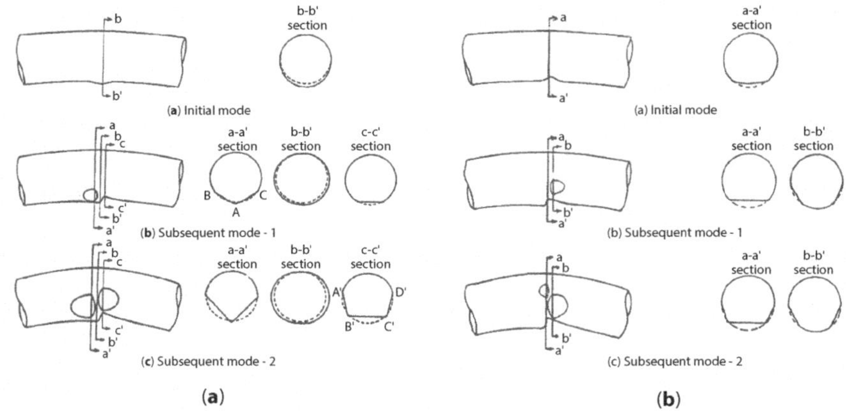


**Figure 9.4** (a) Local buckling in cos mode; (b) local buckling indent mode.

that when specimens buckle in cos mode, the cross-section will become into a more complicated diamond shape. The buckling mode depends on the diameter to thickness ratio, the combination of axial forces and bending moments at the cross-section, and the material properties. All these results are utilized to validate our finite element models.

## 9.3 Numerical Studies

The numerical simulation procedure consists of an eigenvalue buckling analysis, followed by a numerical modeling program using the Riks method in ABAQUS. This section describes the geometry, boundary conditions, material models, and FE modeling of the cylindrical tubes used in this chapter. The initial purpose of this work is to define accurate initial imperfections, which are in accord with the local buckling modes found in the experiments. In order to validate the simulation method, the FE models are used to replicate the obtained full load-deformation histories, ultimate strength, and failure modes from experiments.

### 9.3.1 Model and Material Properties

The commercial software ABAQUS is used to build a series of finite element models. Similar FE models for thin-walled tubular members under different kinds of loads have been established by many researchers. During these FE studies, the four-node reduced integration shell elements, S4R, have been proved capable of simulating the mechanical behavior of tubes with thin walls and therefore are utilized in this chapter. Yadav and Gerasimidis [6] did a mesh convergence analysis to find the accurate element dimensions providing sufficient accuracy and computational efficiency. Based on this research result, the element thickness is chosen as 2 mm in this chapter.

The boundary conditions are shown in Figure 9.5, which are set in accordance with the experiments. Because the deformation of cylinders under combined bending and axial compression is symmetric with respect to the plane perpendicular to the buckling axis, only half-length of the specimen is modeled, while asymmetric boundary condition about Z-axis is applied to the right cross-section. Besides this, a rigid constraint about the reference point RP-1 is imposed at the left end, in order to simulate the specimens fixed by the plug. A concentrated force is applied to the RP-1, of which the offset can be adjusted to realize different eccentricities.

For each load combination of compression and bending moment, the position of the initial neutral surface can be calculated as shown in the right of Figure 9.5. In order to define the difference of the material under compression and tension introduced to form the cold fabrication, two different materials are assigned to two parts divided by the neutral surface. The tensile properties are assigned to part I while the compressive properties are assigned to part II. During the load process, the lower part is always under compression. Stress in some range of the upper half may change from tension to compression but keeps in the elastic stage. As shown in the material tests, compressive and tensile properties are the same in the elastic stage. Therefore, this method can replicate the material properties of test specimens.

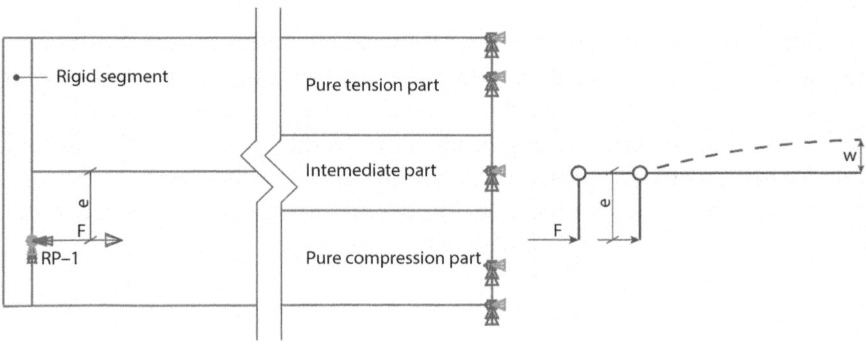


**Figure 9.5** Boundary conditions on the model.

### 9.3.2 Elastic Buckling Analysis

The Riks method calls for an initial imperfection to be introduced into the perfect geometry model so that there will be some response in the buckling mode before the critical load is reached. In previous studies by Zhao [12] and Kshitij Kumar Yadav [6], the initial imperfections were defined as the lowest local buckling mode shape under the considered loading condition. However, what they are concentrated on are stub columns with a small D/t ratio. As the dimensions of the specimen change, this method would be inadequate. Therefore, it is of importance to carry out an elastic buckling analysis to find suitable imperfections.

In this chapter, a parametric investigation was carried out to study the influence of diameter to thickness ratios *D/t*, eccentricity to diameter ratio *e/D*, and slenderness ratio $\lambda$ to the elastic buckling behavior of cylindrical shells. The geometric properties of FE specimens are shown in Table 9.3. For each cross-section, seven different geometric dimensions are adopted, of which the first three groups are in accordance with the experiment specimens.

The results of twenty-eight FE modeling tests are shown in Table 9.4, where buckling modes and eigenvalues are presented. As the parameters are changed, four kinds of buckling modes come up in different combinations. The shapes of them are shown in Figure 9.6. Among these, group (a) is overall buckling mode, (b) is typical column buckling mode, while groups (c) and (d) can be classified into local buckling modes.

As shown in the table, there are three situations. For stub tubes, of which the slenderness is smaller than 25, only local buckling modes occur. However, when the slenderness increases nearly to 50 and eccentricity is big enough, the buckling modes are the column mode and shell buckling mode.

If the eccentricity is small, the COM mode will also appear. Compared to the failure modes found in tests, the local buckling modes will lead to an accurate result; thus, they should be chosen as the initial imperfection rather than the lowest buckling mode. Beyond that, it is interesting to find that for specimens with different slenderness ratios, others being held contrast, the critical stress for local buckling is all the same. In other words, the elastic critical stress is independent of slenderness.

**Table 9.3** Summary of elastic buckling analysis results.

| | No. | Overall buckling | Eigenvalue (N) | COM | Eigenvalue (N) | Shell buckling | Eigenvalue (N) |
|---|---|---|---|---|---|---|---|
| HA | 1 | √ | 2.35E+05 | - | - | DENT | 3.67E+05 |
| | 2 | - | - | - | - | DENT | 5.86E+05 |
| | 3 | - | - | - | - | DENT | 5.86E+05 |
| | 4 | √ | 2.40E+05 | - | - | DENT | 4.89E+05 |
| | 5 | √ | 2.41E+05 | √ | 5.12E+05 | COS | 5.85E+05 |
| | 6 | - | - | - | - | DENT | 3.68E+05 |
| | 7 | - | - | - | - | DENT | 4.89E+05 |
| HB | 1 | √ | 2.90E+05 | √ | 6.77E+05 | DENT | 6.92E+05 |
| | 2 | √ | 2.91E+05 | √ | 7.03E+05 | DENT | 8.28E+05 |
| | 3 | √ | 2.92E+05 | √ | 7.11E+05 | DENT | 9.18E+05 |
| | 4 | √ | 2.86E+05 | - | - | DENT | 5.20E+05 |
| | 5 | - | - | - | - | DENT | 5.19E+05 |
| | 6 | - | - | - | - | DENT | 6.91E+05 |
| | 7 | - | - | - | - | DENT | 8.28E+05 |
| HC | 1 | √ | 3.84E+05 | √ | 1.17E+06 | DENT | 1.21E+06 |
| | 2 | √ | 3.84E+05 | √ | 1.21E+06 | DENT | 1.45E+06 |
| | 3 | √ | 3.84E+05 | √ | 1.23E+06 | COS | 1.61E+06 |
| | 4 | √ | 3.80E+05 | - | - | DENT | 9.13E+05 |
| | 5 | - | - | - | - | DENT | 9.13E+05 |
| | 6 | - | - | - | - | DENT | 1.22E+06 |
| | 7 | - | - | - | - | COS | 1.45E+06 |
| HD | 1 | √ | 6.07E+05 | - | - | DENT | 2.14E+06 |
| | 2 | √ | 6.09E+05 | √ | 2.67E+06 | DENT | 2.85E+06 |
| | 3 | √ | 6.10E+05 | √ | 2.77E+06 | DENT | 3.41E+06 |
| | 4 | √ | 1.97E+06 | - | - | DENT | 2.14E+06 |
| | 5 | - | - | - | - | DENT | 2.15E+06 |
| | 6 | - | - | - | - | DENT | 2.86E+06 |
| | 7 | - | - | - | - | DENT | 3.42E+06 |

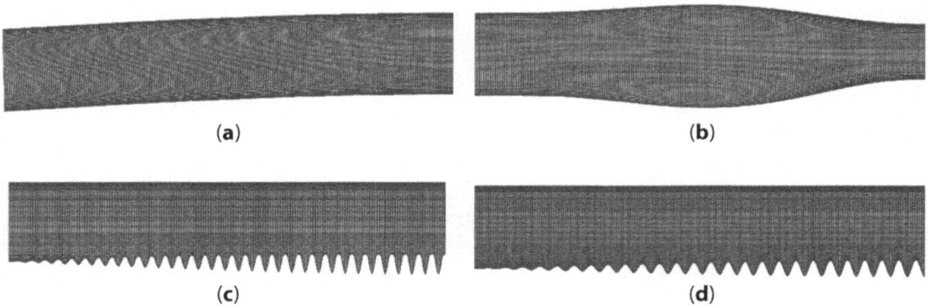


**Figure 9.6** Four kinds of buckling modes: (a) overall mode; (b) COM mode; (c) DENT mode; (d) COS mode.

**Table 9.4** Geometry properties of finite element models.

| (a) HA | | | | | (b) HB | | | | |
|---|---|---|---|---|---|---|---|---|---|
| No. | L | L/i | e | e0 | No. | L | L/i | e | e0 |
| 1 | 1635 | 48.2 | 24.3 | 1.02 | 1 | 1635 | 48.1 | 12.2 | 0.51 |
| 2 | 895 | 25 | 6.1 | 0.26 | 2 | 1635 | 48.1 | 6.1 | 0.26 |
| 3 | 605 | 17.8 | 6.1 | 0.26 | 3 | 1635 | 48.1 | 3.0 | 0.13 |
| 4 | 1635 | 48.2 | 12.1 | 0.51 | 4 | 1635 | 48.1 | 24.4 | 1.02 |
| 5 | 1635 | 48.2 | 6.1 | 0.26 | 5 | 850 | 25 | 24.4 | 1.02 |
| 6 | 605 | 17.8 | 24.3 | 1.02 | 6 | 850 | 25 | 12.2 | 0.51 |
| 7 | 605 | 17.8 | 12.1 | 0.51 | 7 | 850 | 25 | 6.1 | 0.26 |
| **(c) HC** | | | | | **(d) HD** | | | | |
| No. | L | L/i | e | $e_0$ | No. | L | L/i | e | $e_0$ |
| 1 | 1635 | 47.8 | 12.28 | 0.52 | 1 | 1635 | 47.4 | 25 | 1.05 |
| 2 | 1635 | 47.8 | 6.14 | 0.26 | 2 | 1635 | 47.4 | 12.5 | 0.53 |
| 3 | 1635 | 47.8 | 3.07 | 0.13 | 3 | 1635 | 47.4 | 6.25 | 0.26 |
| 4 | 1635 | 47.8 | 24.55 | 1.03 | 4 | 862 | 25 | 25 | 1.05 |
| 5 | 855 | 25 | 24.55 | 1.03 | 5 | 620 | 18 | 25 | 1.05 |
| 6 | 855 | 25 | 12.28 | 0.52 | 6 | 620 | 18 | 12.5 | 0.53 |
| 7 | 855 | 25 | 6.14 | 0.26 | 7 | 620 | 18 | 6.25 | 0.26 |

## 9.4 Validation

### 9.4.1 Induction of Initial Imperfections

The initial geometric imperfections can significantly influence the development of local buckling, the ultimate strength, and the post-buckling response. Therefore, it is important to introduce suitable geometric imperfection amplitudes and modes into the FE models. The amplitude of initial imperfection is one important point that requires additional attention, which will influence the ultimate capacity of members. As many studies have been done to study the suitable amplitude of initial imperfection, t/100 is considered to be capable to obtain accurate results and is used in this chapter. The influence of a different kind of initial imperfection is discussed in this chapter. During the numerical analysis, two different local buckling modes are found. The exact condition of which buckling mode may happen is not clear yet. Therefore, a parametric study is carried out to study the effect of different modes on the buckling behavior of tubes. The relationship between axial load and end displacement for specimens with different initial imperfections is illustrated in Figure 9.7. It can be found that the local buckling modes only affect the post-buckling behavior while the ultimate strength is almost the same. Figure 9.8 shows axial load-end shortening curves. Figure 9.9 presents the failure mode of FE models with different initial imperfections, which shows that the initial imperfection can lead to different failure modes.

By comparison of Figure 9.3 and Figure 9.9, it can also be found that when the suitable initial imperfection is introduced, the failure mode of FE models will be Consistent with experimental results.

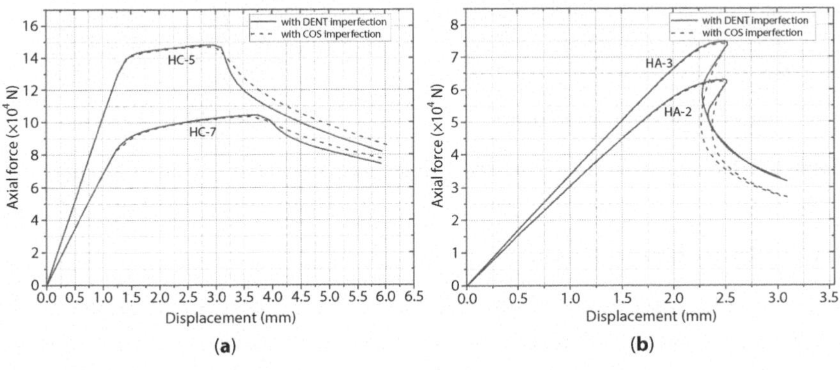


**Figure 9.7** Load-end shortening curves for specimens with DENT and COS imperfections.

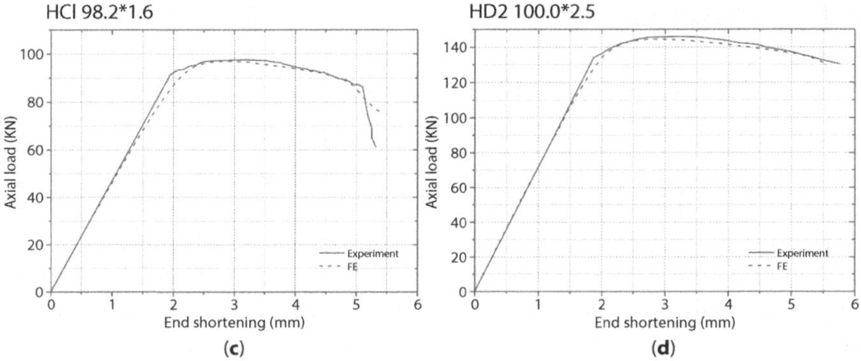


**Figure 9.8** Experimental and numerical axial load-end shortening curves for typically combined loading specimens.

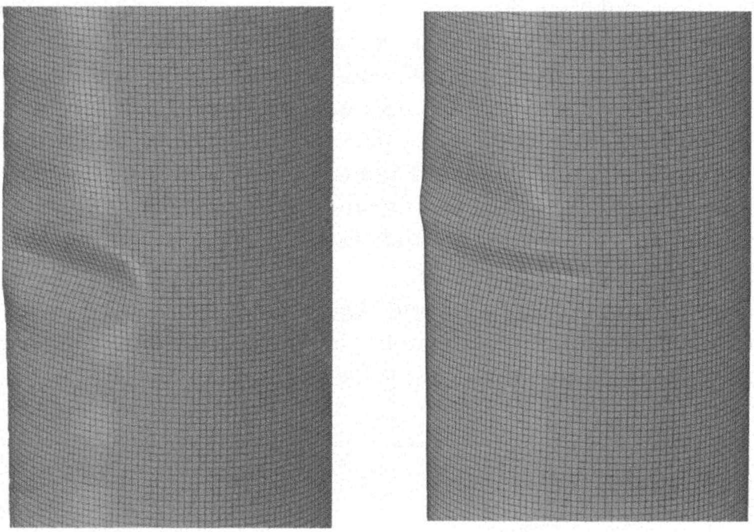

**(a) FE model with DENT mode** **(b) FE model with COS mode**

**Figure 9.9** Failure modes for FE models with different kind of initial imperfections.

## 9.4.2 Simulation Results

After introducing the initial imperfections, the simulation is carried out by the Riks method in ABAQUS. The accuracy of the FE models is validated by comparing the local buckling failure modes and ultimate load obtained from experiments with those from the numerical simulations.

**Table 9.5** Comparison of the test failure loads with FE failure loads.

| Specimen ID | Numerical results $N_{FE}$ (KN) | Experimental results $N_u$ (KN) | Error ratio $N_{FE}/N_u$ |
|---|---|---|---|
| HA1 | 55.0 | 56.4 | 0.98 |
| HA2 | 90.9 | 95.9 | 0.95 |
| HA3 | 103.8 | 105.9 | 0.98 |
| HB1 | 77.9 | 78.0 | 0.99 |
| HB2 | 90.6 | 89.3 | 1.01 |
| HB3 | 97.7 | 97.1 | 1.01 |


| Specimen ID | Numerical results $N_{FE}$ (KN) | Experimental results $N_u$ (KN) | Error ratio $N_{FE}/N_u$ |
|---|---|---|---|
| HA1 | 55.0 | 56.4 | 0.98 |
| HA2 | 90.9 | 95.9 | 0.95 |
| HA3 | 103.8 | 105.9 | 0.98 |
| HB1 | 77.9 | 78.0 | 0.99 |
| HB2 | 90.6 | 89.3 | 1.01 |
| HB3 | 97.7 | 97.1 | 1.01 |


The experimental and numerical results are compared in Table 9.5, which reports the ratios of numerical to experimental ultimate load at the failure point. It can be noticed that the coefficient of variation can be controlled by 5%, which indicates that the FE models can predict the ultimate strength very well.

## 9.5 Buckling Behavior of Thin Tubes Under Combined Loads

Based on the verified models, an extensive parametric investigation was carried out to study the effect of diameter to thickness ratios *D/t*, slenderness ratios $\lambda$, and eccentricity ratios *e/D* on the ultimate strength, deformation capacity, and buckling strain of tube under combined compression and bending loads. The *D/t* of the cross-sections is the main contributor to the buckling behavior of the tubes, and thus, two typical values of 40 and 97 were selected for the study to represent thick and thin shells, respectively. The most extreme range of eccentricity diameter ratio, which is from 1/8 to 1/2, was taken into consideration for the FE models. In order to eliminate the effect of different materials, constant material properties including $E$ = 192658 MPa, $\sigma_y$ = 441.3 MPa, and $n$ = 20 are applied to different simulated tubes. The results are recorded in Figure 9.10, including predicted compression–displacement, ovalization–displacement, and strain–displacement diagrams. Among these diagrams, the abscissa is defined as relative end shortening $x = u/u_0$, where $u_0$ denotes the value when buckling happens.

Some valuable concusions can be obtained here. By comparing two diagrams from Figure 9.10(a), which present the compression–displacement relationship for tubes with $D/t$ ratios of 97 and 40, the $D/t$ ratio has a significant impact on the buckling behavior of tubes. For the thin-walled tubes, once buckling happens, the load decreases dramatically, whereas the thick ones can remain to carry capacity and the load decreases smoothly. It can be obtained that an accurate safety factor should be taken into consideration for the design of thin tubes to prevent the structure from sudden collapses. It can also be concluded that the larger $D/t$ ratio, slenderness ratio $\lambda$, and eccentricity ratio $e/D$ result in a smaller ultimate strength.

Figure 9.10(b) presents the relationship between the deformation of the cross-section and the end displacement. Under eccentric compression, the midspan cross-section of tubes deforms into the shape as shown in the left of Figure 9.10(b), different from the observed elliptical shape in pure bending experiments. However, the ovality $\Delta D/D$ can also be used to evaluate the section deformation. It can be seen that the ovalization of the section is very small in the pre-buckling range, which is generally smaller than 1%. It is to say that the section can be thought of as circular before buckling happens. This assumption is useful for further theoretical study.

The relationship between strain and end displacement is demonstrated in Figure 9.10(c). The directions of $\varepsilon_{11}$ and $\varepsilon_{22}$ are expressed on the left of the figure. The $\varepsilon_{11}$ and $\varepsilon_{22}$ denote hoop and longitudinal strain, respectively. It is clear that the longitudinal strain is much larger than the hoop strain. In conclusion, the hoop strain may be neglected and the longitudinal strain will be used as the sign for buckling, which is in accordance with previous research. As shown by the diagram, the $D/t$ ratio has much more significant effects on the buckling strain than the $e/D$ ratio and slenderness ratio. The normalized compression–moment envelopes are presented in Figure 9.11. The ultimate strength $N_u$ and corresponding $M_u$ are transformed into non-dimensional forms by $N_y = \sigma_y A$ and $M_y = \sigma_y W$ respectively.

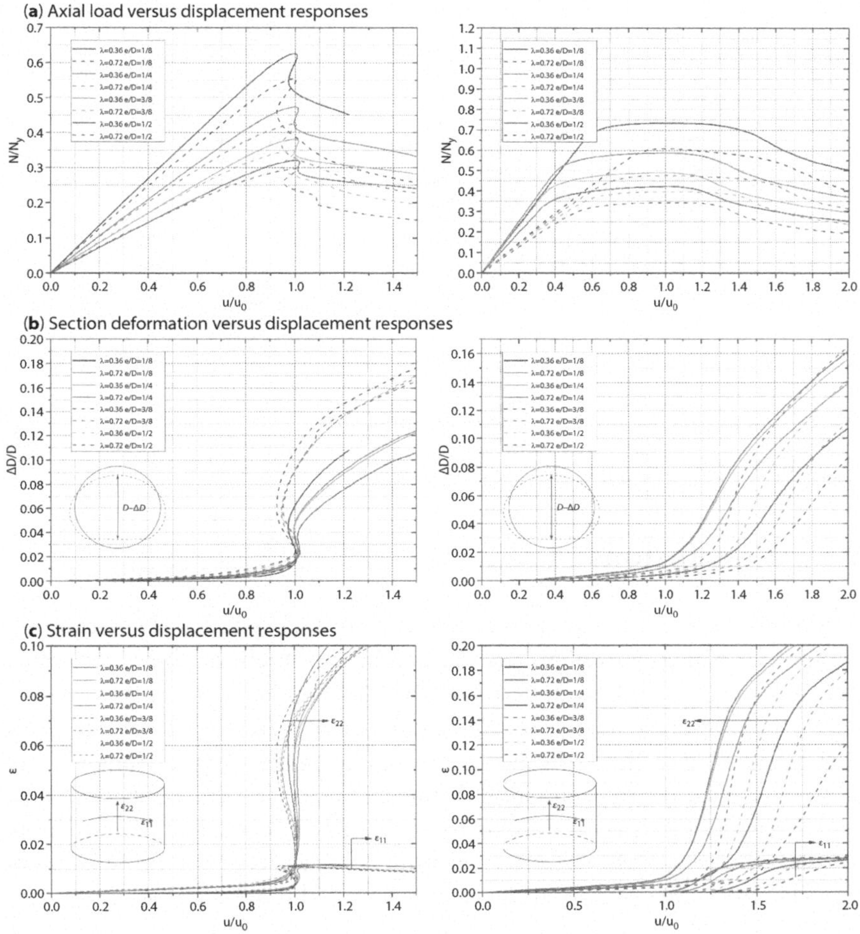


**Figure 9.10** The behavior of tubes under combined compression and bending loads for 97*1 and 100*2.5.

It should be noted that the different points on the same curve represent the simulated models with different eccentricities. The curves for thin-walled tubes are nearly straight lines. And the effect of slenderness ratio to ultimate strength curves of thick tubes is more significant than that to thin-walled ones by comparing the gap between two envelope curves. More simulating tests will be carried out to obtain accurate design codes.

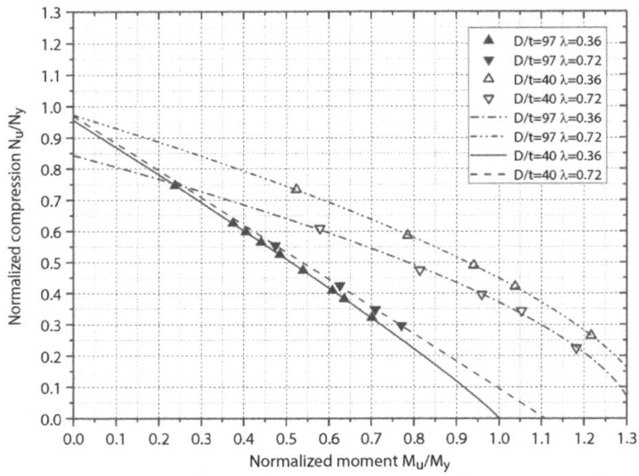


**Figure 9.11** Moment-compression envelopes.

## 9.6 Conclusions

A comprehensive experimental procedure on steel cylindrical shells under combined axial compression and the bending moment has been presented in this chapter in order to investigate the buckling behavior of these members. The currently available experimental data are mostly about thick shells ($D/t \leqslant 25$). The $D/t$ ratio of specimens in this chapter ranges from 40 to 97 so that the experimental results can do a valuable contribution to the experimental data on tubular members under eccentric compression. The experimental results were then utilized to validate the finite element models. The developed FE models were testified to be capable of replicating the test results, load-deformation histories, and local buckling failure modes. The main findings derived from the previous work are summarized as follows:

(1) In the material test, it is interesting to find that the cold-drawn specimens have different tensile and compressive properties. The young's modulus, $E$, increases when the thickness of coupons becomes thinner, whereas the yield stress $\sigma_{0.2}$ increases first and then decreases.

(2) A combined loading test machine is designed to achieve pinned boundary conditions and different eccentricities. All the specimens failed by local buckling because of their relatively large diameter-thickness ratio ($40 \leqslant D/t \leqslant 97$).

However, two different local buckling modes are found, named dent and cos mode, respectively. The first one is the major mode, whereas the second one only happens to HD1, of which the *D*/*t* ratio is small and eccentricity is large. The deformation processes of these two modes are presented by schematic diagrams.

(3) In the numerical simulations, the FE models were developed to simulate the buckling behaviors of the tested specimens. The FE models were verified against the experimental results by comparison of the buckling modes and ultimate strength. The results show that the ratio of error can be controlled in 5%, which is to say the FE models can simulate the experiments very well.

(4) The influence of the two kinds of buckling modes is studied in this chapter. The results show that the buckling mode has no effect on the pre-buckling behavior and ultimate strength. However, different initial imperfections will lead to different buckling deformation and post-buckling behavior. That is to say, it is of importance to introduce the corresponding buckling mode as the initial imperfection when the post-buckling behavior is the focus of research.

(5) A series of parametric studies have been carried out with the verified FE models. The effect of *D*/*t* ratio, slenderness ratio, and eccentricity ratio *e*/*D* was included in the analyses.

The following observations can be made from the results of the present study:

1. The responses of thin tubes under simultaneous axial compression and bending are significantly affected by the three parameters. Larger *D*/*t* ratio and *e*/*D* all result in smaller ultimate strength.
2. The cross-section shape can be considered circular during the pre-buckling range. By this assumption, the classic mechanical formulas are still available before buckling happens.
3. Compared with the longitudinal strain, the hoop strain can be neglected. The buckling strain is significantly affected by the *D*/*t* ratio but is nearly independent of slenderness and eccentricity.

## Bibliography

1. Chen, W.F. and Han, D.J., *Tubular members in offshore structures*, vol. 1, Pitman Publishing, Boston, Massachusetts, USA, 1985.
2. Veritas, D.N., DNV-RP-C202 Buckling Strength of Shells. no. January, 2013.
3. Bulletin API 2U Third Edition, Bulletin on Stability Design of Cylindrical Shells, American Petroleum Institute, Washington, D.C. 20005, 2004.
4. Ju, G.T. and Kyriakides, S., Bifurcation and localization instabilities in cylindrical shells under bending—II. Predictions. *Int. J. Solids Struct.*, 29, 9, 1143–1171, 1992.
5. Kyriakides, S. and Ju, G.T., Bifurcation and localization instabilities in cylindrical shells under bending—I. Experiments. *Int. J. Solids Struct.*, 29, 9, 1117–1142, 1992.
6. Yadav, K.K. and Gerasimidis, S., Instability of thin steel cylindrical shells under bending. *Thin Walled Struct.*, 137, 151–166, 2019.
7. Bardi, F.C. and Kyriakides, S., Plastic buckling of circular tubes under axial compression—part I: experiments. *Int. J. Mech. Sci.*, 48, 8, 830–841, 2006.
8. Bardi, F.C., Kyriakides, S., Yun, H.D., Plastic buckling of circular tubes under axial compression—part II: analysis. *Int. J. Mech. Sci.*, 48, 8, 842–854, 2006.
9. Sohal, I.S. and Chen, W.F., Local buckling and sectional behavior of fabricated tubes. *J. Struct. Eng.*, 113, 3, 519–533, 1987.
10. Sohal, I.S. and Chen, W.F., Local and post-buckling behavior of tubular beam-columns. *J. Struct. Eng.*, 114, 5, 1073–1090, 1988.
11. Sohal, I.S. and Chan, W.F., Moment-curvature expressions for fabricated tubes. *J. Struct. Eng.*, 110, 11, 2738–2757, 1984.
12. Zhao, O., Gardner, L., Young, B., Structural performance of stainless steel circular hollow sections under combined axial load and bending–Part 1: Experiments and numerical modeling. *Thin Walled Struct.*, 101, 231–239, 2016.
13. Zhao, Y., Zhai, X., Wang, J., Buckling behaviors and ultimate strengths of 6082-T6 aluminum alloy columns under eccentric compression–Part I: Experiments and finite element modeling. *Thin Walled Struct.*, 143, 106207, 2019.
14. Skelton, R.P., Maier, H.J., Christ, H.J., The Bauschinger effect, Masing model, and the Ramberg–Osgood relation for cyclic deformation in metals. *Mater. Sci. Eng.: A*, 238, 2, 377–390, 1997.
15. Abaqus, S., 6.10. User Subroutines Reference Manual, 2010.
16. Timoshenko, S.P., Gere, J.M., Prager, W., Theory of elastic stability, 1962.
17. American Bureau of Shipping (ABS). *Guide for Buckling and Ultimate Strength Assessment for Offshore Structures.* American Bureau of Shipping, Houston, TX, 2014.

# 10

# Characterization of Ductile Fracture Criterion for API X80 Pipeline Steel Based on a Phenomenological Approach

***Abstract***
This chapter investigates the ductile fracture behavior of API X80 pipeline steel using both micromechanical and phenomenological damage models. Three fracture models are evaluated: the coupled Gurson--Tvergaard--Needleman (GTN) model and two uncoupled approaches—the Modified Mohr-Coulomb (MMC) and Extended Rice-Tracey (ERT) criteria. The models incorporate the effects of stress triaxiality and Lode angle to account for a wide range of stress states. A comprehensive experimental program, including five notched tensile specimens and a compact tension (CT) specimen, is conducted to calibrate and validate the models. Finite element simulations are implemented in ABAQUS using C3D8R elements and element deletion techniques. A parameter optimization framework using least-squares error minimization is developed to calibrate the models. Mesh size sensitivity is analyzed, with 0.4 mm found optimal for GTN and ERT models, and 0.1 mm for MMC. All three models show good agreement with experimental results; however, the MMC model provides higher resolution in capturing crack shape and initiation in shear-dominated failures. The calibrated models are validated through CT tests, showing accurate predictions of crack propagation and stress state evolution. The chapter concludes by emphasizing mesh size as a critical modeling parameter and confirms the utility of these models for predicting fracture in high-strength pipeline steels.

***Keywords***: Ductile fracture, GTN model, modified Mohr-Coulomb, extended rice-tracey, stress triaxiality, damage modeling, API X80 steel

## 10.1 Introduction

Pipelines made of high-strength steel are the major means of transporting oil and gas (O&G) in both onshore and offshore installations. As the pipeline networks go into more challenging environments such as deeper water

Yong Bai. Plasticity and Ductile Fracture in Structural Engineering, (281–316) © 2026 Scrivener Publishing LLC

or the Arctic region, they might be subjected to large deformation which is far beyond its elastic limit. In recent years, the strain-based design has been introduced and used to guarantee that no failure occurs during the plastic deformation of the pipeline in consideration of strain demand and available strain capacity. Still, it is challenging to capture the actual failure physics which involves localized plastic deformation and material tearing [9].

One common approach to analyzing ductile fracture is based on traditional elastic-plastic fracture mechanics concepts. In this approach, information on the fracture toughness of the material is required, and the fracture toughness is considered in terms of resistance curve, which is normally expressed as J-R curve or CTOD-R curve. Although the dependence on structural geometry and loading mode of resistance curve can be removed by normalization methods proposed by Nyhus *et al.* [18] and Brocks *et al.* [5], this method lacks the ability to account for all the plasticity that occurs in the vicinity of a tearing crack, and it also does not model tearing explicitly [9]. As a result, its application to the actual operation scenario is limited. For example, the reduction of axial plastic deformation of steel pipelines due to the presence of internal pressure is not likely to be considered in this method. Another solution that addressed ductile fracture is by using damage mechanics. Damage mechanics theory is able to describe crack initiation and propagation by allowing the damage evolution to take into account both the influence of local stress and strain variables [13], which removed the limitation of the fracture mechanics model. The micromechanics-based GTN model is one of the famous damage mechanics models. The GTN model is originated from the porous plasticity model proposed by Gurson [11] and later modified by Tvergaard [27] and Tvergaard and Needleman [29]. The failure process in the GTN model can be described as a sequence of void growth, void nucleation, and void coalescence. Over the last decade, the GTN model has been modified by Benzerga *et al.* [3], Nahshon and Huctchison [16] to account for plastic anisotropy and shear failure. The Gurson-type model usually uses void volume fraction as a damage variable, which suggests that the damage evolution is coupled with its plastic behavior. Another type of damage model is called the uncoupled method, in which the damage inner variable has no effect on the material plastic behavior. The uncoupled approach, consisting of a fracture criterion and a damage evolution rule, is often preferred in industrial applications. In the literature, many uncoupled fracture criteria have been proposed, such as the Johnson-Cook criterion [12],

Wilkins criterion [30], modified Mohr-Coulomb (MMC) criterion [2], and extended Rice-Tracey (ERT) criterion [10].

Different kinds of damage mechanics models have been applied to simulate crack initiation and propagations in the pipeline, constraint effect in crack tip region, and etc. Dotta and Ruggieri [7] employ the GTN model to model the ductile crack extension of API X60 steel; the CT specimens are used to calibrate the parameter and the experimental result of burst pressure of pipeline with the longitudinal defect is used to validate the results. Similar works have been done by Oh *et al.* [19–21], who made use of notch tension specimens to obtain the parameters in the GTN model and a power-law fit function fracture criterion of API X65 steel. The full-scale test of defect pipe was also conducted to prove the efficiency of these two models. Mirone and Corallo [15] combined hill'48 yield criterion with Bai-Wizerzbicki failure criterion to model crack propagation in CT specimen made of X100 grade steel; the mesh size effects on the global behavior of CT specimen were investigated. Kofiani *et al.* [13] used a hybrid numerical-experimental calibration method to obtain the X70 parameters of modified Mohr-Coulomb fracture criterion, and the mesh size effect on SENT specimen was also investigated. Paredes *et al.* [22] continued Kofiani's works by introducing a post-initiation softening function to reduce the mesh size effects. Testa *et al.* [26] applied the Bonora damage model to predict ductile failure in X65 steel. Even though the damage model has been extensively used by many researchers to model ductile fracture of API X series pipeline steel, the feasibility of these models still needs to be validated, and there is always debate about the calibration of the parameters.

In this chapter, the GTN model and two recently proposed uncoupled fracture models named MMC criterion and ERT criterion are evaluated using API X80 steel. A comprehensive experimental study including five different notched specimens is conducted, and corresponding finite element models are established. The parameters in these fracture models are calibrated by combining notched tensile specimens test and detailed finite element model. The calibrated fracture models are then applied to simulate the ductile fracture and crack propagation in the CT specimens test. The mesh size that fits the global response of CT specimens best is presented by comparing it with experimental results. The calibrated models can be used for future analysis of the strain capacity of the whole pipe.

## 10.2 Overview of Damage Model

### 10.2.1 Descriptions of the Stress State

Based on the literature on ductile fracture, the hydrostatic pressure and the third deviatoric invariant are two parameters that affect the failure strain. The hydrostatic pressure is represented by stress triaxiality while the third deviatoric invariant can be represented by the lode angle parameter [1]. The failure strain can be defined as a function of these two parameters. The stress triaxiality $\eta$ and the Lode angle parameter $\bar{\theta}$ are defined as follows:

$$\eta = \frac{\sigma_m}{\bar{\sigma}} \tag{10.1}$$

$$\bar{\theta} = 1 - \frac{2}{\pi}\arccos\left(\frac{27J_3}{2\bar{\sigma}^3}\right) \tag{10.2}$$

where $\bar{\sigma}$ is the equivalent von Mises stress, $\sigma_m$ is the mean stress, and $J_3$ is the third invariant of the deviatoric stress.

$$\sigma_m = -p = \frac{1}{3}(\sigma_1 + \sigma_2 + \sigma_3) \tag{10.3}$$

$$\bar{\sigma} = \sqrt{3J_2} = \sqrt{\frac{1}{2}\left[(\sigma_1 - \sigma_2)^2 + (\sigma_2 - \sigma_3)^2 + (\sigma_3 - \sigma_1)^2\right]} \tag{10.4}$$

$$J_3 = (\sigma_1 - \sigma_m)(\sigma_2 - \sigma_m)(\sigma_3 - \sigma_m) \tag{10.5}$$

where $\sigma_1, \sigma_2$, and $\sigma_3$ denote the three principal stresses in the stress tensor.

### 10.2.2 Gurson-Tvergaard-Needleman Model

The Gurson type model was first proposed by Gurson [11] to describe the plastic behavior of the material with the spherical void. The later modification introduced by Tvergaard [28] and Tvergaard and Needleman [29] addressed additional parameters and a modified void volume function.

The yield function of the widely used Gurson-Tvergaard-Needleman (GTN) model has the following form:

$$\phi = \left(\frac{\bar{\sigma}}{\sigma_M}\right)^2 + 2q_1 f^* \cosh\left(-\frac{3}{2}q_2\frac{p}{\sigma_M}\right) - (1 + q_3 f^{*2}) = 0 \quad (10.6)$$

where $q_1$, $q_2$, and $q_3$ are three fitting parameters; the GTN yield function reduces to von Mises yield criterion when $f^*=0$. $\bar{\sigma}$ is von Mises equivalent stress, and $\sigma_M$ is the matrix yield stress. The equivalent void volume fraction $f^*$, which is used to account for the rapid loss of stress capacity due to void coalescence, is shown as

$$f^* = \begin{cases} f & , \ f \le f_c \\ f_c + \dfrac{f_u - f}{f_u - f_c}(f - f_c) & , f_c < f \le f_f, \\ f_u & , \ f > f_f \end{cases} \text{ where } f_u = \frac{q_1 + \sqrt{q_1^2 - q_3}}{q_3} \quad (10.7)$$

where $f$ is the void volume fraction, $f_c$ is the critical void volume fraction corresponding to the onset of coalescence, and $f_f$ is the failure void volume fraction. By specifying $f_c$ and $f_f$, the function $f^*$ describes the acceleration of material failure when $f$ reaches critical value $f_c$ such that the void coalescence occurs.

The void volume fraction $f$ consists of three components: initial void volume fraction $f_0$, void volume fraction due to void growth $f_g$, and void volume fraction due to nucleation $f_n$; therefore, the change of $f$ due to the incremental plastic deformation can be given by the sum of incremental void volume fraction due to void growth and void nucleation:

$$df = df_g + df_n \quad (10.8)$$

Assuming that the matrix material is plastically incompressible, the incremental void volume fraction due to void growth is given by

$$df_g = (1 - f)\,d\varepsilon_{ii}^p \quad (10.9)$$

where $d\varepsilon_{ii}^{p} = \left(d\varepsilon_{11}^{p} + d\varepsilon_{22}^{p} + d\varepsilon_{33}^{p}\right)/3$ is plastic volume dilatation rate. The nucleation of the void is modeled by a statistical approach proposed by Chu and Needleman [6] in this chapter:

$$df_n = \frac{f_N}{s_N\sqrt{2\pi}}\exp\left(-\frac{1}{2}\left(\frac{\bar{\varepsilon}_M^{p} - \varepsilon_N}{s_N}\right)^2\right)d\bar{\varepsilon}_M^{p} \tag{10.10}$$

where $f_N, \varepsilon_N$, and $S_N$ are three parameters controlling void nucleation, the equivalent plastic strain of matrix. According to the principle of equivalent plastic work, the microscopic variables and macroscopic variables can be linked as:

$$\left(1 - f\right)\sigma_M d\bar{\varepsilon}_M^{p} = \sigma_{ij} d\varepsilon_{ij}^{p} \tag{10.11}$$

### 10.2.3 Uncoupled Fracture Model

The uncoupled fracture model usually consists of a fracture criterion and a damage indicator. The damage indicator will increase with plastic deformation. In the following, two recently proposed fracture criteria that account for the effect of the lode angle are introduced.

#### *10.2.3.1 Modified Mohr-Coulomb (MMC) Criterion*

The modified Mohr-Coulomb (MMC) fracture criterion [2] is a phenomenological fracture model modified from the Mohr-Coulomb criterion which has been widely used in rock and soil mechanics. The Mohr-Coulomb criterion assumes that fracture occurs when the combination of normal stress and shear stress in a fracture surface reaches a critical value.

$$\mathrm{Max}\left(C_1\sigma_n + \tau\right) = C_2 \tag{10.12}$$

where $C_1$ and $C_2$ are material constants. By appropriate transformation, the Mohr-Coulomb criterion can be expressed in terms of $\sigma_f, \eta$, and $\theta$ as follows:

$$\bar{\sigma}_f = C_2\left[\sqrt{\frac{1+C_1^2}{3}}\cos\left(\frac{\pi}{6}\bar{\theta}\right)+C_1\left(\eta+\frac{1}{3}\sin\left(\frac{\pi}{6}\bar{\theta}\right)\right)\right]^{-1} \tag{10.13}$$

In order to transform Equation (10.13) into a strain-based criterion which features relatively good resolution in large deformation where the fracture occurs, a modified hardening rule featuring lode angle parameter was used in adjutant with von Mises yield criterion as follows [1]:

$$\bar{\sigma} = A\bar{\varepsilon}^n\left[C_3+\frac{\sqrt{3}}{2+\sqrt{3}}\left(1-C_3\right)\left(\sec\left(\frac{\pi}{6}\bar{\theta}\right)-1\right)\right] \tag{10.14}$$

Substituting Equation (10.14) into Equation (10.13), the final form of the three-parameter modified Mohr-Coulomb criterion can be yielded as follows:

$$\bar{\varepsilon}_f = \left\{C_2\left[C_3+\frac{\sqrt{3}}{2-\sqrt{3}}\left(1-C_3\right)\left(\sec\left(\frac{\bar{\theta}\pi}{6}\right)-1\right)\right]\left[\sqrt{\frac{1+C_1^2}{3}}\cos\left(\frac{\bar{\theta}\pi}{6}\right)+C_1\left(\eta+\frac{1}{3}\sin\left(\frac{\bar{\theta}\pi}{6}\right)\right)\right]\right\}^{-\frac{1}{n}} \tag{10.15}$$

where $\bar{\varepsilon}_f$ is the equivalent strain to fracture; $n$ is the hardening component; and $C_1$, $C_2$, and $C_3$ are three parameters needed to be determined.

### *10.2.3.2 Extended Rice-Tracey (ERT) Criterion*

This chapter follows the extension to the original Rice-Tracey criterion proposed by Gruben *et al.* [10]. Rousselier [23] presented that the Rice-Tracey criterion for void growth may be written as

$$df = 3k_g f(1-f)\exp\left(\psi\eta\right)d\varepsilon_p \tag{10.16}$$

where $f$ denotes void volume fraction, and $k_g$ and $\psi$ are two parameter constants. By following the same lines as Nahshon and Hutchinson [16] in their modification of the Gurson model, the load dependence or damage growth in shear-dominated stress states at low-stress triaxiality can be introduced. Therefore, Equation (10.16) is extended to

$$df = 3k_g f(1-f)\exp(\psi\eta)d\bar{\varepsilon}_p + k_s f \sin^2\left(\frac{\pi}{2}\left(1-\bar{\theta}\right)\right)d\bar{\varepsilon}_p \quad (10.17)$$

Here, $k_s$ is a constant. $f$ should now be regarded as an effective void volume fraction or rather a damage variable since it is no longer uniquely defined by the classical void growth relation. In the special case of axisymmetric stress states, $(1-\bar{\theta})$ is equal to zero, and then $f$ can still be identified as the void volume fraction. As commonly assumed for the original RT fracture criterion, it is here assuming that fracture occurs as $f$ reaches a critical value $f_c$. Assuming proportional loading, the fracture strain can be obtained by the integration of Equation (10.17) as

$$\bar{\varepsilon}_f = \int_0^{\bar{\varepsilon}_f} d\bar{\varepsilon}_p = \int_{f_0}^{f_c} \frac{df}{3k_g f(1-f)\exp(\psi\eta) + k_s f \sin^2\left(\frac{\pi}{2}\left(1-\bar{\theta}\right)\right)}$$

$$= \frac{1}{3k_g\exp(\psi\eta) + k_s\sin^2\left(\frac{\pi}{2}\left(1-\bar{\theta}\right)\right)} \ln\frac{f_c\left[(1-f_0)3k_g\exp(\psi\eta) + k_s\sin^2\left(\frac{\pi}{2}\left(1-\bar{\theta}\right)\right)\right]}{f_0\left[(1-f_c)3k_g\exp(\psi\eta) + k_s\sin^2\left(\frac{\pi}{2}\left(1-\bar{\theta}\right)\right)\right]} \quad (10.18)$$

Generally, $f_0$ and $f_c$ are in the order of $10^{-2}\sim10^{-3}$ which can be taken to be small compared with 1. Equation (10.18) can be reduced to

$$\bar{\varepsilon}_f = \frac{\ln f_c/f_0}{3k_g\exp(\psi\eta) + k_s\sin^2\left(\frac{\pi}{2}\left(1-\bar{\theta}\right)\right)}$$

$$= \frac{1}{R_1\exp(R_2\eta) + R_3\sin^2\left(\frac{\pi}{2}\left(1-\bar{\theta}\right)\right)} \quad (10.19)$$

There are three parameters $R_1$, $R_2$, and $R_3$ needed to be calibrated with respect to the experiment.

### *10.2.3.3 Damage Evolution Rule*

Besides the fracture locus, the rule of damage evolution is also a crucial part of the fracture modeling framework. Assuming monotonic loading conditions, a linear incremental relationship is defined here between damage indicator $D$ and the equivalent plastic strain $\bar{\varepsilon}_p$.

$$D = \int_0^{\bar{\varepsilon}_p} \frac{d\bar{\varepsilon}_p}{\bar{\varepsilon}_f\left(\eta, \bar{\theta}\right)} \tag{10.20}$$

where the stress state parameters $\eta\left(\bar{\varepsilon}_p\right)$ and $\bar{\theta}\left(\bar{\varepsilon}_p\right)$ are unique functions of the equivalent plastic strain, and $\bar{\varepsilon}_f$ is the fracture locus defined by Equation (10.15) or Equation (10.19) depending on the use of MMC or ERT criterion. The above equation suggests that a given increment of the equivalent plastic strain $d\bar{\varepsilon}_p$ contributes to the damage accumulation in a specific way depending on the current fracture strain calculated by a chosen fracture criterion. It is postulated that fracture will initiate at a certain material element when its limit of ductility is reached ($\bar{\varepsilon}_p = \bar{\varepsilon}_f$), and then $D\left(\bar{\varepsilon}_p\right) = 1$.

## 10.3 Experimental Programs

The experimental tests were carried out in Research Labs for Civil and Hydraulic Engineering of Zhejiang University. All specimens were extracted from a 21-inch API X80 UOE pipeline using EDM. All tests are conducted at room temperature on an Instron universal test machine with a maximum load of 250 kN. The measurement precision of the force value is 0.5%.

### 10.3.1 Conventional Tensile Specimens

The specimens of this work are cut from large-diameter UOE pipes; the manufacturing process such as rolling and cold working would inevitably affect the plasticity and anisotropy behavior of the metal structure. Therefore, flat tensile specimens in three different directions are tested.

The standard tensile specimens following ASTM E8 (2004) are cut from different orientations. A schematic representation of the specimens is given in Figure 10.1. In this figure, the specimens cut from the longitudinal direction of the pipeline are shown by "L 0°," parallel to the transverse direction is shown by "T 90°," and between longitudinal and transverse direction is shown by "LT 45°."

Three tests for each direction have been performed to verify the repeatability. Load and gauge section elongation have been measured and recorded by a load cell and extensometer. The loading rate is set to be 0.5 mm/min. Figure 10.2 presents the engineering stress–strain curves up to 5% engineering strain for specimens from three directions. The true stress–strain curves are presented in Figure 10.3. Yield and ultimate tensile strength for three directions are shown in Table 10.1. The API X80 steel

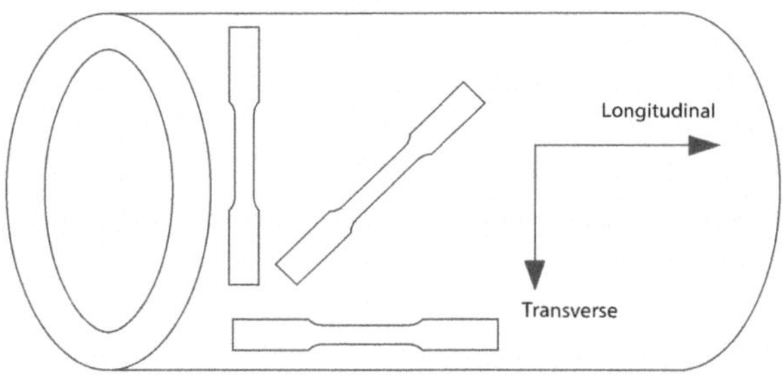


**Figure 10.1** Graphical representation of cutting specimens from the pipe.

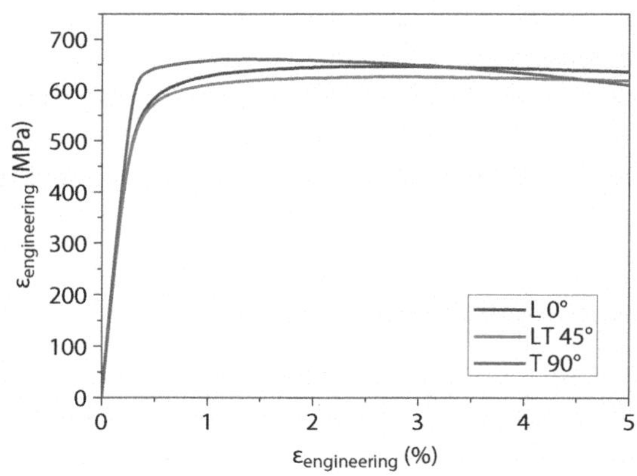


**Figure 10.2** Engineering stress–strain curves of X80 for three material orientation.

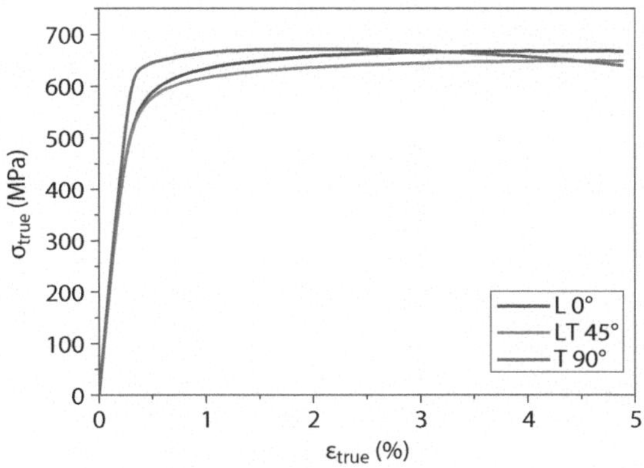


**Figure 10.3** True stress–strain curves of X80 for three material orientation.

**Table 10.1** Mechanical tensile properties at room temperature of the API X80 steel.

| Young's modulus E (GPa) | Poisson's ratio ν | Yield strength $\sigma_y$ (MPa) | | | Ultimate strength $\sigma_u$ (MPa) | | |
|---|---|---|---|---|---|---|---|
| | | **L 0°** | **LT 45°** | **T 90°** | **L 0°** | **LT 45°** | **T 90°** |
| 210 | 0.3 | 580 | 570 | 643 | 650 | 630 | 668 |

does not have an obvious yielding plateau, and therefore, the yield strength is determined by a residual plastic strain of 0.2%.

It can be observed from Figures 10.2 and 10.3 that the pipe material exhibits certain orientation-dependent behavior along with three different directions at the initial yield strength point, but the disparity decreases as deformation increases. The maximum difference between the flow stress in three orientations is 3.3% when the strain exceeds 3%. Since the fracture occurs after strain localization as particle cracking and the difference of stress in large strain is negligible, the material is treated as isotropic in this chapter here. Therefore, the yielding condition is characterized by the conventional $J_2$ flow theory with an associated flow rule as follows:

$$f\left(\sigma,\bar{\varepsilon}_p\right)=\bar{\sigma}-\bar{\sigma}\left(\bar{\varepsilon}_p\right)=0 \tag{10.21}$$

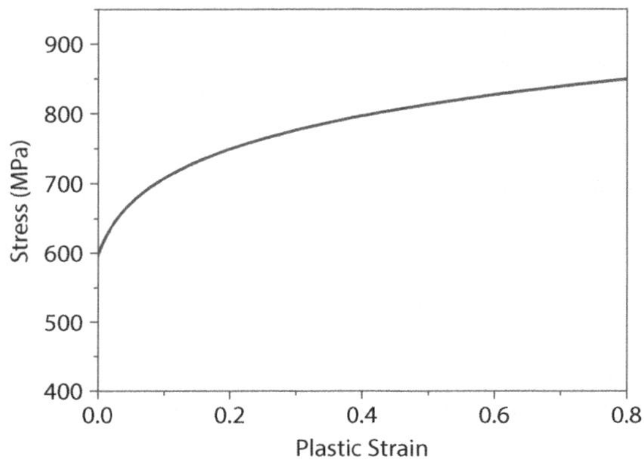


**Figure 10.4** Plastic strain vs. stress with swift law exploitation.

where $\bar{\sigma}$ is the equivalent von Mises stress and $\bar{\varepsilon}_p$ is the corresponding equivalent plastic strain. The hardening law proposed by Swift [25] is used to approximate the flow curve of X80 in the pre-necking and post-necking stages. The Swift law is shown as follows:

$$\bar{\sigma}\left(\bar{\varepsilon}_p\right) = A\left(\bar{\varepsilon}_0 + \bar{\varepsilon}_p\right)^n \tag{10.22}$$

The Swift law is used to fit the pre-necking data, where $A = 865.38MPa$, $\bar{\varepsilon}_0 = 0.02$, and $n = 0.095$ are fitting results. The hardening curve in the post-necking stage is assumed to be the exploitation of Swift law as shown in Figure 10.4.

## 10.3.2 Notched Tensile Specimens

Five different specimens are selected for determining the fracture properties for X80 pipeline steel as shown in Figure 10.5. The five specimens with a different notch are designed to explore a particular stress state. The detailed physical dimension of the five different specimens is shown in Figure 10.6. The five different specimens are all cut out in the longitudinal direction of the pipe only. The first three flat specimens are used for low-stress triaxiality calibration while the high-stress triaxiality is determined by the last two round bar specimens. The specimen in Figure 10.5(a) is named flat shear specimen (FS) and the expectant stress triaxiality is approximately 0. The central hole tension specimen with a hole radius of 4 mm (CH4) is shown

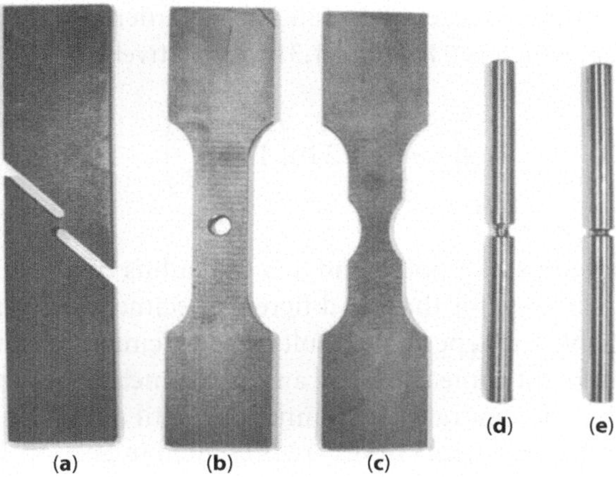


**Figure 10.5** Test specimens: (a) FS, (b) CH4, (c) NT20, (d) NRB3, (e) NRB1.5.

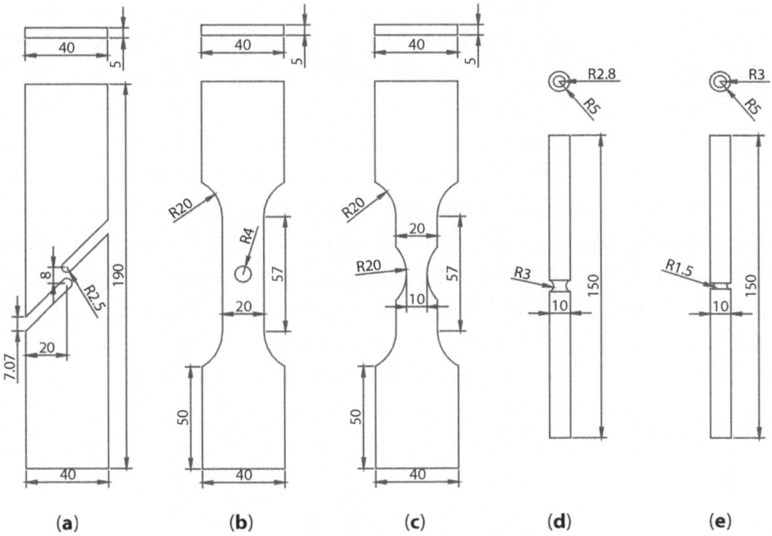


**Figure 10.6** Geometries of specimens (mm): (a) FS, (b) CH4, (c) NT20, (d) NRB3, (e) NRB1.5.

in Figure 10.5(b), and the stress triaxiality is expected to be close to 0.4. Figure 10.5(c) is a flat notched tension specimen (NT20), the notch radius is selected as $R = 20$ mm, and no other notch radius is used in this chapter.

Figures 10.5(d) and (e) is notched round bar tension specimen with a circular notch radius of 3 mm and 1.5 mm respectively (NBR3 and

NBR1.5); the estimated stress triaxiality can be derived from Brigdman (1952) equation, which is 0.875 and 1.314 respectively.

$$\eta = \frac{1}{3} + \sqrt{2}\ln\left(1 + \frac{a}{2R}\right) \quad (10.23)$$

where $R$ is the radius of a notch and a is the radius of a round bar at the notch. The tensile tests for the five different specimens are repeated three times to get reliable and repeatable results. The specimens are fully clamped, and the displacement is measured by an extensometer. The specimens are tensioned with a loading rate of 0.5 mm/min until a fracture occurs. The resulting force-displacement curves are presented in Section 10.5.

### 10.3.3 Compact Tension (CT) Test

The CT specimens were extracted along the longitudinal direction of the pipe. The CT test especimen is shown in Figure 10.7. The geometries of the specimen follow GB/T 21143-2007 [31] as presented in Figure 10.8. As the thickness of the 21-inch pipe is 27 mm, CT specimens with the width $W$=40mm and $B$=20mm are extracted. An initial crack with the size of $a$=18 mm was machined. These dimensions yield the following ratio: $a_0/W$=0.45, $W/B$=2. Subsequent fatigue pre-cracking was introduced and the resulting ratio of final pre-crack was about $a/W$=0.5. The tensile pin loads were applied to the circle hole of the specimen, and stable crack propagation was observed. The applying load is monitored and the crack mouth opening displacement (CMOD) was measured through a clip gauge. The resulting load-CMOD curve is summarized in Section 10.6.

**Figure 10.7** CT test specimen.

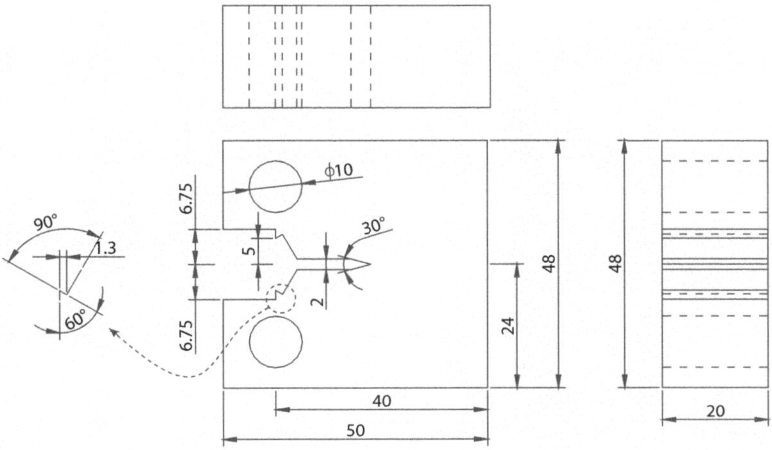


**Figure 10.8** Geometries of CT specimen.

## 10.4 Calibration Procedure

### 10.4.1 Finite Element Model

Five different 3D finite element models were established in ABAQUS. The 8-node solid elements with reduced integration C3D8R were used to establish the model. As shown from Figure 10.9 to Figure 10.13, only one-eighth of the actual specimen was modeled for specimens CH4, NT20, NRB3, and NRB1.5 due to their symmetry features. Three symmetric boundary conditions XSYMM, YSYMM, and ZSYMM were applied on corresponding surfaces, and the displacement load was applied on the upper surface along the Y-axis. Half of the FS specimen was built with only ZSYMM boundary conditions. The lower surface was fixed while the displacement load was applied on the upper surface.

Fine mesh is only assigned near the specimen center while the mesh far from the specimen center is much coarser. Mesh size plays an important role in failure analysis. But for the calibration of the uncoupled model, the only elastic-plastic model with no damage would be used. Therefore, the effect of mesh size tends to be converged as presented by Paredes *et al.* [22]. In this chapter, a 0.2 mm mesh was applied in the notched region of CH4, NT20, and FS specimen while a 0.1-mm mesh was used in the notched region of NRB3, NRB1.5 specimen. All the mesh is converged in the no damage model. As for calibration of the GTN model, even though a smaller mesh size may predict earlier failure initiation, the same mesh is still used and the mesh size seems to be reasonable as shown in Section 10.5.

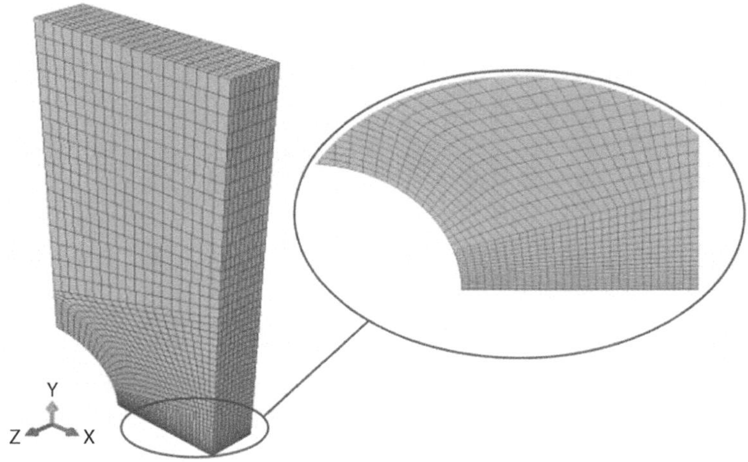


**Figure 10.9** Model and mesh of CH4 specimen.

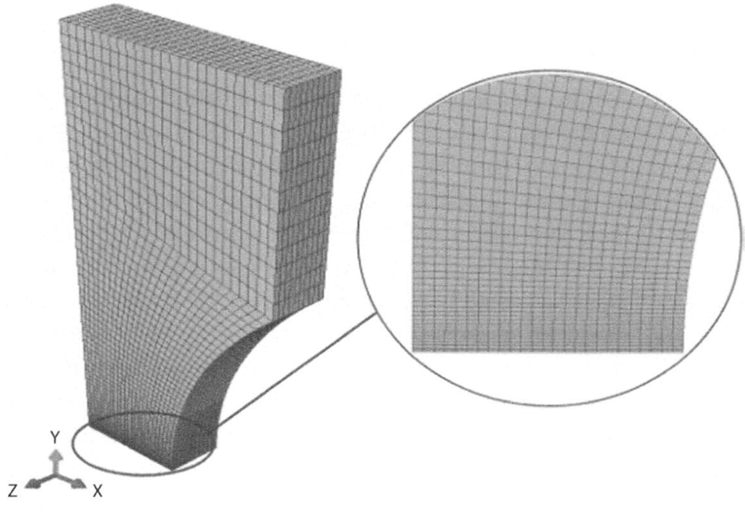


**Figure 10.10** Model and mesh of NT20 specimen.

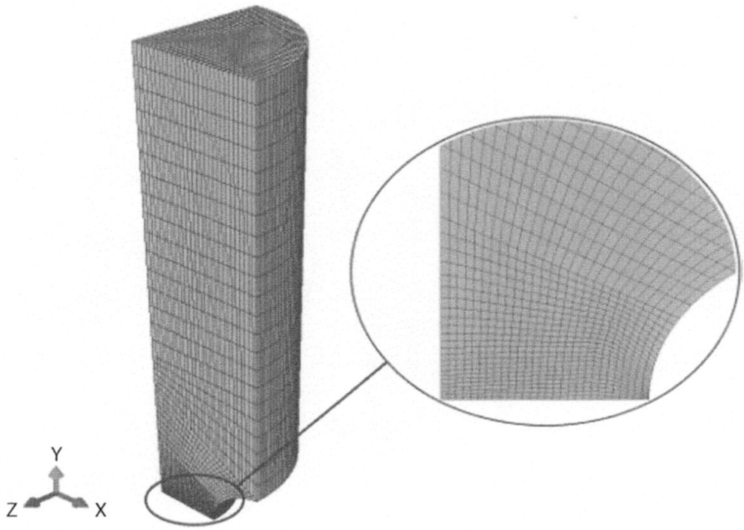


Figure 10.11 Model and mesh of NRB3 specimen.

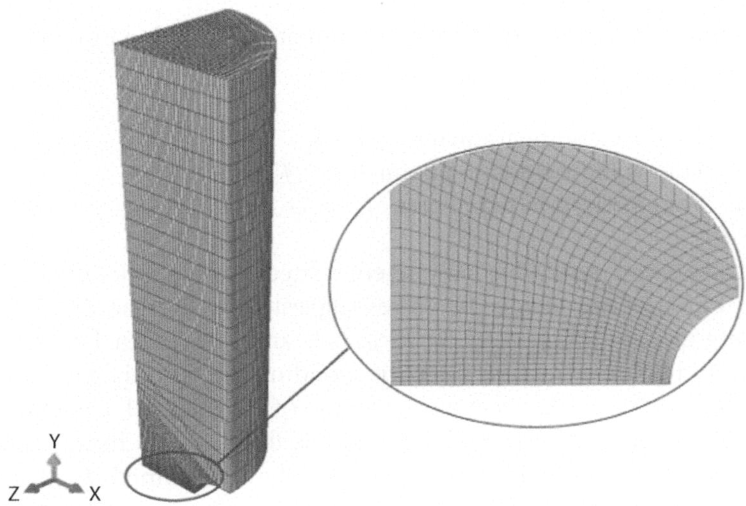


Figure 10.12 Model and mesh of NRB1.5 specimen.

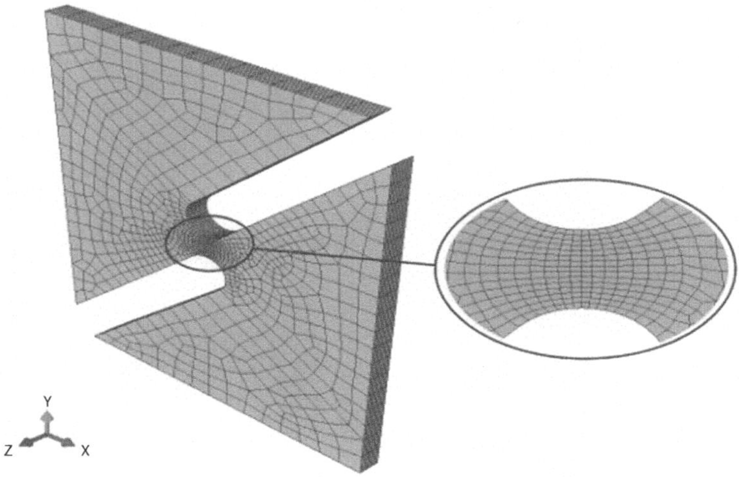


**Figure 10.13** Model and mesh of FS specimen.

### 10.4.2 Parameter Identification for GTN Model

A phenomenal approach was used to determine the parameters in the GTN model. There are a total of nine parameters needed to be determined in the GTN model:

- Yield function parameter: $q_1$, $q_2$, $q_3$
- Void volume fraction parameter: $f_0$, $f_c$, $f_f$
- Void nucleation parameter: $f_N$, $\varepsilon_N$, $s_N$

In the present work, five parameters related to yield function and void nucleation are fixed to specific values suggested by Tvergaard [27, 28] in the literature: $q_1$=1.5, $q_2$=1, $q_3$=$q_1^2$, $\varepsilon_N$=0.3, and $s_N$=0.1. Furthermore, the failure void volume fraction $f_f$ is corresponding to the initial void volume fraction $f_0$ and has little effect on the failure of the material. Generally, the range of $f_f$ is between 0.15 and 0.3, and $f_f$ is fixed as 0.25 in this chapter. Then, only three parameters are needed to be determined: $f_0$, $f_c$, and $f_N$. The different combinations of these three parameters are used in the finite element model; the calibrated values are then obtained by comparing the test results from CH4, NT20, NRB3, and NRB1.5 specimens. After many trials, these three parameters are found to be $f_0$=0.00012, $f_c$ =0.02, and $f_N$ =0.00008, which give the best fits to the test results. The FS specimen is not incorporated in the calibration of the GTN model. The reason is that the

**Table 10.2** Parameters of GTN model.

| $q_1$ | $q_2$ | $q_3$ | $f_0$ | $f_c$ | $f_f$ | $f_N$ | $\varepsilon_N$ | $s_N$ |
|---|---|---|---|---|---|---|---|---|
| 1.5 | 1.0 | 2.25 | 0.00012 | 0.02 | 0.25 | 0.00008 | 0.3 | 0.1 |

GTN model does not account for sheer damage, and therefore, it cannot predict shear-dominated mode failure.

As mentioned by Oh *et al.* [21], it is believed that the unique determination of the parameters in the GTN model would be extremely difficult. Here, we use a phenomenal approach, and after many trials, the parameters listed in Table 10.2 are the best choice to simulate ductile fracture of API X80 steel.

### 10.4.3 Parameter Identification for MMC and ERT Model

The elastic-plastic finite element models without damage are developed to determine the fracture strain. As shown in Figure 10.14 to Figure 10.18, the load-displacement curves obtained from the finite element analysis are compared with the experimental results. The failure points are determined by matching the simulation results at the moment when the critical displacements are reached. The critical displacements are defined as the average displacements of three specimens when a sudden load drop happens.

Once the critical displacements of each model are marked, the histories of equivalent plastic strain with respective to stress triaxiality $\eta$ and lode angle parameter $\bar{\theta}$ in the critical element up to the failure occurs can be obtained, as shown in Figures 10.19 and 10.20. The circular point indicates that failure occurs. For CH4, NRB3, and NRB 1.5 specimens, the stress triaxiality and lode angle parameters remain almost unchanged as equivalent plastic strain increases. It can be observed from Figure 10.19 that the fracture strain does not monotonously decrease with increased stress triaxiality due to the difference of the lode angle parameter. Then, the increment of damage at each equivalent plastic strain increment can be calculated and assembled. Denote the loading history for five different calibration experiments with the equivalent plastic strain $d\bar{\varepsilon}_p$, stress triaxiality $\eta = \eta\left(\bar{\varepsilon}_p\right)$, and lode angle parameter $\bar{\theta} = \bar{\theta}\left(\bar{\varepsilon}_p\right)$. For a given set of MMC model parameters $\chi = \{C_1, C_2, C_3\}$ or ERT model parameters $\chi = \{R_1, R_2, R_3\}$, the strain to fracture $\bar{\varepsilon}_f = \bar{\varepsilon}_f\left(\eta, \bar{\theta}\right)$ at each equivalent plastic strain increment $d\bar{\varepsilon}_p$ can be determined through MMC or ERT criterion. Then, the incremental damage variable can be calculated by $dD = d\bar{\varepsilon}_p / \bar{\varepsilon}_f\left(\eta, \bar{\theta}\right)$.

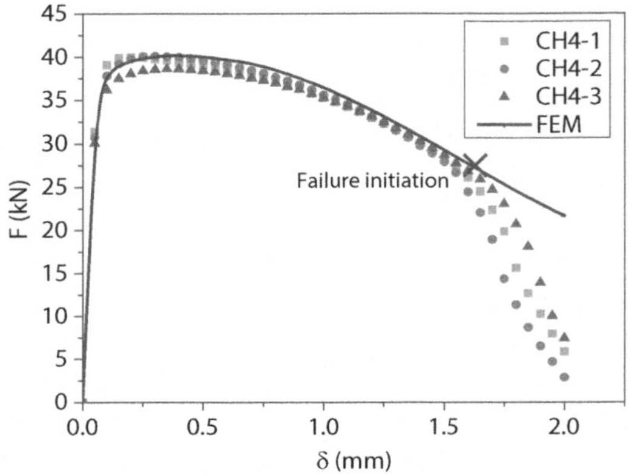


**Figure 10.14** Critical displacement for CH4 specimen.

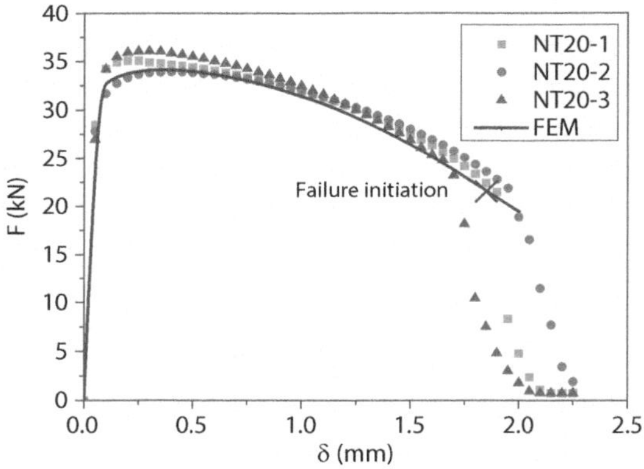


**Figure 10.15** Critical displacement for NT20 specimen.

The incremental damage variable $dD$ in the whole loading history can be assembled as $\int dD$. The average least square error (LSE) between $\int dD$ and constant 1 for five different specimens can be calculated.

The lower the average least-square error means the better fit to the experimental results. In this way, the calibration of the model parameters is transformed into an optimization problem. The objective function is shown as follows:

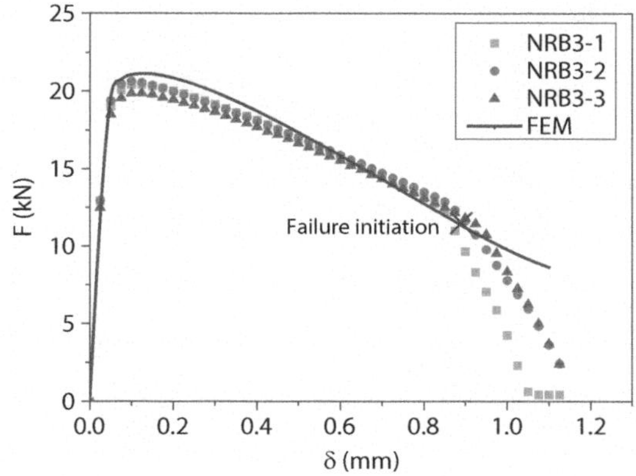


**Figure 10.16** Critical displacement for NRB3 specimen.

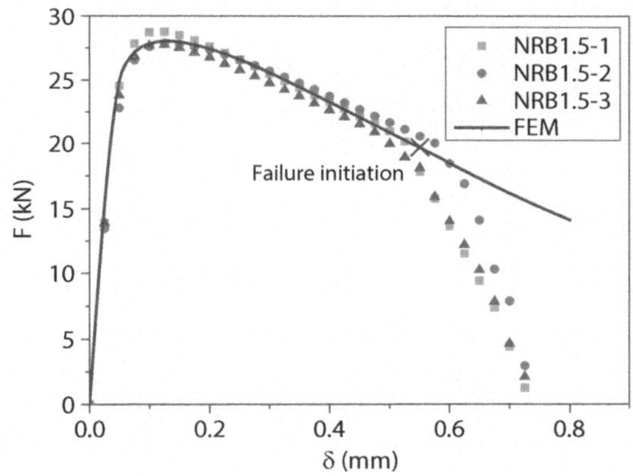


**Figure 10.17** Critical displacement for NRB1.5 specimen.

$$\chi = \arg\min f(\chi)$$

$$f(\chi) = \frac{1}{n}\sum_{i=1}^{n}\left(\int dD - 1\right)^2 = \frac{1}{n}\sum_{i=1}^{n}\left(\int_0^{\bar{\varepsilon}_f} \frac{d\bar{\varepsilon}_p}{\bar{\varepsilon}_f(\eta,\bar{\theta})} - 1\right)^2 \quad (10.24)$$

where $n$ is the number of calibration experiments, in this chapter, $n$=5. $\bar{\varepsilon}_f = \bar{\varepsilon}_f(\eta,\bar{\theta})$ is determined based on Equation (10.15) or Equation (10-19) depending on which model is used. A MATLAB code using the quasi-newton method is developed to find the minimum of this

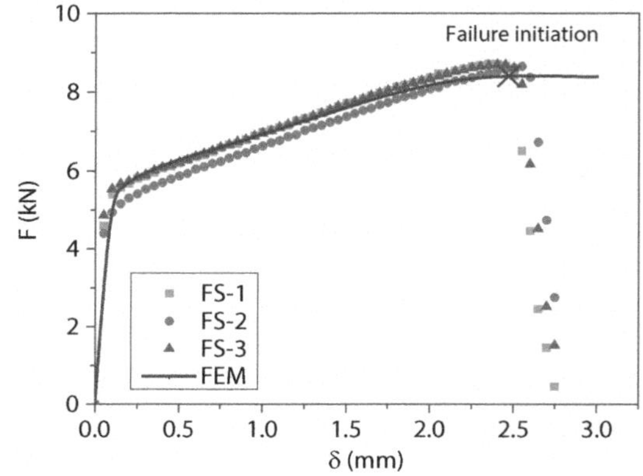


**Figure 10.18** Critical displacement for FS specimen.

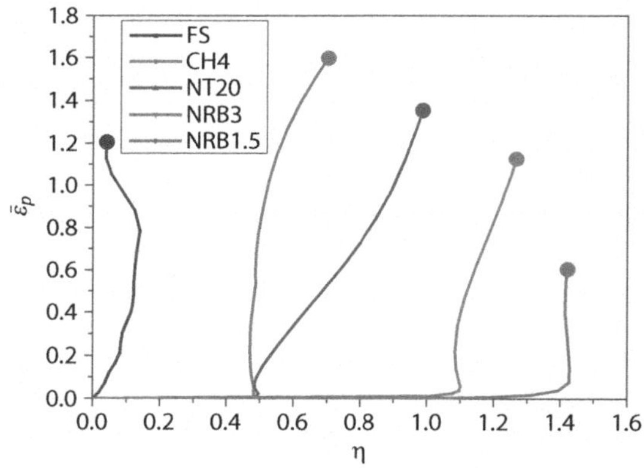


**Figure 10.19** Stress triaxiality vs. equivalent plastic strain.

unconstrained multivariable function. Since the quasi-newton method may find local minimization, the value of the start point must be chosen carefully. In this chapter, the initial value for the MMC model is set as $C_1 = 0.05, C_2 = 1.86, C_3 = 0.91$ as the calibration values for the X70 pipeline steel from Kofiani *et al.* [13]. The initial value for the ERT model is set as $R_1 = R_2 = R_3 = 1$. The optimized parameters for MMC and ERT criteria are given in Table 10.3. The least-square error of the MMC and ERT model is about $3\times10^{-2}$ which shows that both models are in good ability to fit the results of all five calibration experiments.

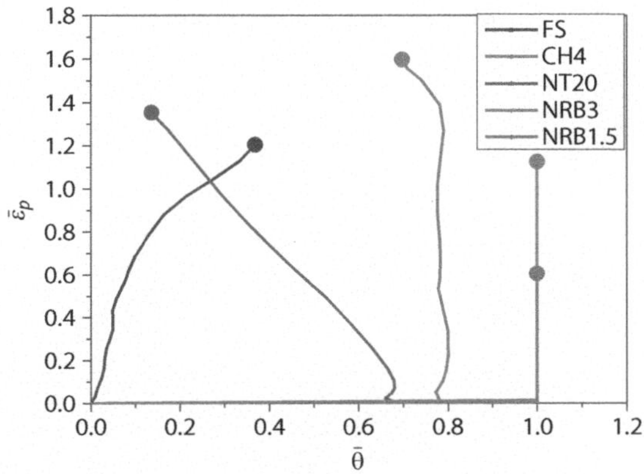


**Figure 10.20** Lode angle parameter vs. equivalent plastic strain.

**Table 10.3** Optimized parameters for MMC and ERT model.

| | C1 | C2 | C3 | LSE |
|---|---|---|---|---|
| **MMC** | 0.0291 | 1.8497 | 0.8867 | 3.6e-2 |
| ERT | $R_1$ | $R_2$ | $R_3$ | LSE |
| | 0.2818 | 1.0743 | 0.4082 | 2.9e-2 |

The calibration fracture locus in the $\left(\eta, \bar{\theta}, \bar{\varepsilon}_f\right)$ space for MMC and ERT criterion is shown in Figures 10.21 and 10.22 respectively. The ERT criterion predicts an axisymmetric fracture locus while the MMC criterion gives a relatively larger fracture strain in generalized tension than generalized compression. The fracture strain predicted by MMC and ERT criterion differs significantly at the negative stress triaxiality region.

## 10.5 Comparison with Notched Tensile Specimen Test

The MMC and ERT failure criteria are implemented into ABAQUS subroutine VUMAT in conjunction with $J_2$ flow theory using radial return algorithm, respectively. The GTN model is implemented by using the ABAQUS material model "porous metal plasticity." The ductile crack

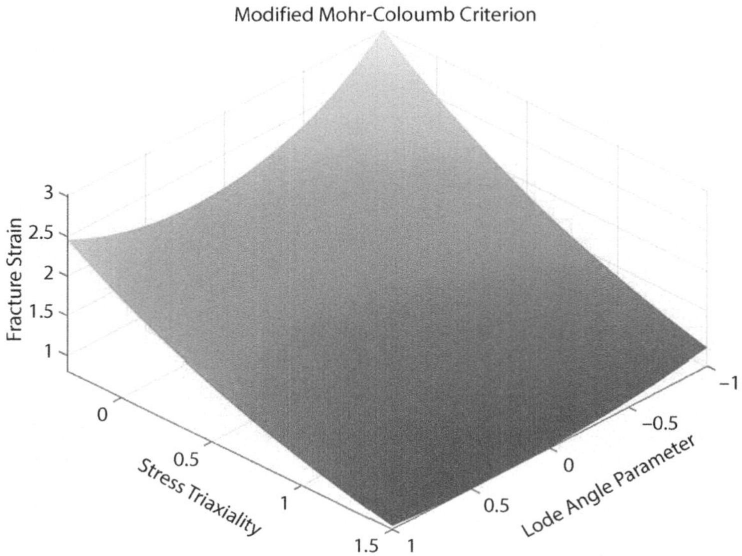


**Figure 10.21** Calibrated fracture surface for MMC criterion.

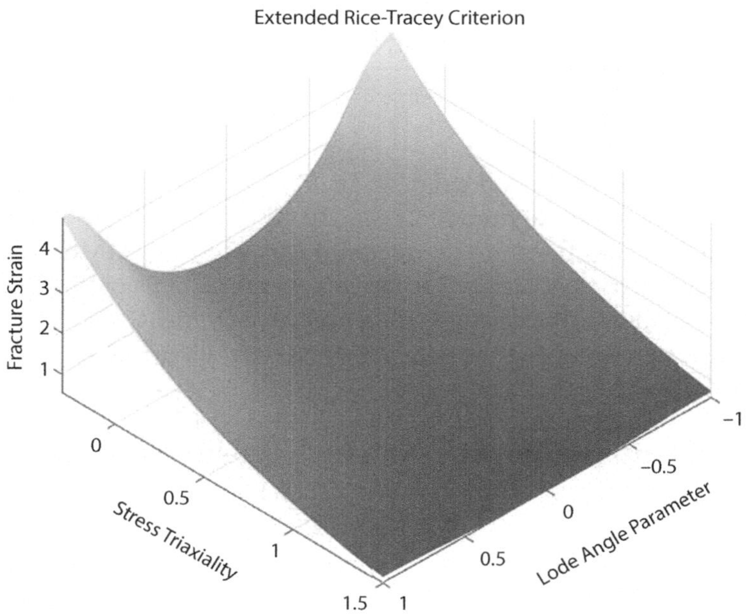


**Figure 10.22** Calibrated fracture surface for ERT criterion.

growth is described through the element deletion technique. The FE load-displacement curves of CH4, NT20, NRB3, NRB1.5, and FS by including the elastic-plastic model, GTN model, MMC model, and ERT model are reported in Figure 10.23 to Figure 10.27 respectively together with corresponding experimental results.

This comparison provides a first assessment of the damage model parameters. It can be found that there are small differences between predicted and measured load-displacement curves before a fracture occurs. This can be explained due to material anisotropy or section heterogeneity. But the differences are relatively small, and the numerical results follow closely the experimental results up to the failure initiation point, which also suggests that the assumption of ignoring the anisotropy behavior is applicable in this chapter. It can be observed that the elastic-plastic model cannot predict the onset of failure, and therefore, the force increases with displacement and no-load drop happens. As shown in Figure 10.23 for the CH4 specimen, the MMC model predicts slightly early onset of failure while the GTN and ERT model almost matches the experimental results.

For NT20 and NRB3 specimens in Figure 10.24 and Figure 10.25, three damage models exhibit good agreement with experimental results. The MMC and ERT model predicts a late onset of failure while the GTN model

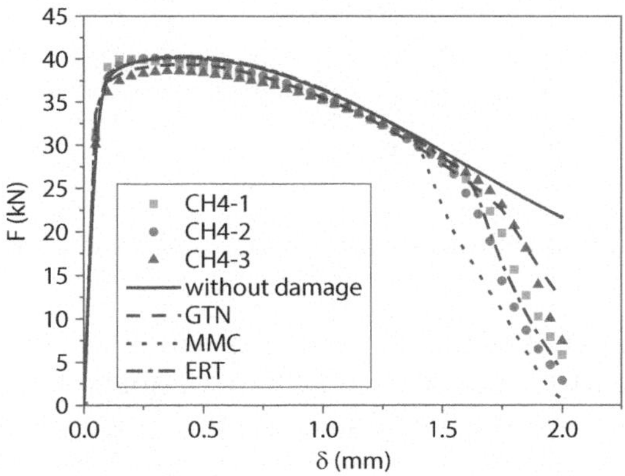


**Figure 10.23** Numerical and experimental comparison for CH4 specimen.

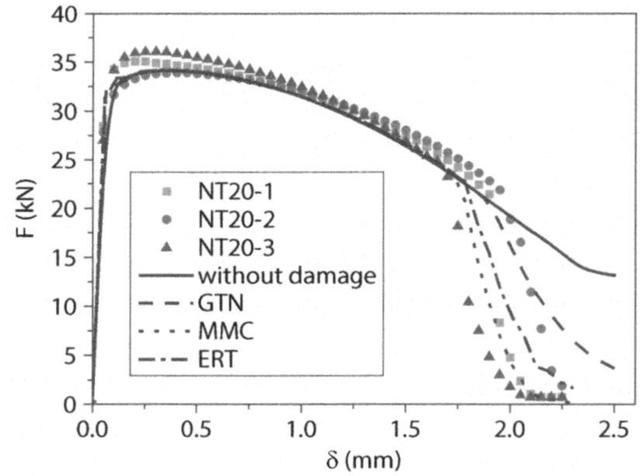


**Figure 10.24** Numerical and experimental comparison for NT20 specimen.

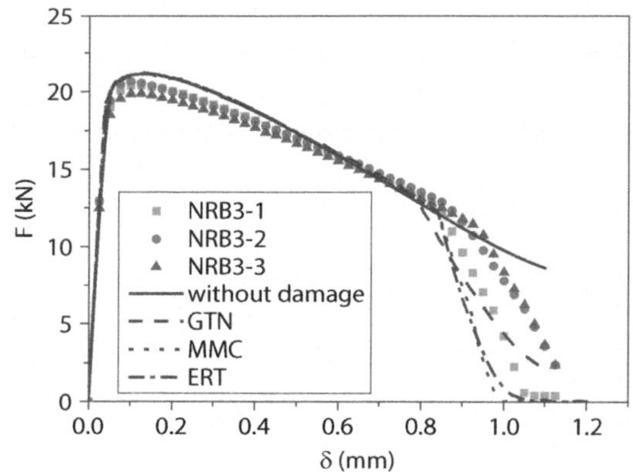


**Figure 10.25** Numerical and experimental comparison for NRB3 specimen.

gives an early onset of failure for NRB1.5 specimen as presented in Figure 10.26. In Figure 10.27, the GTN model shows a similar force-displacement curve as the elastic-plastic model since the GTN model is not able to capture the damage in shear-dominated failure mode. The MMC and ERT models are able to predict shear failure due to the lode angle parameter or third stress invariant introduced in these two models. The MMC model gives a more accurate onset of failure as compared with the ERT model.

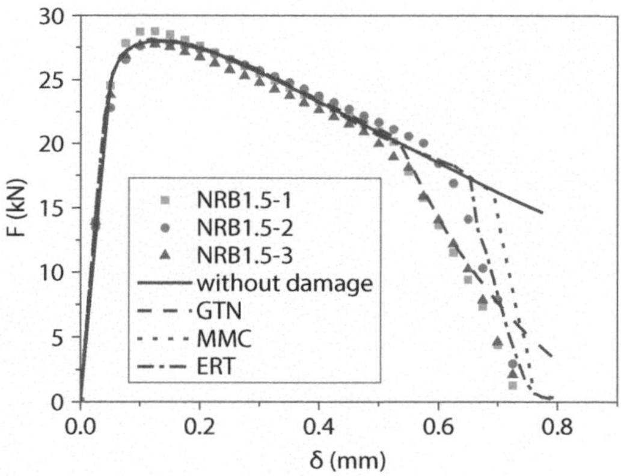


**Figure 10.26** Numerical and experimental comparison for NRB1.5 specimen.

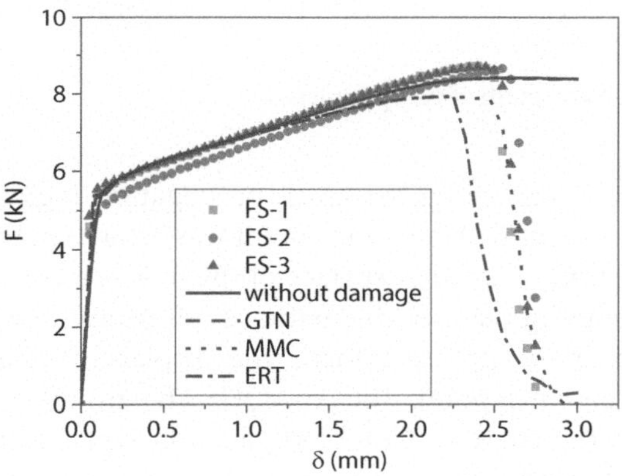


**Figure 10.27** Numerical and experimental comparison for the FS specimen.

# 10.6 Validation through CT Test

## 10.6.1 Finite Element Model

The 3D finite element model in accordance with the CT test was established in ABAQUS. Only a quarter of the CT test model was built, and the symmetry boundary conditions ZSYMM and YSYMM were applied on its corresponding surface. As shown in Figure 10.28, the mesh in the notch region is locally refined.

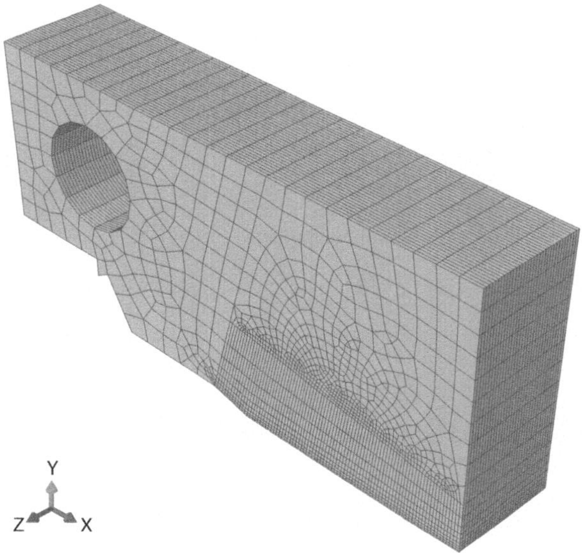

**Figure 10.28** Model and mesh of CT specimen.

The exact mesh size along the crack propagation region will be illustrated in the next section. A reference point was located in the center of the circular loading hole. The inner surface of the hole is a constraint with the related reference point using a kinematic coupling, and the rotation is not restricted in order to apply pin load. The displacement along Y-axis is applied to this reference point. The ABAQUS Dynamic/Explicit scheme is used to conduct the analysis. Crack propagation was simulated using the element removal technique.

### 10.6.2 Mesh Size Sensitivity Analysis

The mesh size effect on the simulation of crack propagation using the element removal technique has been reported by many researchers. Different mesh size is assigned in the crack extension region to investigate the mesh size effect. The mesh size of 0.3mm means that in the crack propagation direction and thickness direction, the mesh length is 0.3 mm, and in the direction perpendicular to the crack surface, the mesh length is assigned as 0.15 mm. Figures 10.29, 10.30, and 10.31 show the mesh size effect on the global response of the CT specimen.

The crack mouth opening displacement (CMOD) versus applied force is presented. It can be observed that increasing mesh density does not

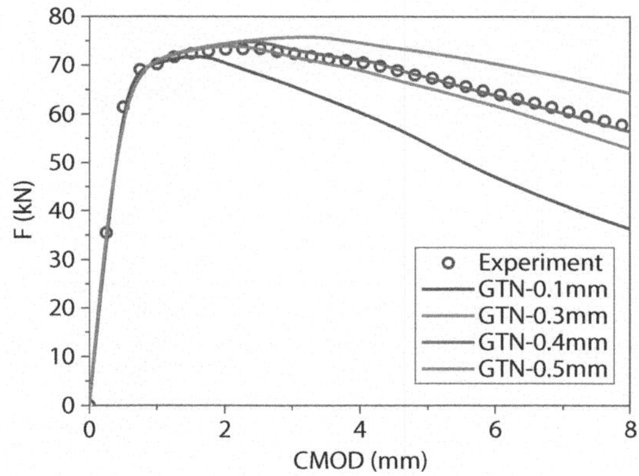


**Figure 10.29** Force-CMOD curves for GTN model.

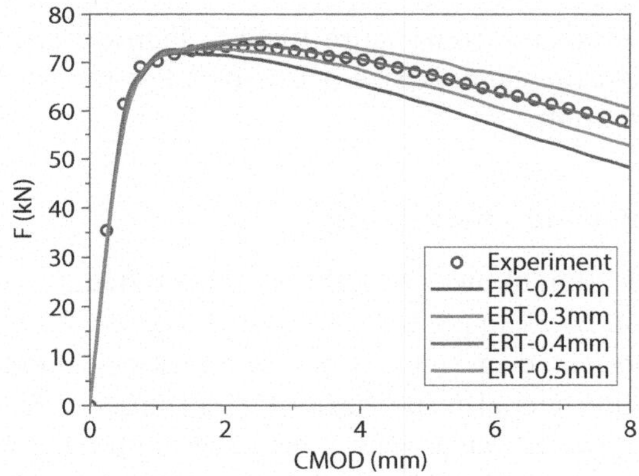


**Figure 10.30** Force-CMOD curves for ERT model.

enhance the convergence of the results. On the contrary, it shows a tendency that the fine mesh model will predict early crack initiation and quick crack propagation as well. It suggests that the fracture resistance or fracture toughness may be underestimated with fine mesh. The reason is that dissipated mechanical work is related to mesh size. By increasing the element height, more mechanical work is dissipated or energy "released" per increment of crack extension. The larger the elements are, the steeper the fracture resistance curve. As the results are mesh size-dependent, the element size becomes an additional material parameter [17].

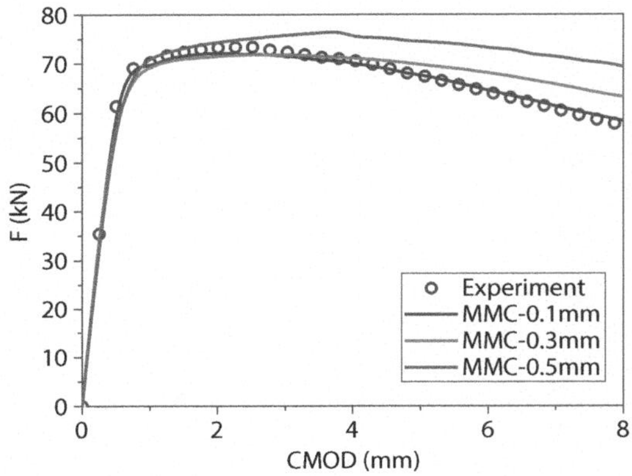


**Figure 10.31** Force-CMOD curves for MMC model.

The best-fitting curve to the experimental result for different damage models can be obtained from Figure 10.29 to Figure 10.30. For the GTN and ERT models, the element size is 0.4 mm. For the MMC model, the element size is 0.1mm.

### 10.6.3 Results and Discussions

With proper selection of mesh size, three different damage models namely GTN, MMC, and ERT can predict the accurate Force-CMOD curves of CT specimen, as presented in Figure 10.32. Even though these three models can predict similar global responses of CT specimens, the smaller mesh size in the MMC model can provide a better resolution for the crack shape and crack propagation. The calibrated GTN, ERT, and MMC models with a mesh length of 0.4 mm, 0.4 mm, and 0.1 mm respectively can be used for future fracture analysis of the whole pipe.

The state of stresses developed in the crack extension region plays an important role in crack initiation and propagation. The stress state in the MMC model is extracted since the mesh length in the MMC model is 0.1 mm, and fine mesh provides higher resolution in the crack front. The evolution of three fracture variables $\left(\bar{\varepsilon}_p, \eta, \bar{\theta}\right)$ with respect to crack mouth opening displacement (CMOD) is monitored in the gauss point.

The stress states of ten elements along the crack extension direction are presented in Figure 10.33. The first element is the element in the crack tip, after that, each element is 0.5 mm away from the previous one. It can be found in Figure 10.33 that the equivalent plastic strain at failure decreases from element 1 to element 3, and then it seems to be saturated after element 4.

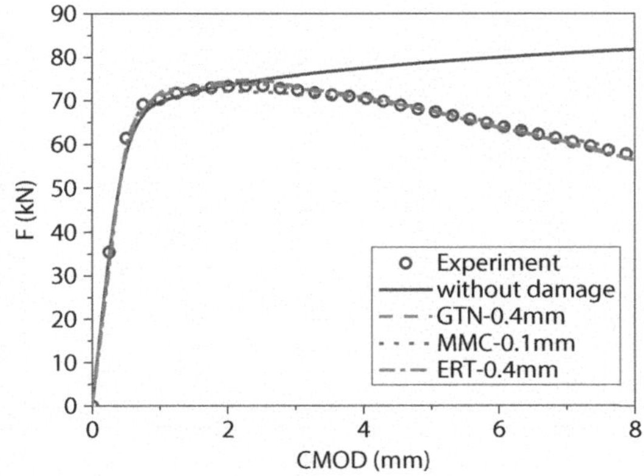


**Figure 10.32** Force-CMOD curves for a different model.

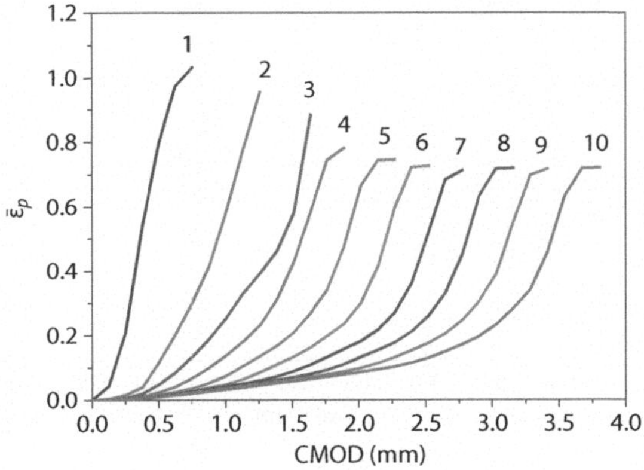


**Figure 10.33** Equivalent plastic strain vs. CMOD.

It suggests that the model can capture the transition from initial crack propagation to stable crack propagation.

From Figure 10.34 and Figure 10.35, the reason for different failure strains in the initial crack propagation and stable crack propagation is due to the difference of stress triaxiality. When the crack propagation went into a stable phase, the elements in the crack extension path follow similar stress states up to fracture occurs.

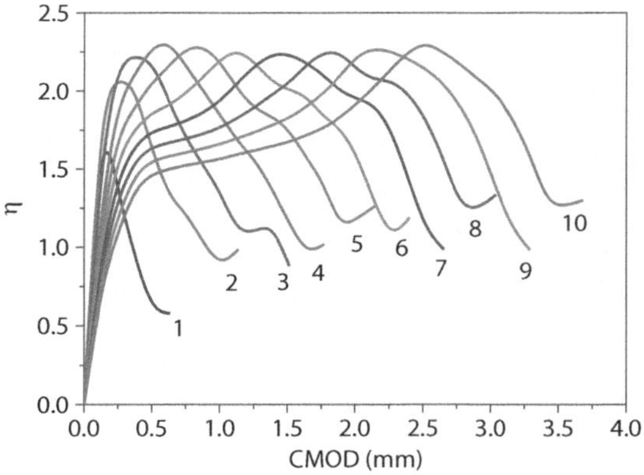


**Figure 10.34** Stress triaxiality vs. CMOD.

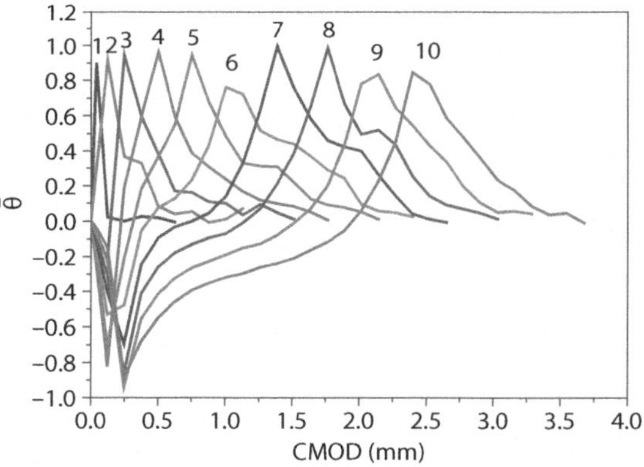


**Figure 10.35** Lode angle parameter vs. CMOD.

## 10.7 Conclusions

This chapter presents the characterization of three ductile fracture criterion for API X80 steel. A micro-based coupled fracture model: Gurson–Tvergaard–Needleman (GTN) model and two macro-based uncoupled fracture models: modified Mohr-Coulomb (MMC) and extended Rice-Tracey (ERT) criterion are employed in this chapter.

A comprehensive experimental test including five different notched specimens is conducted, and corresponding detailed finite element models are established. The parameters in these three models are determined through a hybrid numerical-experimental approach. The proposed parameters are checked and validated by comparing simulated results with experimental ones from notched tensile specimen tests and compact tension (CT) tests.

The presented ductile fracture models would be quite useful to predict the ductile failure of API X80 steel. The mesh size, which has an evitable effect on the fracture responses using the element removal technique, should be included in the interpretation of the results. The mesh size can be regarded as an extra parameter for ductile fracture simulation. The mesh length of 0.4 mm is found to be suitable for the GTN and ERT models, and 0.1 mm is reasonable for the MMC model. Even though these three models can predict similar results, the smaller mesh size used in the MMC model provides a better resolution of crack shape. The calibrated parameters for the three-fracture model with respect to their specific mesh length can be used for future simulations.

The models can predict the transition from initial crack growth to stable crack growth. The difference in stress triaxiality is responsible for the different failure strains at these two-crack growth phases. The failure of CT specimen is not a shear-dominated failure, and therefore, lode angle parameter-free GTN model can predict accurate results as the MMC and ERT models.

## Bibliography


1. Bai, Y. and Wierzbicki, T., A new model of metal plasticity and fracture with pressure and Lode dependence. *Int. J. Plast.*, 24, 6, 1071–1096, 2008.
2. Bai, Y. and Wierzbicki, T., Application of extended Mohr-Coulomb criterion to ductile fracture. *Int. J. Fract.*, 161, 1, 1, 2010.
3. Benzerga, A.A., Besson, J., Pineau, A., Anisotropic ductile fracture: Part II: theory. *Acta Mater.*, 52, 15, 4639–4650, 2004.
4. Bridgman, P.W., *Studies in Large Plastic Flow and Fracture*, McGraw-Hill, New York, USA, 1952.
5. Brocks, W., Anuschewski, P., Hellmann, D., A concept for scaling J R-curves by plastic constraint factors. *Int. J. Fract.*, 130, 1, 455–469, 2004.
6. Chu, C.C. and Needleman, A., Void nucleation effects in biaxially stretched sheets. *J. Eng. Mater. Technol.*, 102, 3, 249–256, 1980.
7. Dotta, F. and Ruggieri, C., Structural integrity assessments of high-pressure pipelines with axial flaws using a micromechanics model. *Int. J. Press. Vessels Pip.*, 81, 9, 761–770, 2004.

8. Dunand, M. and Mohr, D., Hybrid experimental–numerical analysis of basic ductile fracture experiments for sheet metals. *Int. J. Solids Struct.*, 47, 9, 1130–1143, 2010.
9. Fairchild, D.P., Crapps, J.M., Cheng, W., Tang, H., Shafrova, S., Full-scale pipe strain test quality and safety factor determination for strain-based engineering critical assessment, in: *2016 11th International Pipeline Conference*, American Society of Mechanical Engineers, pp. V002T06A003–V002T06A003, 2016.
10. Gruben, G., Hopperstad, O.S., Børvik, T., Evaluation of uncoupled ductile fracture criteria for the dual-phase steel Docol 600DL. *Int. J. Mech. Sci.*, 62, 1, 133–146, 2012.
11. Gurson, A.L., Continuum theory of ductile rupture by void nucleation and growth: Part I—Yield criteria and flow rules for porous ductile media. *J. Eng. Mater. Technol.*, 99, 1, 2–15, 1977.
12. Johnson, G.R. and Cook, W.H., Fracture characteristics of three metals subjected to various strains, strain rates, temperatures, and pressures. *Eng. Fract. Mech.*, 21, 1, 31–48, 1985.
13. Kofiani, K., Nonn, A., Wierzbicki, T., New calibration method for high and low triaxiality and validation on SENT specimens of API X70. *Int. J. Press. Vessels Pip.*, 111, 187–201, 2013.
14. Liu, T., Gao, Y.F., Li, Y., GB/T 21143—2007 Metallic materials—unified method of test for determination of quasistatic fracture toughness, 2008.
15. Mirone, G. and Corallo, D., Stress-strain and ductile fracture characterization of an X100 anisotropic steel: Experiments and modeling. *Eng. Fract. Mech.*, 102, 118–145, 2013.
16. Nahshon, K. and Hutchinson, J.W., Modification of the Gurson model for shear failure. *Eur. J. Mech. A. Solids*, 27, 1, 1–17, 2008.
17. Nègre, P., Steglich, D., Brocks, W., Crack extension in aluminum welds: a numerical approach using the Gurson–Tvergaard–Needleman model. *Eng. Fract. Mech.*, 71, 16-17, 2365–2383, 2004.
18. Nyhus, B., Zhang, Z.L., Thaulow, C., Normalization of material crack resistance curves by the T stress, in: *ECF14*, 2002.
19. Oh, C.K., Kim, Y.J., Baek, J.H., Kim, Y.P., Kim, W.S., Ductile failure analysis of API X65 pipes with notch-type defects using a local fracture criterion. *Int. J. Press. Vessels Pip.*, 84, 8, 512–525, 2007a.
20. Oh, C.K., Kim, Y.J., Baek, J.H., Kim, W.S., Development of stress-modified fracture strain for ductile failure of API X65 steel. *Int. J. Fract.*, 143, 2, 119, 2007b.
21. Oh, C.K., Kim, Y.J., Baek, J.H., Kim, Y.P., Kim, W., A phenomenological model of ductile fracture for API X65 steel. *Int. J. Mech. Sci.*, 49, 12, 1399–1412, 2007c.
22. Paredes, M., Wierzbicki, T., Zelenak, P., Prediction of crack initiation and propagation in X70 pipeline steels. *Eng. Fract. Mech.*, 168, 92–111, 2016.

23. Rousselier, G., Ductile fracture models and their potential in local approach of fracture. *Nucl. Eng. Des.*, 105, 1, 97–111, 1987.
24. Standard A. E8, Standard Test Methods for Tension Testing of Metallic Materials. Annual book of ASTM standards, vol. 3, pp. 57–72, 2004.
25. Swift, H., Plastic instability under plane stress. *J. Mech. Phys. Solids*, 1, 1, 1–18, 1952.
26. Testa, G., Bonora, N., Gentile, D., Carlucci, A., Madi, Y., Ductile fracture assessment of X65 steel using damage mechanics. *Procedia Struct. Integrity*, 3, 508–516, 2017.
27. Tvergaard, V., Influence of voids on shear band instabilities under plane strain conditions. *Int. J. Fract.*, 17, 4, 389–407, 1981.
28. Tvergaard, V., On localization in ductile materials containing spherical voids. *Int. J. Fract.*, 18, 4, 237–252, 1982.
29. Tvergaard, V. and Needleman, A., Analysis of the cup-cone fracture in a round tensile bar. *Acta Metall.*, 32, 1, 157–169, 1984.
30. Wilkins, M.L., Streit, R.D., Reaugh, J.E., Cumulative-strain-damage model of ductile fracture: simulation and prediction of engineering fracture tests (No. UCRL-53058), Lawrence Livermore National Lab., CA (USA); Science Applications, Inc., San Leandro, CA (USA), 1980.
31. G. 21143-2007, *Metallic Material sunified Method of Test for Determination of Quasistatic Fracture Toughness*, 2007.

# 11

# Fracture Response of Steel Pipelines Under Combined Tension and Torsion

***Abstract***
This chapter presents a detailed investigation into the fracture behavior of API X80 steel pipelines containing axial surface cracks under internal pressure-induced hoop stress. The study employs three fracture assessment models: the Gurson–Tvergaard–Needleman (GTN) damage model and two uncoupled criteria, the Modified Mohr-Coulomb (MMC) and Extended Rice–Tracey (ERT) models. A hybrid experimental–numerical calibration procedure is used to extract model parameters from tensile and notched specimens. Finite element simulations in ABAQUS, validated against compact tension tests and benchmark design codes (ASME B31G, DNV RP-F101), demonstrate the effectiveness of damage models in predicting ductile fracture and burst pressure. The pipe is modeled with a semi-elliptical axial crack, and crack propagation is simulated using element deletion. Parametric studies indicate that defect depth has a greater influence on burst pressure than defect length. Axial compressive forces and bending moments further reduce burst resistance, while tensile stresses marginally enhance it. The results affirm the capability of advanced fracture criteria to accurately simulate failure processes in pipelines and highlight the need to move beyond conventional standards when assessing structural integrity under realistic service conditions.

***Keywords:*** X80 steel pipeline, hoop stress, internal pressure, axial surface crack, GTN model, MMC criterion, finite element modeling

## 11.1 Introduction

In the oil and gas industries, offshore pipelines play the first role. The installation of pipelines in deep waters is a significant matter, which induces severe loads to the structure. The most demanded pipes are steel pipelines; the main reason is economic issues. Besides, steel pipelines are very compatible with severe environmental conditions, especially in deep waters, for conveying the crude oil and gas from subsea to vessel or coast. Offshore

Yong Bai. Plasticity and Ductile Fracture in Structural Engineering, (317–338) 

pipelines under both installation and operation conditions are inevitably confronted tension and torsion. The steel pipes have a good response to tension, but torsion affects the integrity and safety of the steel pipes during installation and application [1]. Nowadays, offshore pipelines are being used for long-distance transmissions. Hence, applying long steel pipes causes some defects, particularly in material and welding. In addition, due to corrosion, pipes could be damaged or cracked. Accordingly, studying the cracks in pipelines is vital. The types and amounts of the cracks are based upon the manufacturing process [2]. The pipes are under high external and internal pressure; these pressures are also accompanied by torsion and tension. In this way, considering the plastic strain, the pipelines are vulnerable to fracture, which may cause the collapse of the pipes and lead to a huge economic loss. In the normal reeling condition, the plastic strain can be up to 3% [3]. There are two main standards used to evaluate the fracture resistance for calculating the rate of strain, BS 7910 and R6 [4, 5]. However, these standards are established for elastic response, and they are limited to small plastic strains. Therefore, a precise fracture evaluation method that is compatible with the deformation capacity of offshore pipelines under tension and torsion is required. Østby [6] proposed a design based on the strain that was called SBD. The permissible strain is determined by this method which is not dependent on material properties or crack geometry. The tensile strain capacity of steel pipes could be validated by a numerical model which is presented by Tang *et al.* [7]. The response of fracture mechanics depends on CTOD (crack-tip opening displacement) and J-integral (energy release rate). This mono parametric approach describes the severity of crack growth at initiation and stable states and assesses the unstable state of ductile materials as follows [8]. Nevertheless, if the crack starts to expand, the J is not applicable any more [9]. Therefore, Paris *et al.* [10] presented the J-integral R-curve approach for unstable cracks. Moreover, the CTOD and J parameters are well known for their validation in the crack-tip fields characterized in certain cases [11]. In recent years, several types of researches have been done about the buckling response of the pipes under different load combinations. Finally, the theoretical model which assumed the tube surface as a thin shell was developed to analyze the buckling and collapse of the tubes, by using the Newton-Raphson method for solving the equilibrium equations [3]. However, in the literature, less attention has been paid to the fracture response of the pipes. Rahman suggested a new method to calculate the J-integral for cracked pipes, based upon the classical beam theory and deformation theory of plasticity. This method assumes that the plane defect of pipe has limited length and depth persistent, so they are applicable for both elastic and plastic conditions [12]. The approach of this method is load based, and this is its advantage since loads have more effect on the fracture parameters than strains [13]. The primary J-estimation method (GE/EPRI procedure) was studied by Kumar and German [14] for the elastic-plastic fracture analysis of cracked pipes. The fracture response is increased when the internal pressure in the pipes induces hoop stresses; this was an issue that no one had considered until Foxen and Rahman [15]. They also proved the hoop stresses increase J and CTOD for high-hardening materials. However, reaching precious answers for fracture response was not possible, so three dimensions of finite element simulations of surface cracked pipes were performed. Jayadevan *et al.* [16–18] did many types of research on the fracture response of the pipelines. They employed a strain-based fracture mechanics approach for pipelines in different loading conditions. Bending and tension were the loading type of their model. Likewise, the parameters which influence the fracture response of the pipe (crack depth, crack length, radius-to-thickness ratio, and material hardening) were identified. These parameters also have a strong effect on the slope of the CTOD-strain curve. They perceived that a simple linear CTOD–strain relationship is approached to large plastic strain levels for both tension and bending. The CTOD of a line-pipe section with external surface crack was developed, and a comprehensive 3D elastic-plastic finite element model was achieved. The pipelines with girth welds under biaxial loadings and large plastic deformation were studied by Dake *et al.* [19, 20]. In order to improve their study, Zhang *et al.* [21–25] proposed a 3D nonlinear elastic-plastic FE method to determine the effects of girth welds in the failure of offshore pipelines. They also studied in girth-welded pipelines how corrosion defect parameters affect the remaining strengths. Nevertheless, the torsion was not considered in any of these studies. Hence, Xie and Wang [26] used a theoretical model to obtain the crack initiation angle of pipes under tension and torsion. They only applied the S-theory to calculate the critical load of circumferential periodic cracks, so experimental or numerical methods were not applied to reach more accurate results. For isotropic materials, the plasticity model was presented by Gao *et al.* [27], whereas the second and third invariants of the stress deviator and also the hydrostatic stress were its function. Therefore, recent studies had many deficiencies in this field. Accordingly, to use with ductile metals, Bai and Wierzbicki [28] presented a modified version of the Mohr-Coulomb criterion which had been used before as an uncoupled method for rock and soil [29]. Chang-Sik Oh *et al.* [30] used finite element analysis based upon the stress-modified fracture strain to propose a new model for ductile failure simulation. However, their model was generally based on two-dimension space representation, and the fracture strain was taken into account as a function of stress triaxiality level. Thus, to determine the behavior of the crack under tension and torsion, Haltom *et al.* [31] did some experiments on Al-6061-T6 tubes subjected to tensile load and torque; the state of stress and strain was determined from these experiments. They used a wide and thin cylindrical specimen, and they concluded that once the triaxiality is decreased, failure strain will be increased. In order to assess the ductile failure response of pipes, a double-notched tube (DNT) specimen was examined under a fixed ratio of tension and torsion by Faleskog and Barsoum [32]. Then, their study was developed to obtain the effective void volume fraction and a shear damage coefficient by Xue *et al.* [33]. Recently, Damjanovic *et al.* [34] used the pipe ring-notched specimen (PRNB) to analyze the torsion effect on the fracture behavior of the pipe. The maximum torsion effect takes place in the circumferences of the crack region, so they compared the response of the pipe in the presence and absence of torsion in the crack region. MMC criterion obtained the alteration in fracture strain at the small variety of stress triaxiality and predicted the fracture strain by way of a stress triaxiality function in-plane stress. Pack *et al.* [35] found that in order to predict the start of failure and crack extension in smaller components and metal sheet troubles, the MMC criterion is operative. In fact, the unique feature of this criterion in describing the state of stresses in 3D space, where strain to fracture is considered as a function of stress triaxiality, provided the estimate of the start of failure for pressure sensibility rates and various loading conditions. It is noted that the aforementioned studies mainly focused on axial forces and experimental methods, so they are not comprehensive enough. Therefore, the fracture response of offshore pipelines under various load combinations still deserves additional study. The fracture model which has been used in this paper is based on the modified Mohr-Coulomb (MMC) criterion. Meanwhile, a 3D FE model is established under the different values of axial deformations and rotational strains for the same loading conditions by using ABAQUS. The influences of pipe misalignment and weld mismatch on strain capacity have not been investigated in this study. The calibration procedure and fracture criterion have been explained in previous chapters.

## 11.2 Theoretical Equations

In this section, firstly, intersection equations are proposed in the form of tension–torsion interaction. The interaction equations have been obtained

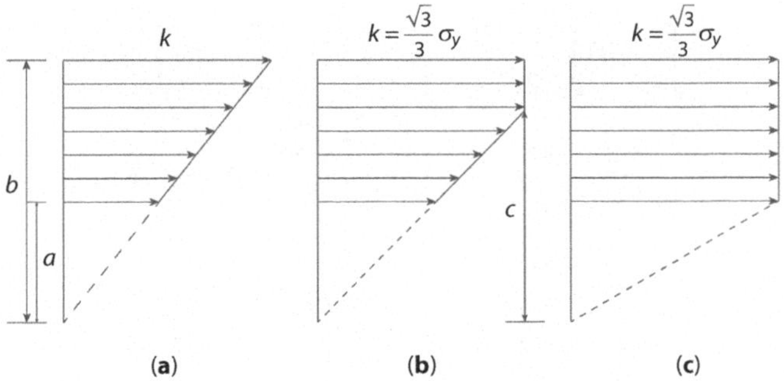


**Figure 11.1** Shear stress distribution along the radius.

through curve fitting to the finite element simulations. For the sake of completeness, equations for tubes under single loads and coupled loads are presented. The collapse torsion strength is considered as the fully plastic torque moment determined using elastic-perfectly plastic assumptions. The shear stress distribution for pipes under pure torque is displayed in Figure 11.1. The torsion moment ($T$) and the ultimate torque strength ($T_u$) can be calculated by

$$T = \int_0^{2\pi} d\theta \int_a^b \tau r^2 dr \tag{11.1}$$

$$T_u = \frac{2\sqrt{3}\pi}{9}\left(b^3 - a^3\right)\sigma_y \tag{11.2}$$

The deformation process under the torsion is divided into three parts, elastic, elastic-plastic, and perfectly plastic state. The related equations to these three states are given in the following, respectively:

$$T = \int_0^{2\pi} d\theta \int_a^b \frac{E}{2(1+v)} r \frac{\varphi}{L} r^2 dr = \frac{E\pi}{4(1+v)} \frac{\varphi}{L}\left(b^4 - a^4\right) \tag{11.3}$$

$$T = \int_0^{2\pi} d\theta \int_a^c \frac{E}{2(1+\nu)} r \frac{\varphi}{L} r^2 dr + \int_0^{2\pi} d\theta \int_c^b kr^2 dr$$

$$= \frac{E\pi}{4(1+\nu)} \frac{\varphi}{L}\left(c^4 - a^4\right) + \frac{2\sqrt{3}\pi}{9}\left(b^3 - c^3\right)\sigma_y \tag{11.4}$$

$$T = \int_0^{2\pi} d\theta \int_a^b kr^2 dr = \frac{2\sqrt{3}\pi}{9}\left(b^3 - a^3\right)\sigma_y \tag{11.5}$$

In Equation (11.4), variable c is defined as

$$c = \frac{\sqrt{3}\sigma_y L}{3\varphi G} \tag{11.6}$$

where G is the plastic modulus of the material. The tension–torsion interaction based on a curve fitting to the numerical data can be expressed as

$$\left(\frac{N_c}{N_u}\right)^c + \left(\frac{T_c}{T_u}\right)^d = 1 \tag{11.7}$$

where Nc and Tc denote collapse tension and torque moment under combined loads, respectively.

## 11.3 Finite Element Procedure

### 11.3.1 The Pipe Model

In order to simulate the fracture response of a steel offshore pipeline under simultaneous loading, an FE model is developed in the framework of the software ABAQUS. The pipe is modeled as a solid element in the way that is the best suited for this method [36]. A surface cracked pipe which is under combined tension and bending is considered in this study. As shown in Figure 11.2, the outer diameter and thickness of the pipe are shown by (D) and (t), and also the (a) and (c) indicate the crack depth and crack length, respectively. The shape of the crack is assumed elliptical since it is not a significant matter. However, the external surface of the crack is assumed with

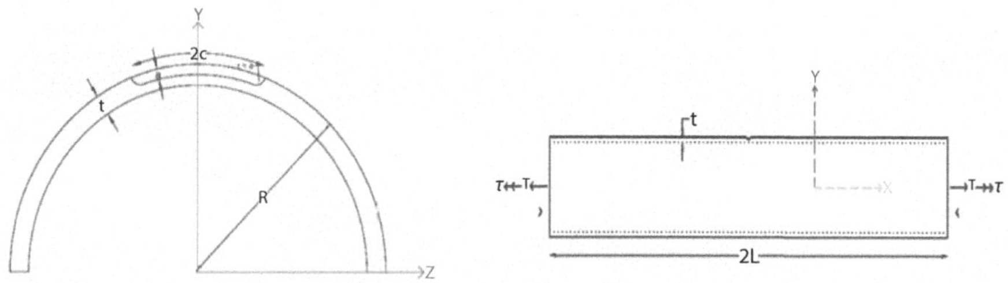


**Figure 11.2** The section and cross-section of the pipe with the details of crack.

uniform depth along the peripheral crack length, and the crack end radius is taken into account as same as the crack depth. In addition, regarding the three axes, the crack is positioned on the pipe symmetrically. For the simulation, the outer diameter and thickness of the pipe are chosen as 200 and 10 mm. In this way, the D/t is going to be 20. The crack depth to thickness ratio (a/t) and the crack length to perimeter ratio (2c/πD) are considered 0.5 and 0.15, respectively. Besides, the total length of the pipe is assumed as six times of the pipe diameter (2L = 6D).

## 11.3.2 Finite Element Modeling

ABAQUS/Explicit ABAQUS/Explicit is used to analyze the fracture behavior of the pipe since it has the ability to include the failure criterion through a user-defined subroutine VUMAT. Based on these codes, when the elements reach a critical amount, calculated by previous equations, they will be removed and the stresses become zero. Then, the ductile crack growth is characterized by an incremental element removal method. Since the pipe is subjected to torsion, the whole of the pipe has been modeled. A typical finite element mesh used in the analyses is demonstrated in Figure 11.3. The pipe model is meshed into different parts around the circumference and along the length, as well as the thickness of the pipe (Figure 11.3(a)). Besides, as can be seen in Figure 11.3(b), the crack surface is meshed into additional elements in comparison with the other sections of the model. The mesh study was carried out through trial analysis and was found to be adequate. As shown in Figure 11.4, the boundary conditions are prescribed on the left end of the pipe model and restrict this end to all directions (U1 = U2 = U3 = UR1 = UR2 = UR3 = 0). The reference point (RP-2) is chosen at the center of the unrestricted end of the pipe for considering the coupling relationship, as displayed in Figure 11.5. It is assumed that the right end remains plane under loading but the cross-section is free to deform.

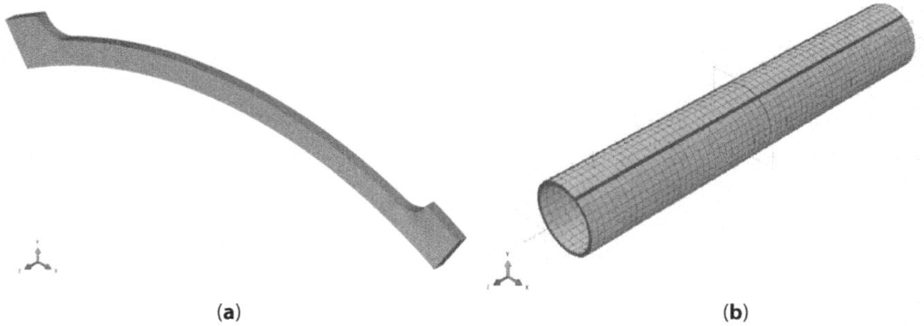


**Figure 11.3** Finite element mesh.

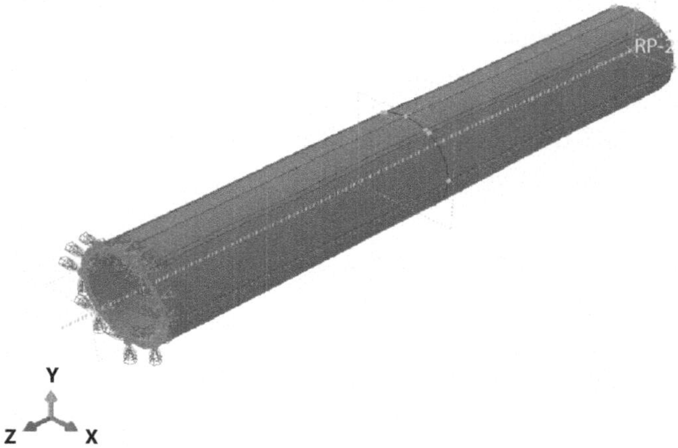


**Figure 11.4** The applied boundary condition surface.

The displacement (ΔL) and rotation angle around the Z-axis (ψ'), which present tension and torsion, are applied to the reference point to obtain the axial strain (ε) and twist angle (φ), and they are given as

$$\varepsilon = \frac{\Delta L}{L} \tag{11.8}$$

$$\varphi = \frac{\psi'}{L} \tag{11.9}$$

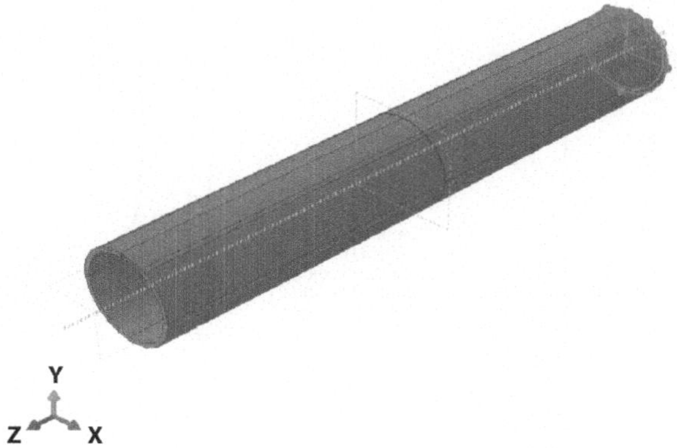


**Figure 11.5** The reference point and the related surface.

The pipe is placed under five types of loadings. In the first step, to determine the ultimate strain of the pipe, the model is loaded in pure tension by specifying a displacement ($\varepsilon/\varphi = +\infty$). Then, the pipe model is subjected to pure torsion by considering the rotation angle ($\varepsilon/\varphi = 0$), to obtain the ultimate twist. Subsequently, three different types of combined loading of tension and torsion (A: $\varepsilon/\varphi = 0.25$m, B: $\varepsilon/\varphi = 0.5$m, and C: $\varepsilon/\varphi = 1$m) are applied to the model for analyzing the fracture response of multiple loadings. The model which is employed, with rational limits, can precisely predict the fracture parameters. Besides, the chosen pipe length is long enough to eliminate the end effects in the calculations.

## 11.4 Results

The responses of the pipe subjected to two types of pure loading (tension or torsion) and three types of combined tension and torsion (combined load type A, B, and C) are computed using the numerical simulation. Based on the results obtained from the numerical simulation, the behavior of the pipe under different loading conditions can be analyzed and compared. As demonstrated in Figures 11.6 and 11.7 the comparisons corresponding to different types of loading are performed. As you can see in Figure 11.6, tension increases so fast while strain has a small change. The reason is that the whole pipe was modeled, and the displacement was applied to one end of the pipe. Therefore, a high-tension force is required to make change to the length of the pipe. Then, by reaching the tension to a specific amount, the strain increased intensely; from this value onward, the strain has no

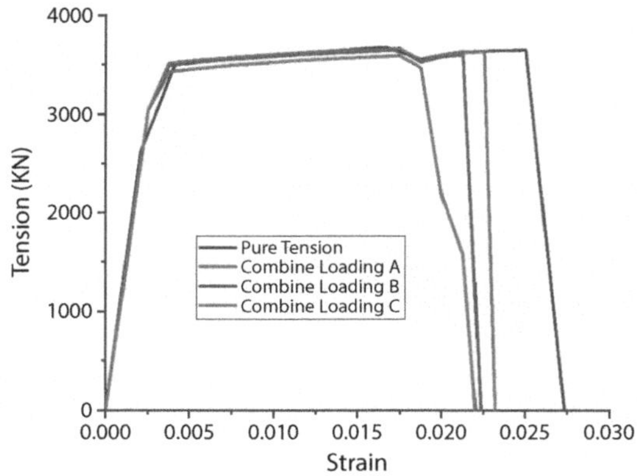


**Figure 11.6** The strain–tension relationship.

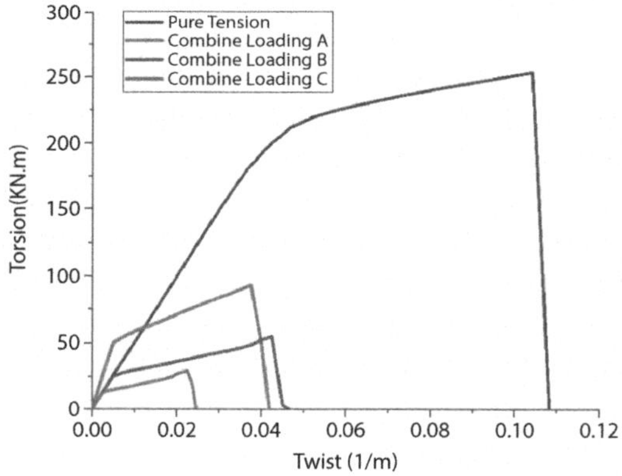


**Figure 11.7** Twist–torsion relationship.

noteworthy effect on the tension. The tension remained almost constant until the pipe model entirely lost its load-carrying capacity, resulting in circumferential crack propagation. Eventually, the pipe failed at the values which are displayed in Table 11.1. Afterwards, adding displacement does not affect the strain, since the pipe has lost its efficiency. The plastic zone of the pipe has a higher degree of resolution than the elastic zone, and before the curves collapse sharply, they exhibit some misbehavior, as the pipe has lost some element first and then begins to collapse completely. Figure 11.7 demonstrates that torsion and twist have a direct relationship at first, but it depends on the tension value. In fact, when the pipe is under pure torsion, the relationship between torsion and twist remains more constant in

**Table 11.1** The parameter values for different types of loading.

| Loading type | Displacement (mm) | Rotation angle (rad) | Strain | Twist (1/m) |
|---|---|---|---|---|
| Pure tension | 30 | 0 | 0.025 | 0 |
| Combine loading A | 22.5 | 0.09 | 0.01875 | 0.075 |
| Combine loading B | 25.5 | 0.051 | 0.02125 | 0.0425 |
| Combine loading C | 27 | 0.027 | 0.0225 | 0.0225 |
| Pure torsion | 0 | 0.125 | 0 | 0.1042 |

comparison with different combined loading. As far as the first stage of the elastic region goes, all three curves (black, blue, and green curves) are well agreed upon, while the red curve differs from the first. This is due to the fact that the applied strain is much smaller than the applied torque, which means that the pipe's behavior is more affected by torsion than tension. At specific values, the effect of torsion on twist will be decreased dramatically.

Lastly, this relationship continues until the process of failing the crack is started. At the points which are mentioned in Table 11.1, the loads decreased dramatically to zero, which means no more twist amount was added to the model. The four points where the curves suddenly go down in Figure 11.6 are called critical displacements, and in Figure 11.7, they are defined as critical rotation angles. Consequently, the final relationship between strain and twist for all types of loading is derived in Table 11.1 and shown in Figure 11.8. The extracted curve can be used in engineering applications; if the amount of strain and twist is below it, the pipe can continue to perform, but if it is beyond it, the pipe will no longer function. The torsion process generates the shear force, and the tension force causes stress in the circumferential of the cracked surface. Therefore, once the pipe model approaches failure, the primary crack blunts notably before advances needing a significant value of strain energy for the process.

In addition, Figure 11.9 illustrates the tension–torsion relationship for all three types of combined loading and shows that the influence of tension and torsion on crack failure resoundingly depends on the ratio of $\varepsilon$ to $\varphi$. It could be observed that when tension increased instantly, torsion did not change much. However, both of them are increased gradually at different stages for all types of loading until they reach certain quantities. At these values, the condition is reversed. Tension changes just a little,

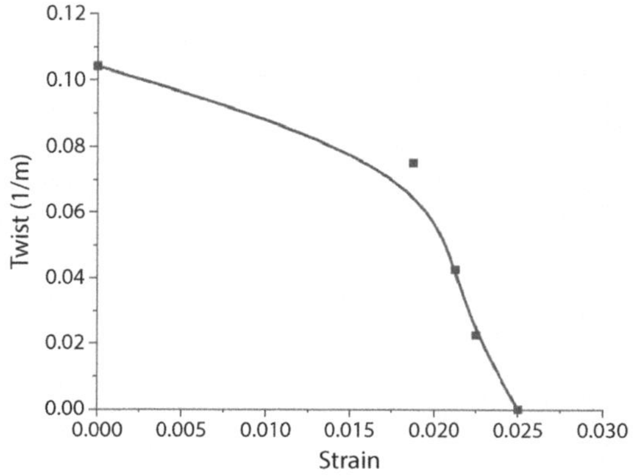


**Figure 11.8** The strain–twist relationship.

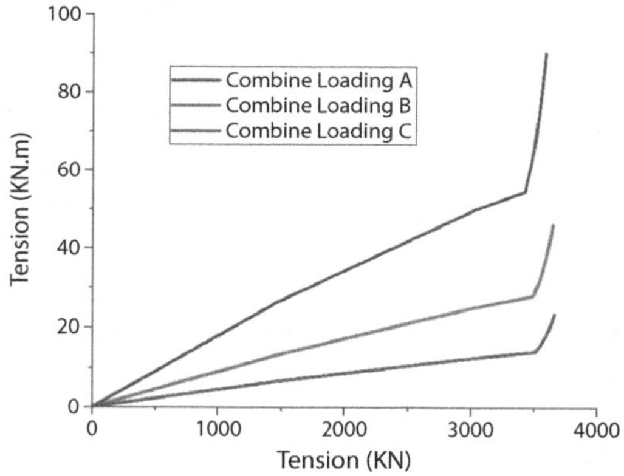


**Figure 11.9** The tension–torsion relationship of different combine loadings.

whereas torsion is rising fast. This rule continues until the tube collapses. Figure 11.10 displays the numerical result demonstration of crack growth in the period of time. Figure 11.10(a) shows that at certain stress, the crack abruptly starts to extend around the crack tip section, which causes a stable crack growth through the thickness (Figure 11.10(b) is an upside view of the pipe). Then, by increasing the load, the crack growth rate increases rapidly, and as long as the complete separation of the pipe is achieved, the crack grows fast in the circumferential direction, as shown in Figure 11.10(c). (Figure 11.10(d) is an upside view of the pipe.) It should be noted that, because the whole pipe was modeled for this simulation, it was not

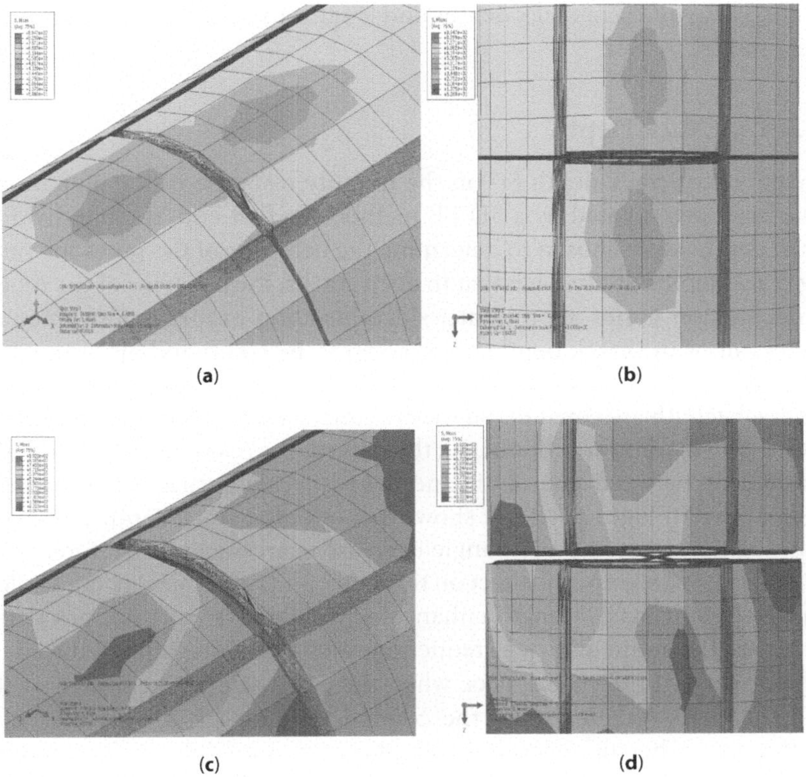


**Figure 11.10** The crack behavior during the time in two views.

possible to observe crack behavior in detail. This led to the failure point of the pipe, which was considered in the calculation, being determined by seeing a time-based plot.

## 11.5 Parametric Studies

In this section, for a better understanding of crack behavior under tension and torsion, the effects of some important parameters are examined based on the proposed finite element model. Since the crack is the most important part of the pipe to analyze in this study, the various crack depths (*a*) and lengths (*2c*) are modeled to investigate the crack size influence on the fracture behavior of the X80 steel pipeline. This study aims to understand the impact of crack size on the fracture behavior of the X80 steel pipeline by analyzing different crack depths and lengths using a proposed finite element model.

### 11.5.1 Crack Depth

The effects of crack depth (a) on the fracture response of the pipe model have been investigated by a 3D FE method. To that end, three individual crack depths were chosen to determine the behavior of the pipes for varying crack depths. It was seen that in deep cracks, the behavior of the model under torsion is not appropriate, since buckling occurs before fracture. Hence, all of the crack depths were taken to be less than half of the pipe thickness (a/t = 0.45, a/t = 0.4, a/t = 0.25). The properties of the pipe and the crack length are equal for all the specimens (2c/πD = 0.15). Five different types of loading are conducted, the same as the main model. Finally, the relationship between axial strain and twist angle with various crack depths is illustrated in Figure 11.11. It shows that once the crack depth grows, the critical axial strain and twist angle decrease aggressively. Therefore, crack geometry has a significant effect on the strain and twist capacity of the pipe, deprived of ultimate strength enhancement. Moreover, as the crack depth increases, the strain and twist reduction rates are significantly reduced. In comparison with deeper cracks, when the crack size is one-fourth of the wall thickness (green curve), the combined loading has a greater effect on the pipe behavior. It becomes significantly more difficult for a pipe to withstand strain and torque when the crack depth reaches half the wall thickness (black curve). Besides, when the depth of the crack decreases for 0.05 mm more than the half of the pipe thickness, the strain and twist

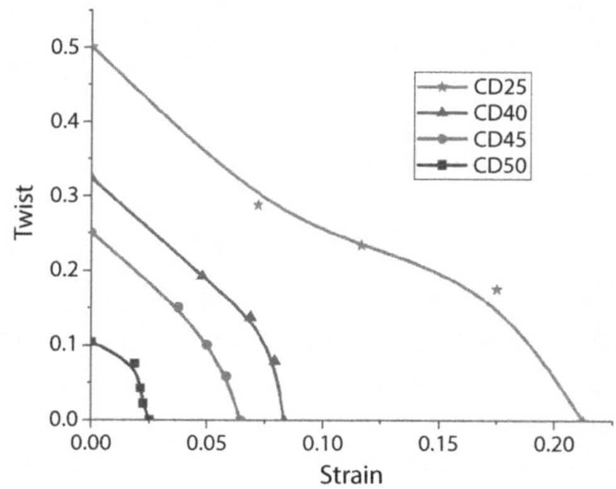


**Figure 11.11** The comparison between the different crack depths.

capacity increase larger compared to when the crack depth reduces from 0.45 mm (red curve) to 0.40 mm (blue curve). It shows that 0.5 mm is a critical crack depth for our model. Besides, the relationship between axial strain and tension force, as well as twist angle and the torsion moment of the pipe for the different crack depths, is demonstrated in Figures 11.12 and 11.13, respectively. The general behavior of all graphs is similar to the main model. Figure 11.12 shows that the influence of crack depth in the elastic zone is negligible, unlike the plastic behavior of pipe that is affected by crack depth significantly. The pipe that behaves elastically experiences a sharp increase in tension without affecting strain increments noticeably. The decreasing rate of strain has a direct relation to reducing the depth of the crack. It is clear that the crack depth has a more significant effect on the strain capacity when the pipe is under pure tension (black curves).

However, because torque plays such a significant role in combined loading, the strain resolution becomes less, especially in the first combination of loadings (red curves), which has the lowest displacement when combined. The response of the pipe model in Figure 11.13 is also similar. The values of torsion moment and twist angle increase simultaneously until reaching the specific points, then they fail sharply to below zero. The influence of crack depth on the twist capacity of the pipelines can be observed in this graph, as well. In the elastic region, crack depth does not significantly influence the tube behavior, and in the plastic region, they have much less of an effect, since the applied loading type influences the pipe behavior more. The crack depth, in fact, only affects the pipe model's failing point.

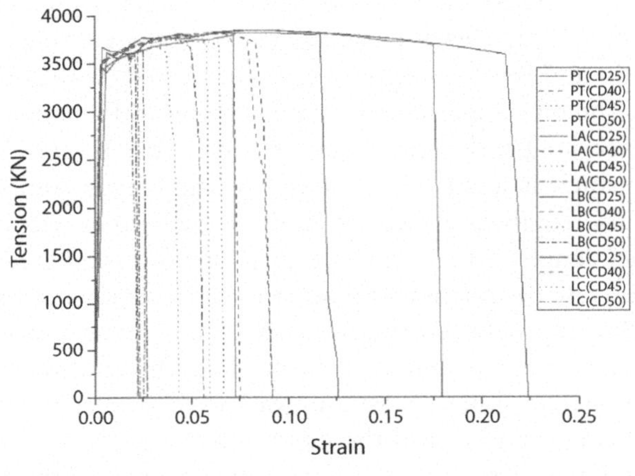


**Figure 11.12** The strain–tension relationship.

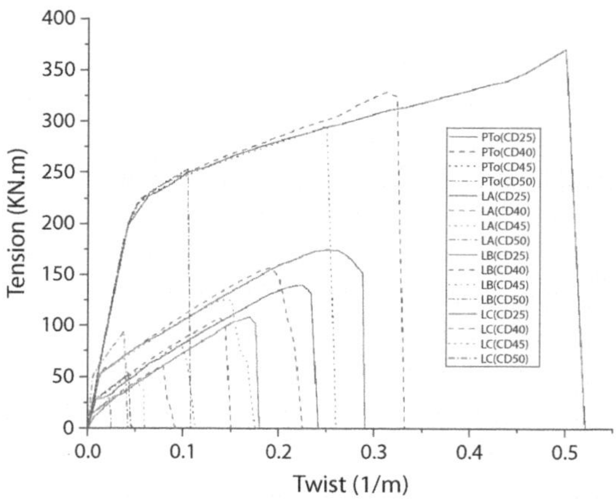


**Figure 11.13** Twist–torsion relationship.

## 11.5.2 Crack Length

In order to determine the fracture response of the pipe for various crack lengths, a 3D FE model was established by considering the circumferential crack length of the pipe. Two pipes with different crack lengths were adopted, one with shorter ($2c/\pi D = 0.13$) and one with larger ($2c/\pi D = 0.20$) crack lengths. The other crack features and the crack depth ($a/t = 0.5$) are the same for this model, as well. The model is considered to be under different types of loading, equal to the main model. As shown in Figure 11.14, by increasing the crack length, the strain and twist capacity are not decreased sharply. This graph shows that although the pipe collapses over larger cracks, they do not decrease much as the crack length increases from $0.15\pi D$ (black curve) to $0.2\pi D$ (blue curve). On the contrary, the strain and twist capacity raised significantly with shortening the crack length. The pipe experiences applied torque and tension much more when the crack length decreases from $0.15\pi D$ (black curve) to $0.13\pi D$ (red curve), which is a smaller variation in the length compared to the increasing crack length case. As a result, there is a kind of critical length for the pipe's crack length, which can be $0.15\pi D$ for our model. Figures 11.15 and 11.16 illustrate the relationship between axial strain and tension force and the relationship between the twist angle and torsion moment of the pipe, respectively. Approximately in both plots, the relationship between strain and tension, as well as twist and torsion, is similar for the different lengths and loading types until the yield points. Figure 11.15 shows that the crack length has a greater influence on the behavior of the pipe when it

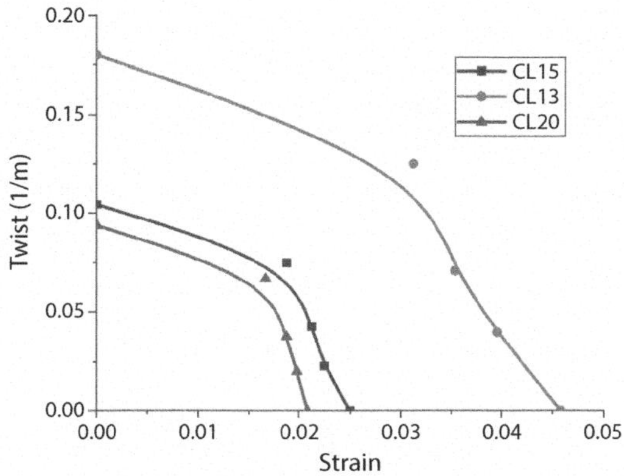


**Figure 11.14** The comparison between the different crack lengths.

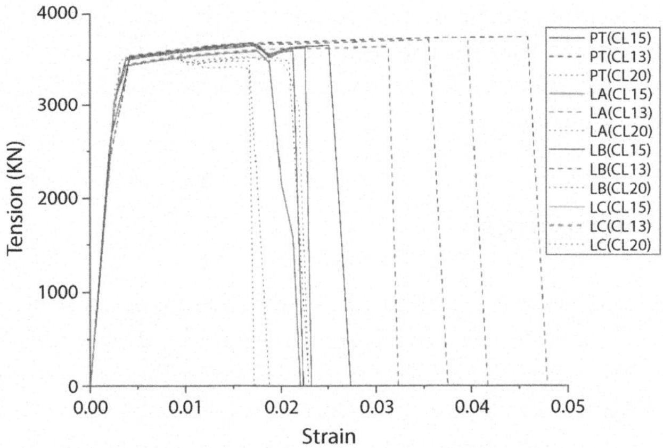


**Figure 11.15** The strain–tension relationship.

is under pure tension (black curves). As the amount of tension is reduced in combined loading types, the pipe responds less to variations in crack length. Figure 11.16 demonstrates that all graphs show good agreement at the elastic range and that the behavior of the pipe model in the plastic zone is similar for a given loading type applied without accounting for the crack length. With the exception of loading type A (red curves), which is caused by a relatively high applied torque in this combined loading, the pipe's behavior varies depending on the length of the cracks. This phenomenon is also visible when the tube is under pure torsion (black curves).

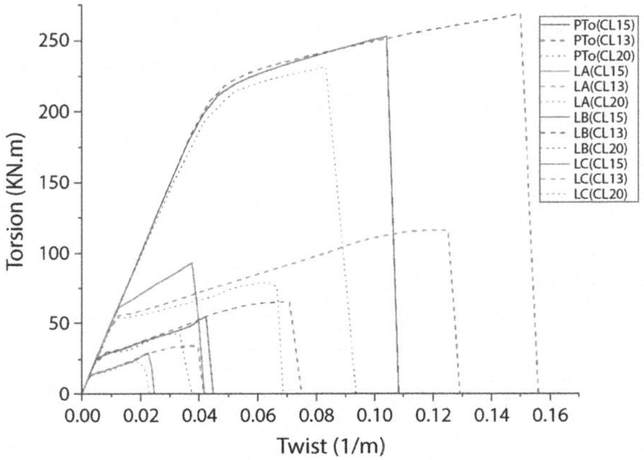


**Figure 11.16** Twist–torsion relationship.

In fact, crack length affects the failure point of the pipe. It could be pointed out, in both figures, that for different crack lengths, the elastic and plastic zones of the pipe model, as well as the yield points for different types of loading, are not closed.

## 11.6 Conclusions

Tension and torsion are applied to the X80 steel pipe simultaneously. The variations of twist with strain under different loading types have been carried out. The pipe can continue its service in the presence of a crack till a specific value, and these values are extracted and the related curve is drawn. Since the crack depth and length have the largest effect on the fracture response of the pipe, the models with different crack dimensions are analyzed. It was seen that the crack depth has a large influence on the strain and twist capacity of the steel pipes. Likewise, the strain and twist reduction rates have a direct relation with reducing the depth of the crack. When the crack size is one-fourth of the wall thickness, the combined loading has a greater effect on the pipe behavior. It becomes significantly more difficult for a pipe to withstand strain and torque when the crack depth reaches half the wall thickness. Although the pipe collapses over larger cracks, they do not decrease much as the crack length increases from 0.15πD to 0.2πD. On the contrary, the strain and twist capacity raised significantly with shortening the crack length. The pipe experiences applied torque and tension much more when the crack length decreases from 0.15πD to 0.1πD, which is the same variation that the length was increased previously. As a result, there is a kind of critical length for the pipe's crack length, which can be 0.15πD for our model.

## References


1. Fang, P., Yuan, S., Cheng, P., Bai, Y., Xu, Y., Mechanical responses of metallic strip flexible pipes subjected to pure torsion. *Appl. Ocean Res.*, 86, 13–27, 2019, https://doi.org/10.1016/j.apor.2019.02.009.
2. Mousselli, A., *Offshore Pipeline Design, Analysis, and Methods*, Pennwell Books, Tulsa, Oklahoma, 1981.
3. Kyriakides, S. and Corona, E., *Mechanics of Offshore Pipelines*, vol. 1, Buckling and Collapse, Burlington, Massachusetts, Oxford, UK, 2007.
4. British Standards Institution, BS 7910, Guide on methods for assessing the acceptability of flaws in metallic structures, London, UK, 2000.
5. British Energy, Assessment of the integrity of structures containing defects, revision 4, London, UK, 2001.
6. Østby, E., Fracture control - Offshore pipelines: New strain-based fracture mechanics equations including the effects of biaxial loading, mismatch and misalignment. *Proc. Int. Conf. Offshore Mech. Arct. Eng. - OMAE*, 3, 649–658, 2005, https://doi.org/10.1115/OMAE2005-67518.
7. Tang, H., Fairchild, D., Panico, M., Crapps, J., Cheng, W., Strain capacity prediction of strain-based pipelines, in: *Proc. 2014 10th Int. Pipeline Conf.*, Calgary, Alberta, Canada, pp. 1–12, 2014.
8. Hutchinson, J.W., Fundamentals of the phenomenological theory of nonlinear fracture mechanics. *J. Appl. Mech. Trans. ASME*, 50, 1042–1051, 1983, https://doi.org/10.1115/1.3167187.
9. Hutchinson, J.W., Singular behaviour at the end of a tensile crack in a hardening material. *J. Mech. Phys. Solids*, 16, 13–31, 1968.
10. Paris, P., Tada, H., Zahoor, A., Ernst, H., The Theory of Instability of the Tearing Mode of Elastic-Plastic Crack Growth. Elastic-Plastic Fract., pp. 5-5–32, 2009, https://doi.org/10.1520/stp35825s.
11. O'Dowd, N.P. and Shih, C.F., Family of crack-tip fields characterized by a triaxiality parameter-I. Structure of fields. *J. Mech. Phys. Solids*, 39, 989–1015, 1991, https://doi.org/10.1016/0022-5096(91)90049-T.
12. Rahman, S. and Brust, F.W., Approximate methods for predicting J-integral of a circumferentially surface-cracked pipe subject to bending. *Int. J. Fract.*, 85, 111–130, 1997, https://doi.org/10.1023/A:1007322018722.
13. Anderson, T., *Fracture Mechanics - fundamentals and applications*, 2nd ed, CRC Press, London, 1995.

14. Kumar, V. and German, M., Elastic-Plastic Fracture Analysis of Through-Wall and Surface Flaws in, Gen. Electr. Co., Schenectady, NY (USA). Corp. Res. Dev. Cent., 1988.
15. Foxen, J. and Rahman, S., Elastic-plastic analysis of small cracks in tubes under internal pressure and bending. *Nucl. Eng. Des.*, 197, 75–87, 2000, https://doi.org/10.1016/S0029-5493(99)00258-7.
16. Jayadevan, K.R., Ostby, E., Thaulow, C., Strain based fracture mechanics analysis of pipelines. *Proc. Int. Conf. Adv. Struct. Integrity*, Bangalore, India, pp. 14–17, 2004.
17. Jayadevan, K.R., Østby, E., Thaulow, C., Fracture response of pipelines subjected to large plastic deformation under tension. *Int. J. Press. Vessel. Pip.*, 81, 771–783, 2004, https://doi.org/10.1016/j.ijpvp.2004.04.005.
18. Østby, E., Jayadevan, K.R., Thaulow, C., Fracture response of pipelines subject to large plastic deformation under bending. *Int. J. Press. Vessel. Pip.*, 82, 201–215, 2005, https://doi.org/10.1016/j.ijpvp.2004.08.012.
19. Dake, Y., Sridhar, I., Zhongmin, X., Kumar, S.B., Fracture capacity of girth welded pipelines with 3D surface cracks subjected to biaxial loading conditions. *Int. J. Press. Vessel. Pip.*, 92, 115–126, 2012, https://doi.org/10.1016/j.ijpvp.2011.10.019. URL10.1016/j.ijpvp.2011.10.019.
20. Yi, D., Xiao, Z.M., Idapalapati, S., Kumar, S.B., Fracture analysis of girth welded pipelines with 3D embedded cracks subjected to biaxial loading conditions. *Eng. Fract. Mech.*, 96, 570–587, 2012, https://doi.org/10.1016/j. engfracmech.2012.09.005. URL 10.1016/j.engfracmech.2012.09.005.
21. Zhang, Y.M., Xiao, Z.M., Zhang, W.G., On 3-D crack problems in offshore pipeline with large plastic deformation. *Theor. Appl. Fract. Mech.*, 67–68, 22–28, 2013, https://doi.org/10.1016/j.tafmec.2014.01.001. URL 10.1016/j.tafmec.2014.01.001.
22. Zhang, Y.M., Yi, D.K., Xiao, Z.M., Huang, Z.H., Kumar, S.B., Elastic-plastic fracture analyses for pipeline girth welds with 3D semi-elliptical surface cracks subjected to large plastic bending. *Int. J. Press. Vessel. Pip.*, 105–106, 90–102, 2013, https://doi. org/10.1016/j.ijpvp.2013.03.009.
23. Zhang, Y.M., Ariffin, M.Z., Xiao, Z.M., Zhang, W.G., Huang, Z.H., Nonlinear elastic-plastic stress investigation for two interacting 3-D cracks in offshore pipelines. *Fatigue Fract. Eng. Mater. Struct.*, 38, 5, 540–550, 2015, https://doi.org/10.1111/ ffe.12259.
24. Zhang, Y.M., Yi, D.K., Xiao, Z.M., Huang, Z.H., Engineering critical assessment for offshore pipelines with 3-D elliptical embedded cracks. *Eng. Fail. Anal.*, 51, 37–54, 2015, https://doi.org/10.1016/j.engfailanal.2015.02.018.
25. Zhang, Y.M., Tan, T.K., Xiao, Z.M., Zhang, W.G., Ariffin, M.Z., Failure assessment on offshore girth welded pipelines due to corrosion defects. *Fatigue Fract. Eng. Mater. Struct.*, 39, 4, 453–466, 2016, https://doi.org/10.1111/ffe.12370.

26. Xie, Y.J. and Wang, X.H., Crack initiation and direction for circumferential periodic cracks in pipe under tension and torsion. *Theor. Appl. Fract. Mech.*, 40, 153–159, 2003, https://doi.org/10.1016/S0167-8442(03)00043-0.
27. Gao, X., Faleskog, J., Shih, C.F., Dodds, R.H., Ductile tearing in part-through cracks: experiments and cell-model predictions. *Eng. Fract. Mech.*, 59, 761–777, 1998, https://doi.org/10.1016/S0013-7944(97)00174-4.
28. Bai, Y. and Wierzbicki, T., Application of extended Mohr-Coulomb criterion to ductile fracture. *Int. J. Fract.*, 161, 1–20, 2010, https://doi.org/10.1007/s10704-009-9422-8.
29. Wylllie, D., *Foundations on rock*, Second ed, E & FN Spon, London, 1999.
30. Oh, C.S., Kim, N.H., Kim, Y.J., Baek, J.H., Kim, Y.P., Kim, W.S., A finite element ductile failure simulation method using stress-modified fracture strain model. *Eng. Fract. Mech.*, 78, 124–137, 2011, https://doi.org/10.1016/j.engfracmech.2010.10.004.
31. Haltom, S.S., Kyriakides, S., Ravi-Chandar, K., Ductile failure under combined shear and tension. *Int. J. Solids Struct.*, 50, 1507–1522, 2013, https://doi.org/10.1016/j.ijsolstr.2012.12.009.
32. Faleskog, J. and Barsoum, I., Tension-torsion fracture experiments - Part I: Experiments and a procedure to evaluate the equivalent plastic strain. *Int. J. Solids Struct.*, 50, 4241–4257, 2013, https://doi.org/10.1016/j.ijsolstr.2013.08.029.
33. Xue, Z., Faleskog, J., Hutchinson, J.W., Tension-torsion fracture experiments - Part II: Simulations with the extended Gurson model and a ductile fracture criterion based on plastic strain. *Int. J. Solids Struct.*, 50, 4258–4269, 2013, https://doi.org/10.1016/j.ijsolstr.2013.08.028.
34. Damjanović, D., Kozak, D., Gelo, I., Gubeljak, N., The influence of torsion effect on fracture behavior of Pipe Ring Notched Bend specimen (PRNB). *Theor. Appl. Fract. Mech.*, 103, 2019, https://doi.org/10.1016/j.tafmec.2019.102286.
35. Pack, K., Luo, M., Wierzbicki, T., Sandia fracture challenge: Blind prediction and full calibration to enhance fracture predictability. *Int. J. Fract.*, 186, 155–175, 2014, https://doi.org/10.1007/s10704-013-9923-3.
36. Hibbitt, H.D., Karlsson, B.I., Sorensen, P., ABAQUS Theory Manual, 2006.

# 12

# Fracture Response of Steel Pipelines Under Combined Tension and Bending

***Abstract***

This chapter investigates the mechanical behavior and fracture response of X80 steel pipelines subjected to simultaneous axial tension and bending, a common loading condition during offshore installation and operation. A 3D finite element model is developed in ABAQUS using the Modified Mohr-Coulomb (MMC) fracture criterion through a user-defined VUMAT subroutine to simulate ductile crack propagation. The pipe, containing a circumferential surface crack, is analyzed under pure tension, pure bending, and three combined loading cases. Results reveal that as the pipe transitions from the elastic to plastic regime, fracture initiates and propagates rapidly once strain or curvature thresholds are exceeded. Parametric studies highlight that both crack depth and length significantly reduce the pipe's strain and curvature capacity, with crack depth exerting a greater influence. Validation is performed by comparing finite element results with theoretical models under elastic conditions, showing strong agreement. Theoretical and numerical comparisons also emphasize the limitations of existing standards and the need for advanced criteria like MMC in capturing fracture behavior under multi-axial plastic deformation. The study offers practical insight into structural integrity limits and failure mechanisms in offshore pipeline applications.

***Keywords*:** Steel pipelines, combined loading, crack geometry, MMC criterion, finite element analysis, strain capacity, ductile fracture

## 12.1 Introduction

In the oil and gas industries, offshore pipelines play the first role. The most demanded pipes are steel pipelines; the main reason is economic issues. Besides, steel pipelines are very compatible with severe environmental conditions, especially in deep waters, for conveying the crude oil and gas from sub-sea to vessel or coast. Without paying attention to the installation method (J-lay or S-lay), pipelines have to endure the combined loading

Yong Bai. Plasticity and Ductile Fracture in Structural Engineering, (339–372) © 2026 Scrivener Publishing LLC

of tension and bending [1]. Nowadays, o shore pipelines have been used for long-distance transmits. Hence, applying long steel pipes cause some defects, particularly in material and welding. In addition, due to corrosion, pipes could be damaged or cracked. Accordingly, studying the cracks in pipelines is vital. The types and amounts of the cracks are based upon the manufacturing process [2]. Occasionally, the pipelines in operation are under simultaneous tension and bending. In this way, considering the plastic strain, the pipelines are vulnerable to fracture, which may cause the collapse of the pipes and lead to a huge economic loss. There are two main standards used to evaluate the fracture resistance for calculating the rate of strain, BS 7910 and R6 [3, 4]. However, these standards are established for elastic response, and they are limited to small plastic strains. Therefore, a precise fracture evaluation method that is compatible with the deformation capacity of o shore pipelines under tension and bending is required. Ostby [5] proposed to design based on the strain that was called SBD. The permissible strain is determined by this method which is not dependent on material property or crack geometry. The tensile strain capacity of steel pipes could be validated by a numerical model which is presented by Tang *et al.* [6]. The response of fracture mechanics depends on CTOD (crack-tip opening displacement) and J-integral (energy release rate). This mono parametric approach describes the severity of crack growth at initiation and stable states and assesses the unstable state of ductile materials as follows [7]. Nevertheless, if the crack starts to expand, the J is not applicable any more [8]. Therefore, Paris *et al.* [9] presented the J-integral R-curve approach for unstable cracks. Besides, the CTOD and J parameters are well known for their validation in the crack-tip fields characterized in certain cases [10]. The primary J-estimation method (GE/EPRI procedure) was studied by Kumar and German [11] for cracked pipes under pure tension and by Ahamed *et al.* [12] for cracked pipes under pure bending. These studies were not useful enough for the application since they did not consider combined loading. Hence, the responses of thick tubes under combined tension and bending have been studied by Bai *et al.* [13]. They presented an FEM analysis of the collapse behavior of pipes. Rahman and Burst [14] suggested a new method to calculate the J-integral for cracked pipes, based upon the classical beam theory and deformation theory of plasticity. This method assumes that the plane defect of pipe has limited length and depth persistence, so they are applicable for both elastic and plastic conditions. The approach of this method is loading base, and this is its advantage since loads have more effect on the fracture parameters than strains [15]. The fracture response is increased when the internal pressure in the pipes induces hoop stresses; this was an issue that no one had considered until

Foxen and Rahman [16]. They also proved that hoop stresses increase J and CTOD for high-hardening materials. However, reaching precious answers for fracture response was not possible, so three-dimension finite element simulations of surface cracked pipes were performed. Hence, Shimanuki and Inoue [17] applied an equivalent CTOD concept to the model simulation of brittle fracture in girth weld joints of pipelines. Their model assumed that the surface cracks have elliptical form, and the pipes are under bending and axial forces, in the presence of internal pressure. They assessed the X80 steel pipe and understood that in the elastic region, the fracture persistence is fixed; but in the plastic region, the CTOD surpasses the critical level. Nevertheless, their work has some weaknesses. For example, loading on finite CTOD or J is uncertain and the effects of crack depth, crack length, radius-to-thickness ratio, and material hardening are not considered. Jayadevan and Ostby did many types of research on the fracture response of the pipelines [18–20]. They employed a strain-based fracture mechanics approach for pipelines in different loading conditions. Bending and tension were the loading type of their model. Likewise, the parameters which influence the fracture response of the pipe (crack depth, crack length, radius-to-thickness ratio, and material hardening) were identified. These parameters also have a strong effect on the slope of the CTOD-strain curve. They perceived that a simple linear CTOD–strain relationship is approached large plastic strain levels for both tension and bending. The CTOD of a line-pipe section with external surface crack was developed, and comprehensive 3D elastic-plastic finite element model was achieved. Rozumek and Marciniak [21] studied the fatigue properties of different notched specimens. They found fatigue weakening of the material and an increase in strain. Chang-Sik Oh *et al.* [22] used finite element analysis based upon the stress-modified fracture strain to propose a new model for ductile failure simulation. However, their model was generally based on two-dimension space representation, and the fracture strain was taken into account as a function of stress triaxiality level. For isotropic materials, the plasticity model was presented by Gao *et al.* [23], whereas the second and third invariant of the stress deviator and also the hydrostatic stress were its function. Therefore, recent studies had many deficiencies in this field. Accordingly, Bai and Wierzbicki [24] to use ductile metals presented a modified version of the Mohr-Coulomb criterion which had been used before as an uncoupled method for rock and soil [25]. MMC criterion obtained the alteration in fracture strain at the small variety of stress triaxiality and predicted the fracture strain by way of a stress triaxiality function in-plane stress. Pack *et al.* [26, 27] found that to predict the start of failure and crack extension in smaller components and metal sheet troubles, the MMC criterion is operative. The unique feature of this criterion in describing the state of stresses in 3D space, where strain to fracture is considered as a function of stress triaxiality, provided the estimate of the start of failure for pressure sensibility rates and various loading conditions. It is noted that the aforementioned studies for obtaining the fracture response of the pipes mainly focused on J-integral and CTOD, and does not consider the simultaneous loading, so they are not comprehensive enough. These methods were not able to capture the biaxial effects and the plastic region was assumed small. Therefore, assessing the ductile behavior of o shore pipelines under various load combinations still deserves additional study. The fracture model which has been used here is based on the modified Mohr-Coulomb (MMC) criterion. Meanwhile, a 3D FE model is established under the different values of displacements and rotation angles by using ABAQUS. Besides, the parametric studies are conducted to comprehend the effects of crack depth and crack length which have the largest influence on the fracture response of the steel pipes. The influences of pipe misalignment and weld mismatch on strain capacity have not been investigated in this study. It should be noted that the calibration process of the pipe model has been done in the previous chapters.

## 12.2 Finite Element Procedure

A finite element model is established in the ABAQUS software to simulate the fracture response of a steel offshore pipeline under combined loading, which in this section is tension and bending. Based on Hibbitt *et al.* [31], the most suited element for the applied method is solid, so the pipe is modeled as a solid element. The side section of the pipe under tension and bending is displayed in Figure 12.1. The pipe is located along the x-axis and is perpendicular to the y-axis. The strain and curvature are applied on

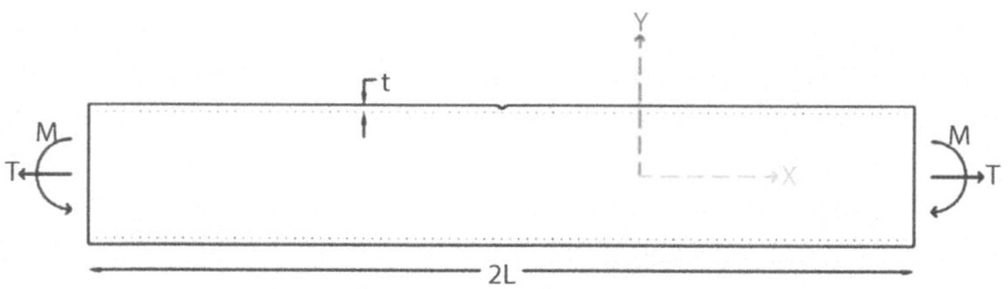


**Figure 12.1** The section of the pipe under tension and bending.

both ends of the pipe model, and the location of the crack is exactly at the middle of the pipe.

### 12.2.1 The Pipe Model

As shown in Figure 12.2, the outer diameter and thickness of the pipe are shown by (D) and (t), respectively. Besides, a and 2c indicate the crack depth and crack length, respectively. The shape of the crack is assumed elliptical since it's not a significant matter. However, the external surface of the crack is assumed with uniform depth along the peripheral crack length, and the crack end radius is taken into account as the same as the crack depth. The crack is positioned on the pipe symmetrically with regard to all three axes. For the simulation, the outer diameter and thickness of the pipe are chosen as 200 and 10 mm, respectively. In this way, D/t ratio is going to be 20. The crack depth to thickness ratio (a/t) and the crack length to perimeter ratio (2c/πD) are considered 0.5 and 0.15, respectively. The crack length angle is shown by 2θ, and it can be calculated as C = Rθ. Besides, the total length of the pipe is assumed as six times the pipe diameter (2L = 6D).

### 12.2.2 Finite Element Modeling

In order to analyze the fracture behavior of the pipe, the ABAQUS software is used, because it can include the failure criterion through a user-defined subroutine VUMAT, which was established through the explained process in previous chapters. As a matter of fact, when the elements reach a critical amount, calculated by previous equations, based on these codes, they will

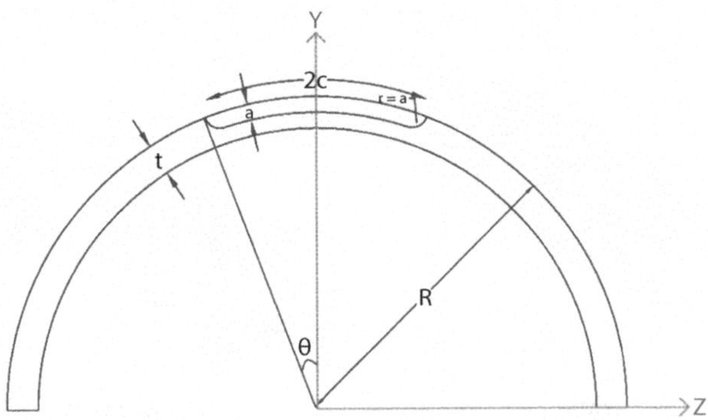


**Figure 12.2** The section of the pipe with the details of the crack.

be removed and the stresses become zero. At that point, the ductile crack growth is characterized by an incremental element removal method. As shown in Figure 12.3, a quarter of the pipe has been modeled, because of the symmetry deformations and loads. The axial symmetric boundary condition (ZSYMM) along the z-axis is set at the left end surface of the model (defected surface), as displayed in Figure 12.3(a), and the axial symmetry boundary condition along the x-axis (XSYMM) is set at the pipe edges at the top and bottom of the model, as illustrated in Figure 12.3(b). As demonstrated in Figure 12.4, a reference point (RP-1) is chosen at the center of the non-cracked end of the pipe (non-defected surface), and kinematic coupling is set for all degrees of freedom at this end with the reference point in order to conveniently control the boundary conditions of the model. As can be seen, the reference point's degrees of freedom are all constrained, with the exception of the z-axis direction. It is assumed that the right end plane remains plane under loading, but the cross-section is free to deform. In order to obtain strain (ε), the displacement (ΔL) is applied to the reference point, and by proposing the rotation angle (ψ), the curvature (κ) is applied at this point, as well. The strain and curvature of the pipe can be calculated as

$$\varepsilon = \frac{\Delta L}{L} \tag{12.1}$$

$$\kappa = \frac{\psi}{L} \tag{12.2}$$

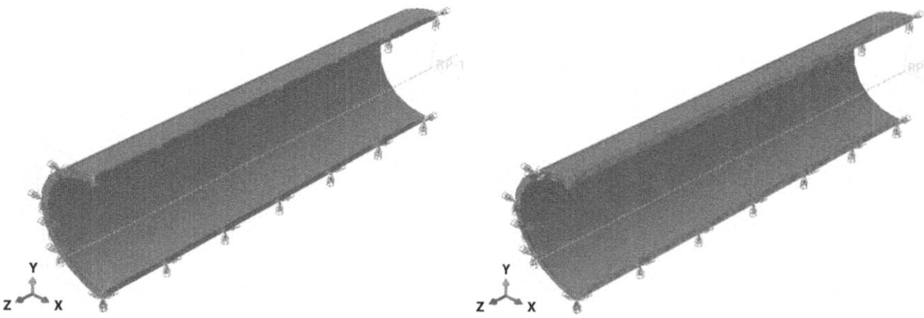


(a) ZSYMM (b) XSYMM

**Figure 12.3** Symmetric boundary conditions of the model.

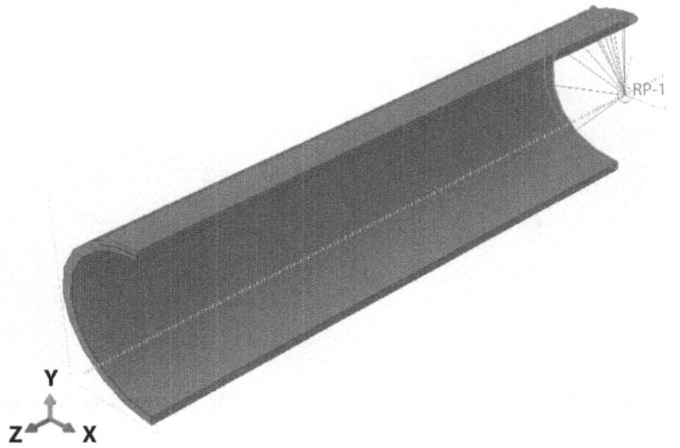


**Figure 12.4** Reference point and kinematic coupling of the model.

The pipe model is meshed into different parts around the half circumference and along the length, also the thickness of the pipe, especially the crack surface, is meshed into additional elements in comparison with the other sections of the model. The mesh study was carried out through trial analysis and was found to be adequate. A typical finite element mesh used in the analyses is demonstrated in Figure 12.5. The model's mesh is constructed using the C3D8R element, which is an eight-node continuum linear brick element with reduced integration and hourglass control. The mesh near the right end of the model is sparse, with a maximum mesh size of about 20 mm, as shown in Figure 12.5(a), and a relatively dense mesh is used at the crack section, as displayed in Figure 12.5(b). The mesh that was selected is appropriate, and the mesh size for the MMC model is also set to 0.1 mm. Finally, the pipe model is placed under the pure and combined loadings. In the first step, to determine the ultimate strain of the pipe, the model is loaded in pure tension by specifying a displacement ($\varepsilon/k = +\infty$). Then, the pipe model is subjected to pure bending by considering the rotation angle ($\varepsilon/k = 0$), to obtain the ultimate curvature. Subsequently, three different types of simultaneous loading of tension and bending which are A: $\varepsilon/k = 0.05$m, B: $\varepsilon/k = 0.1$m, and C: $\varepsilon/k = 0.2$m are applied to the model for analyzing the fracture response of multiple loading. The model which is employed, with rational limits, can precisely predict the fracture parameters. Besides, the chosen pipe length is long enough to eliminate the end effects in the calculations.

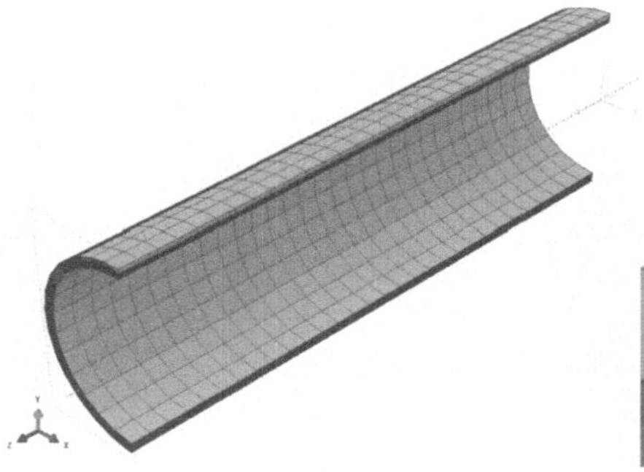

(a) Circumferential directions mesh.

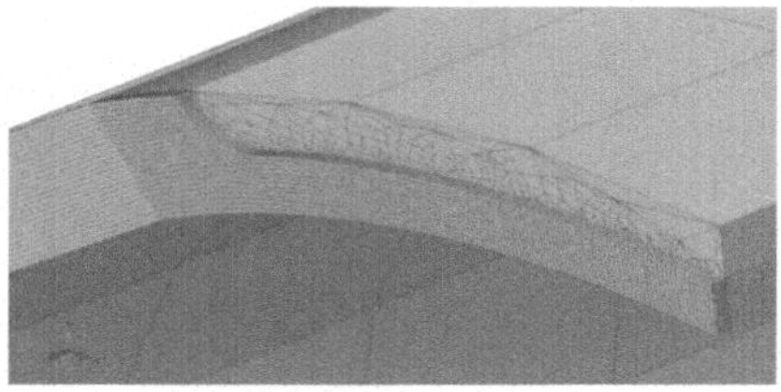

(b) Cracked surface mesh.

**Figure 12.5** The finite element mesh.

## 12.3 Results

The comparisons corresponding to different types of loading are performed. Figure 12.6 demonstrates the relationship between strain and tension for all types of applied loadings. It can be seen in the elastic zones that all graphs overlap at initiation; when they reach the yield point, they commence to enter the plastic zone, individually. In the phase of nonlinearity, the tension increment is almost parallel to the axial strain. As much as the tension is increased, the elastic zone takes place for a longer time. The bending moment makes the model enter the plastic zone much sooner, so there is a large gap between black and red curves. The green curve shows that when the model is tolerating less curvature, with the constant tension, the strain increases more. Before complete failure, a short unstable curve appears which is because the pipe lost an element at this time. Figure 12.7 illustrates the relationship between the curvature angle and the bending moment. When the moment is raised, the elastic zone is increased. The reason is that once the strain is at its lowest, at the same time the curvature is at its highest, and vice versa. By comparing the black curve with the green one, it can be found that the strain has a large influence on the bending failure point of the pipe. When the pipe is under a large amount of axial strain, the shape of curves becomes less smooth after their yielding points. It happens because when the model is under large strain and curvature at the same time, the material enters its plastic zone suddenly. In both figures, tension and moment remain almost constant, but before the pipe model reaches the ultimate strength, all the graphs are dropped down shortly. These are the values that the pipe loses its efficiency, resulting in

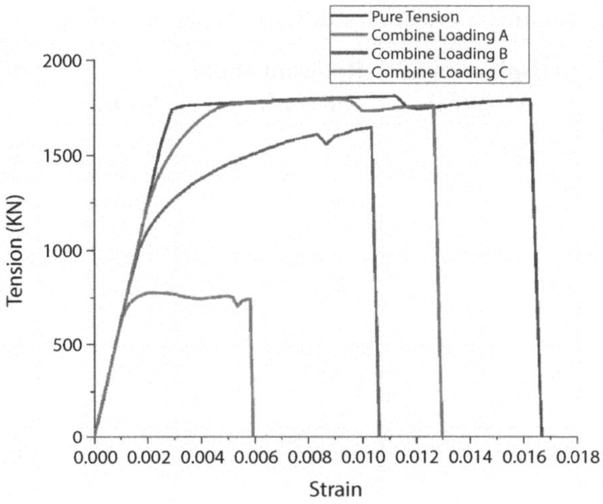


**Figure 12.6** The strain–tension relationship.

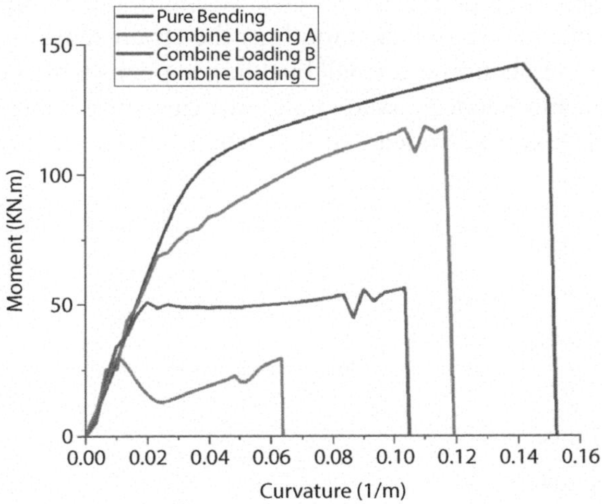


**Figure 12.7** The curvature–moment relationship.

circumferential crack propagation. The quantity of these points is shown in Table 12.1, and the relationship between strain and curvature for all types of loading is illustrated in Figure 12.8. Finally, at ultimate strength, the model entirely lost its load-carrying capacity, so tension and moment are decreased dramatically to zero. Since the pipe is cracked, the graphs do not

**Table 12.1** The parameters values for different types of loading.

| **Loading type** | **Displacement ΔL (mm)** | **Rotation angle $\theta$ (rad)** | **Strain $\varepsilon$** | **Curvature $\kappa$ (1/m)** |
|---|---|---|---|---|
| Pure tension | 7.25 | 0 | 0.12 | 0 |
| Combine loading A | 3.2 | 0.064 | 0.0054 | 0.107 |
| Combine loading B | 5.2 | 0.052 | 0.0087 | 0.087 |
| Combine loading C | 6 | 0.03 | 0.01 | 0.05 |
| Pure bending | 0 | 0.0875 | 0 | 0.146 |

go down gradually, instead, they are sharply dropped to zero. It should be noted that when the pipe reaches its yield strength, the graphs in the curvature-moment plot are not as smooth as the strain-tension plot. As the pipe is subjected to bending, a small buckling happens which is negligible.

The relationship between axial strain and curvature angle is displayed in Figure 12.8. It can be concluded that when the amounts of strain and

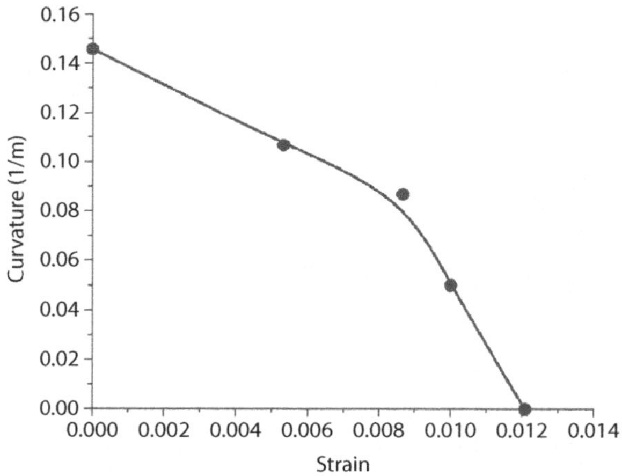


**Figure 12.8** The strain–curvature relationship of all types of loading.

curvature are under the curve, the model is in the safe zone, and fracture still has not taken place in the pipe. This curve can be used in engineering applications to verify the applicability of the cracked pipe at specific tension and bending amounts. Figure 12.9 displays the tension–moment relationship for all three types of combined loading. Both moment and tension are increased gradually at the different stages for all types of loading. It shows that the influence of tension and moment on crack failure resoundingly depends on the ratio of ε/k.

The behavior of the crack during the loading in the period of time is demonstrated in Figure 12.10. It could be observed that once the pipe model approaches failure, the primary crack blunts notably before advances needing a significant value of strain energy for the process (Figure 12.10(a)). At certain stress, the crack abruptly starts to extend around the crack tip section, which causes a stable crack growth through the thickness (Figure 12.10(b)). The values which are used in the calculations are based on the point that the crack surface lost the elements. As a matter of fact, this is the time that the pipe has not had its last efficiency anymore. Then, by increasing the load, the crack growth rate increases rapidly, as well (Figure 12.10(c)). Eventually, as long as the complete separation of the pipe is achieved, the crack grows fast in the circumferential direction (Figure 12.10(d)).

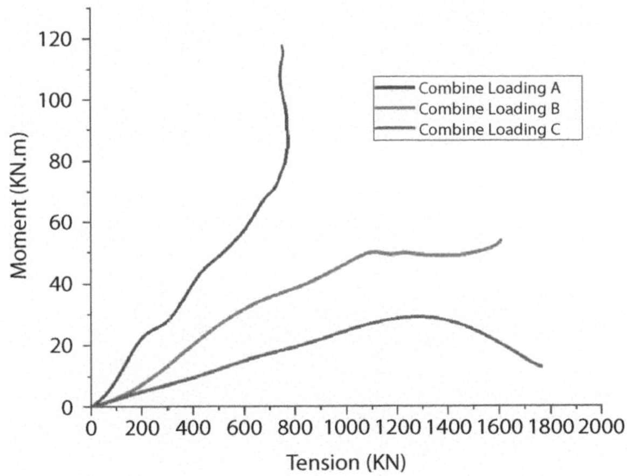


**Figure 12.9** The tension–moment relationship of different combined loading.

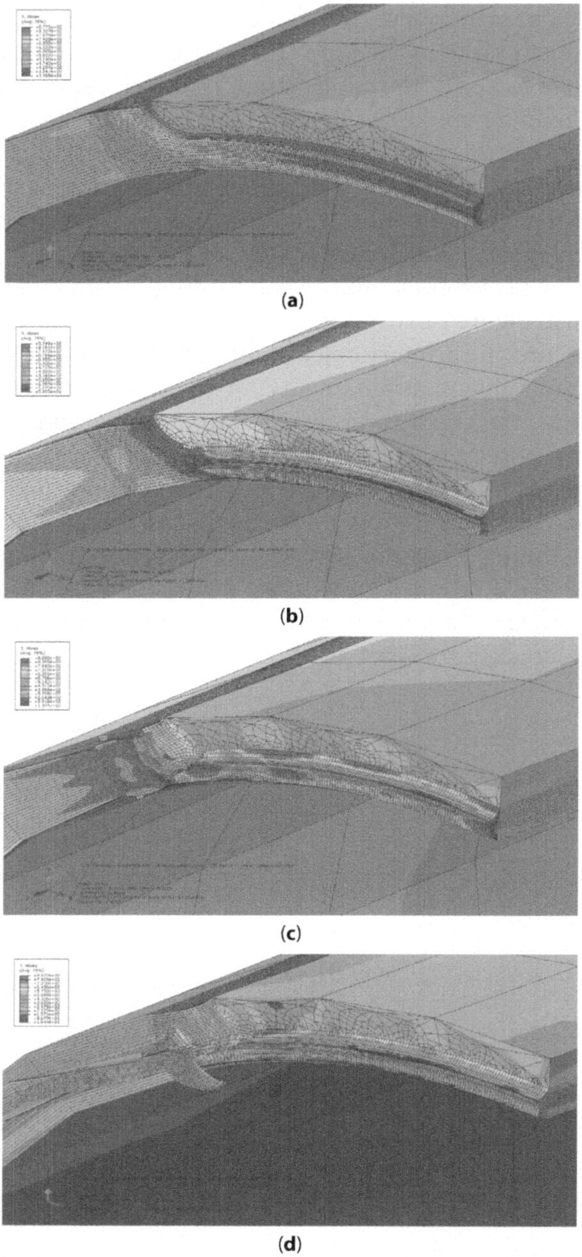

Figure 12.10 The crack behavior under simultaneous tension and bending during the time.

## 12.4 Parametric Studies

This section uses a numerical model to investigate the effects of two parameters that have the biggest bearing on how steel pipes behave mechanically when bent and under tension. In actuality, the crack depth (*a*) and length (2*c*) of the pipe are related to the fracture response. As a result, eight pipe models in total are subjected to the same loading as the previous section, with four distinct crack depths and four different crack lengths. It should be noted that everything else remains the same, including the diameter-to-thickness ratio and the characteristics of the material.

### 12.4.1 Crack Depth

Understanding the relationship between crack geometry and the strain and curvature capacity of the pipe is crucial for assessing the integrity and potential failure modes of the pipe. By analyzing the crack geometry and its impact on the strain and curvature capacity of the pipe, engineers can better evaluate the potential failure modes and overall integrity of the pipe and make informed decisions regarding maintenance and repair strategies. A three-dimensional finite element model is adopted to study the effects of crack depth (a), on the fracture response of offshore pipelines. Three various crack depths were taken to obtain the behavior of the pipe model for varying crack depths. In order to prevent the pipe model from being subjected to buckling, all of the crack depths were chosen to be more than half of the pipe thickness ($a/t = 0.55$, $a/t = 0.6$, and $a/t = 0.75$). All other characteristics of the crack and the crack length are the same for all the specimens ($2c/\pi D = 0.15$). Then, the comparisons corresponding to five different types of loading are conducted, the same as the main model.

Lastly, the relationship between strain and curvature with dissimilar crack depths is displayed in Figure 12.11. It shows that once the crack depth grows, the critical axial strain and curvature angle decrease aggressively. Therefore, crack geometry has a significant effect on the strain and curvature capacity of the pipe, deprived of ultimate strength enhancement. It should be noticed that when the deepness of the crack is going to be more than half of the pipe thickness, the strain and curvature capacity of the pipe decrease sharply, but after that the decreasing, rate becomes slow. Moreover, as the crack depth increases, the strain and curvature reduction rates are significantly reduced. If the pipe experiences applied curvature or strain within its limits, it can still maintain its operational effectiveness. Therefore, understanding the behavior of crack depth in relation to strain

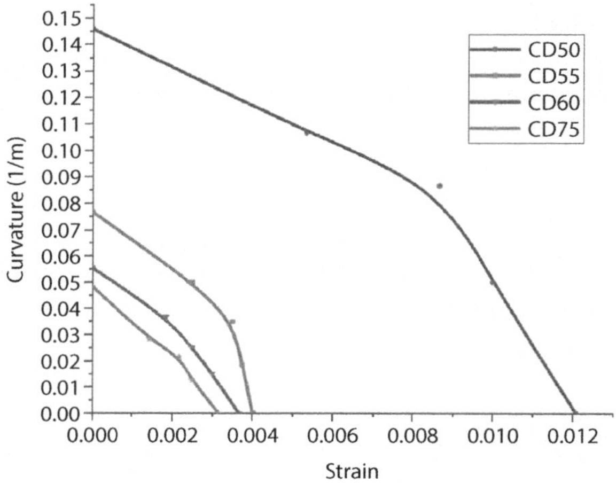


**Figure 12.11** The comparison between the different crack depths.

and curvature is crucial for determining the threshold at which the pipe's performance is compromised and appropriate. The relationship between axial strain and tension force, as well as the curvature angle, and the bending moment of the pipe for the different crack depths are demonstrated in Figure 12.12 and Figure 12.13, respectively. The general behavior of all graphs is similar to the main model. Figure 12.12 demonstrates when in the applied combined loads tension is relatively low, the crack depths effect becomes less significant, and the pipe behavior is relevant to the loading type.

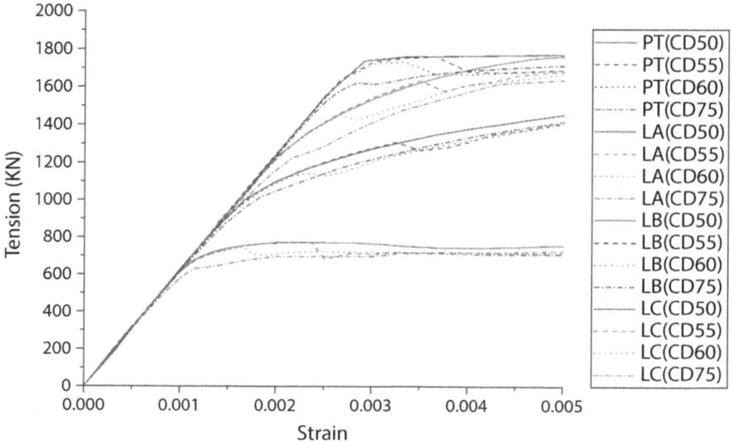


**Figure 12.12** The strain–tension relationship of different crack depths.

As can be seen, the red and blue curves are in good correspondence with each other. However, the crack depth of the pipe plays a much more significant role in its tensile capacity when it is under pure tension (black curves) or a fairly large tension in the applied combined load (green curves). This suggests that the influence of crack depth on tensile capacity is more pronounced in scenarios with pure tension or larger tension in combined loads, while its effect becomes less.

Figure 12.13 shows the curvature–moment relationship of the model is affected by crack depth in all types of loadings more significantly. Therefore, the crack depth significantly affects the moment capacity of the pipe model, particularly under combined loadings with a higher degree of bending. Even so, the crack depth effect on the moment capacity of the pipe is less when there is a low degree of bending in the combined load (green curves). This loading type has a large application of tension, so the pipe model behaves in a less smooth manner in the plastic zone due to the high applied tension. The influence of the crack depth in other curves is quite obvious. Besides, when the pipe is under pure bending (black curves), increasing the crack depth more than half of the pipe thickness does not noticeably affect the collapse capacity of the pipe. These figures show that the crack depth has much more effect on the bending strength than the tensile strength of the pipe. The buckling effect can also be observed here. The results indicate that the depth of cracks in the pipe has a significant influence on its capacity to withstand bending and tension loads.

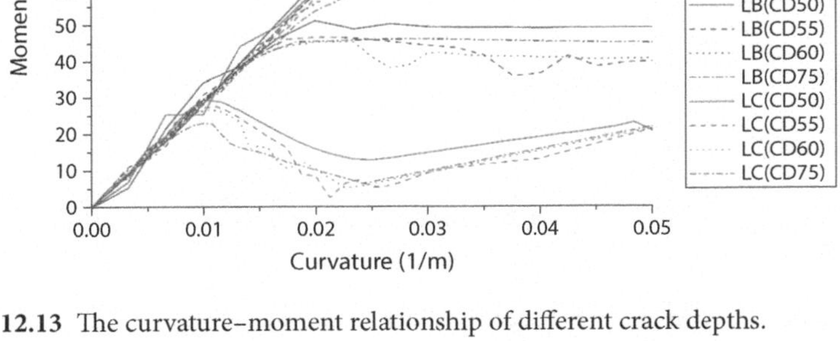


**Figure 12.13** The curvature–moment relationship of different crack depths.

### 12.4.2 Crack Length

A study on the circumferential crack length of the pipe was performed by using the 3D FE method. With the intention of determining the fracture response of the pipe for dissimilar crack lengths, three different crack lengths were adopted. Hereupon, these lengths are chosen to one be less than the main crack length ($2c/\pi D = 0.10$) and two be more ($2c/\pi D = 0.17$ and $2c/\pi D = 0.20$). The other crack features and the crack depth ($a/t = 0.5$) are equal for the models. The model is considered to be under different types of loading, identical to the previous parts. As shown in Figure 12.14, as the crack length (2c) increases from approximately 60 to 120 mm, the strain and curvature capacity are decreased significantly. It should be noted that the fracture behavior of the pipe to different crack lengths is not as expected. This unexpected fracture behavior suggests that further investigation is needed to understand the factors influencing strain and curvature capacity in relation to crack length. As can be observed, for the all-various crack lengths, the ratio of strain to curvature is similar. However, the green, blue, and black curves are closed to each other, while there is a noticeable distance from black to red curve. This leads to a crucial conclusion; if the crack length decreases more than a specific amount which here is $2c/\pi D = 0.15$, the strain and curvature capacity of the pipe increases significantly. Additionally, when the crack length becomes larger than this amount (blue and green curves), both the tension and bending capacity of the tube are not reduced noticeably. Conversely, everything changes for smaller crack

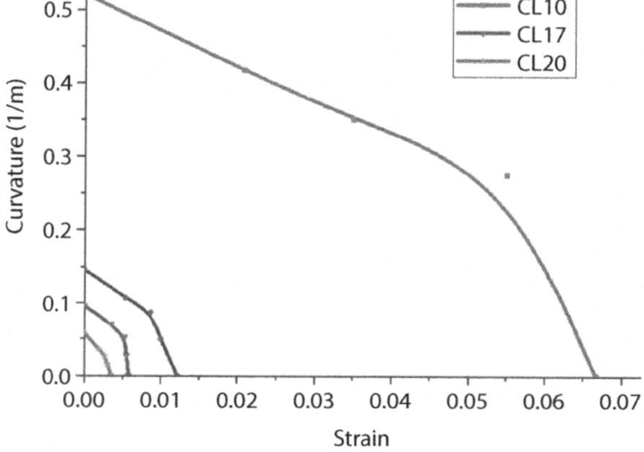


**Figure 12.14** The comparison between the different crack lengths.

lengths; as seen in the case of the red curve where pipes with this feature have a much greater collapse capacity. These findings suggest that the relationship between strain, curvature, and crack length plays a crucial role in determining the failure capacity of the pipe. For the various crack lengths, Figure 12.15 and Figure 12.16 illustrate the relationship between axial strain and tension force and the relationship between the curvature angle and bending moment of the pipe, respectively. Approximately in both plots, the relationship between strain and tension, as well as curvature and moment, is similar for all different lengths and loading types until the yield points. It could be pointed out, in both figures, that for different crack lengths, the elastic and plastic zones of the pipe model are so nearby. However, the yield points for different types of loading are so far. These results indicate that the pipe's behavior remains relatively consistent under various crack lengths and loading conditions until it reaches the yield point. However, there is a notable difference in the pipe's response beyond this stage. Figure 12.15 shows that the variety of crack lengths does not have a very large influence on the pipe when tension is low in the applied combined loading (red curve). By increasing the strain, the crack length effect becomes more noticeable (blue and green curves). However, when the pipe is under pure tension (black curve), the crack length plays a more significant role especially when it is in the smallest tested form ($2c/\pi D = 0.10$).

Figure 12.16 displays that the various crack lengths have a significant impact on the bending moment capacity of the tube. This issue is visible in all curves and becomes more pronounced when the curvature is at its

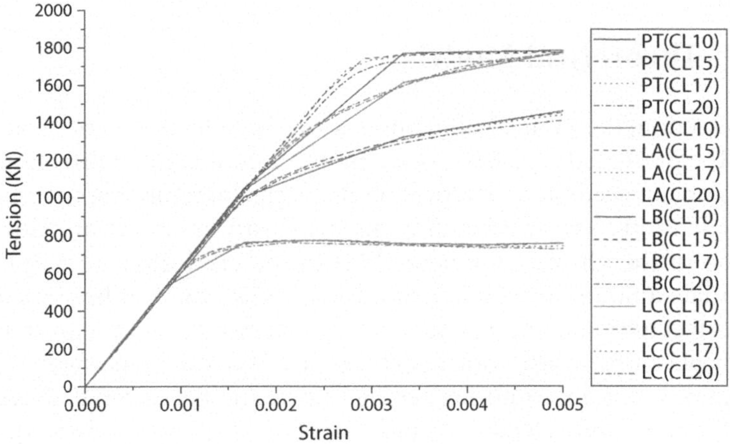


**Figure 12.15** The strain–tension relationship of different crack lengths.

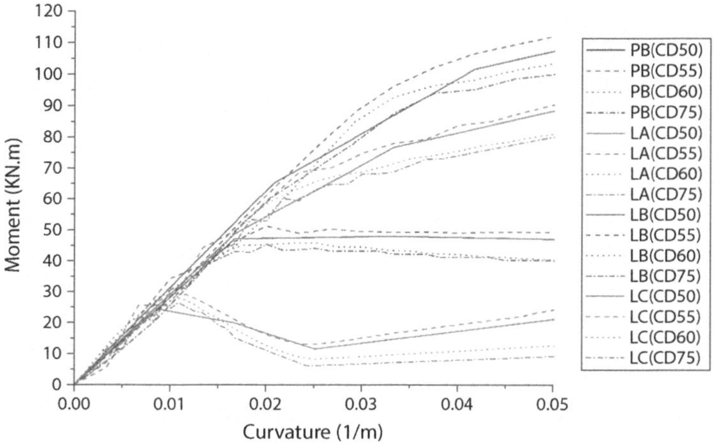


**Figure 12.16** The curvature–moment relationship of different crack lengths.

lowest applied amount (green curve). The decrease in bending moment capacity is more evident as the crack length increases, particularly when the curvature is at its minimum level. It could be concluded that more influence of the crack length on the bending strength than the tensile strength of the pipe can be observed. When considering the structural integrity of the pipe, it is crucial to prioritize addressing the impact of crack length on bending strength. It may be valuable to explore methods for minimizing the effects of crack propagation in the pipe's structure, particularly in areas where bending forces are prominent.

## 12.5 Theoretical Models

In this section, the numerical method is validated by the theoretical solutions. The presented pipe has a crack in the circumferential direction, and there is not an accurate theoretical model in the literature for determining a pipe's tension force and bending moment with imperfection. However, the ratio of the ultimate moment (Mu) to the yield moment (My) is an appropriate amount for comparison. Hence, this ratio can be similar to a perfect steel tube and our pipe model. The collapse strength is greater than the fully plastic bending moment determined from elastic-perfectly-plastic assumptions due to strain-hardening effects in the material. Based on curve fitting to finite element results, Bai *et al.* [13] proposed an equation to calculate the collapse moment for a cylinder under pure bending to determine the ultimate moment to the yield moment (Mu/My).

$$\frac{M_u}{M_y}=\left[\begin{array}{l}1.0094-0.00184\left(\frac{D_m}{t}\right)+0.0000333\left(\frac{D_m}{t}\right)^2+\left(\frac{0.15}{n}\right)\left(0.14835-0.00275\left(\frac{D_m}{t}\right)\right)\\+\left(\frac{0.15}{n}\right)^2\left(0.03913-0.001742\left(\frac{D_m}{t}\right)+0.0000258\left(\frac{D_m}{t}\right)^2\right)\end{array}\right]*$$
$$\left[0.971+15.5\left(\frac{\sigma_y}{E}\right)\right]*\left[0.861+0.139S\right] \tag{12.3}$$

where $S$ is the yield anisotropy parameter which is the ratio of the yield stress in circumferential direction and longitudinal direction. Accordingly, the ratio of the ultimate curvature ($\kappa_u$) to the yield curvature ($\kappa_y$) under a single load is calculated as

$$\frac{\kappa_u}{\kappa_y}=1.255\left[0.5+0.586\left(\frac{0.15}{n}\right)-0.0939\left(\frac{0.15}{n}\right)^2\right]\left[1.19S-0.185\right] \tag{12.4}$$

Besides for considering pipe under pure tension, by applying the flow stress equation [32], the ratio of the ultimate tensile strength ($N_u$) to the yield tensile strength ($N_y$) of steel pipes is given below:

$$\frac{N_u}{N_y}=\frac{1}{2}\left[1+\frac{(0.2n)^{-\frac{0.01}{n}}}{\exp\left(\frac{0.01}{n}\right)}\right] \tag{12.5}$$

By using Equations (12.3), (12.4), and (12.5), the ratios of Mu/My, κu/κy, and Nu/Ny are calculated, respectively. Due to the fact that the ratio chosen for comparison is for a pipe without imperfections, it is expected they will be in good agreement with the ratios obtained from the finite element model. However, determining the tension forces and bending moments from the numerical method should be accurate enough. Therefore, a MATLAB code has been written to extract the most possible accurate amounts for yield and ultimate moments. First, the curves that are obtained from ABAQUS results have to be smoothed, so more values between the current amounts are determined. Then, the first derivative, second derivative, and third derivative of each curve are determined, and their graphs are compared with each other. It has been seen when the sign of the curve changes from positive to negative in the third derivative graph; the curve at that moment has a reasonable relationship with the first and second derivatives as well. Hence, these points are chosen as the model yield tension and moment and ultimate tension and moment. The extracted ratios from numerical analysis are compared with the theoretical results, as shown in Table 12.2. The results obtained from the finite element model closely align with the theoretical expectations. Besides, for a better understanding, the relationship between the $N_u/N_y$, $M_u/M_y$, and $\kappa_u/\kappa_y$ are illustrated in Figure 12.17.

Based on Figure 12.17, we can see that the values extracted from the numerical results for different crack lengths and depths are well correlated with each other. Therefore, their variation is reasonable from theoretical results. Additionally, consistent numerical outcomes are observed across different crack depths and lengths. Pipe fractures are larger when it is under pure tension as opposed to when the pipe is subjected to pure bending, as observed. The comparison between finite element and theoretical results demonstrates the reliability and accuracy of the proposed numerical analysis approach in predicting and analyzing the behavior of X80 steel pipe with a circumferential crack. Moreover, the analysis also considers the effect of different loading conditions on crack behavior, providing a

**Table 12.2** The comparison between models under pure bending and pure tension.

| **Method** | **Model** | $N_u/N_y$ | ***Mu/My*** | ***κu/κy*** |
|---|---|---|---|---|
| *Theoretical results* | | 1.183 | 1.210 | 1.698 |
| *Numerical analysis results* | CD50/CL10 | 1.084 | 1.190 | 1.509 |
| | CD50/CL15 | 1.072 | 1.178 | 1.476 |
| | CD50/CL17 | 1.055 | 1.164 | 1.348 |
| | CD55/CL15 | 1.049 | 1.152 | 1.228 |
| | CD50/CL20 | 1.041 | 1.118 | 1.184 |
| | CD60/CL15 | 1.033 | 1.091 | 1.124 |
| | CD75/CL15 | 1.030 | 1.072 | 1.113 |

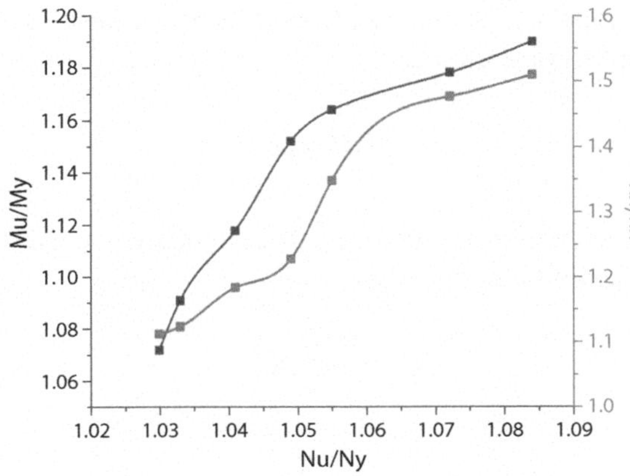


**Figure 12.17** The relationship between various crack sizes.

comprehensive understanding of fracture mechanics in X80 steel pipes with circumferential cracks.

Since the pipe model is subjected to simultaneous tension and bending in this chapter, there are some theories which consider the pipe under combined tension and bending. Therefore, it is important to analyze the effects of varying loads on different types of pipes. Firstly, to consider a perfect pipe without a crack under combined tension and bending, the below equation based on the ultimate capacity problem is suggested by Bai *et al.* [33]:

$$\frac{M_c}{M_u} = 1 - \left(\frac{N_c}{N_u}\right)^{2.4} \tag{12.6}$$

where $N_c$ and $M_c$ are collapse tension and collapse bending, respectively. It should be noted that $N_u$ and $M_u$ indicate ultimate tension and bending under single load. Besides, this equation is valid under below circumstances:

- Diameter to thickness ratio: $10<D/t<40$
- Ratio between yield parameter and Young's modulus: $0.001<\sigma_y/E<0.003$
- Strain-hardening parameter: $5<n<25$
- Yield anisotropy parameter: $0.8<S<1.2$

The studied model has all the mentioned features. Besides, the yield tensile strength can be computed as

$$N_y = 2\pi\sigma_y R_m t \tag{12.7}$$

In addition, the yield moment and subsequently the yield curvature of a perfect steel tube can be deduced as

$$M_y = 4\sigma_y R_m^2 t \tag{12.8}$$

$$\kappa_y = \frac{t}{4R_m^2} \tag{12.9}$$

However, the proposed equations do not consider the effect of the crack on the fracture response of the pipe. Currently, most plastic load limits for pipes are derived either empirically or analytically using a simple yield criterion. These approaches underestimate the actual limit load with unknown conservatism [34]. In order to consider the effect of the crack on the loading capacity of a pipe to those for the corresponding limit loads of a through-wall cracked pipe under pure bending and pure tension, respectively, the below equation has been presented by Foxen and Rahman [16]:

$$N_0 = 2\pi\sigma_o R_m t\left(1 - \frac{\theta}{\pi} - \frac{2}{\pi}\sin^{-1}\left(\frac{1}{2}\sin\theta\right)\right) \tag{12.10}$$

$$M_0 = 4\sigma_o R_m^2 t\left(\cos\frac{\theta}{2} - \frac{1}{2}\sin\theta\right) \tag{12.11}$$

$$\theta = \frac{c}{R_m} \tag{12.12}$$

where $\sigma_0$ is the collapse of the flow stress of the material. In Equations (12.10) and (12.11), the decreasing coefficients are added to Equations (12.7) and (12.8) to consider the crack effect of the bending moment and tension force of the pipe. The below equation is proposed by Foxen and Rahman [16] to those for the corresponding limit loads of a through walled-crack pipe under combined tension and bending:

$$N_0' = \frac{1}{2}\left( -\frac{\lambda N_0^2 R_m}{M_0} + \sqrt{\left(\frac{\lambda N_0^2 R_m}{M_0}\right)^2 + 4N_0^2} \right) \tag{12.13}$$

where $\lambda$ is a dimensionless load factor and can be utilized to describe a proportionality relationship between tension load ($N$) and bending moment ($M$) according to Kumar and Germen [11] as below:

$$\lambda = \frac{M}{NR_m} \tag{12.14}$$

By solving Equations (12.13) and (12.14), the relationship between the ratio of the bending to the tension and the combined load of tension and bending is determined and illustrated in Figure 12.18. The various crack lengths are compared while the crack depth for all models is the same, and it is equal to the pipe thickness. There is a complex relationship between the equivalent combined tension and bending load and the bending to tension ratio, indicating potential non-linear behavior of the material or structure

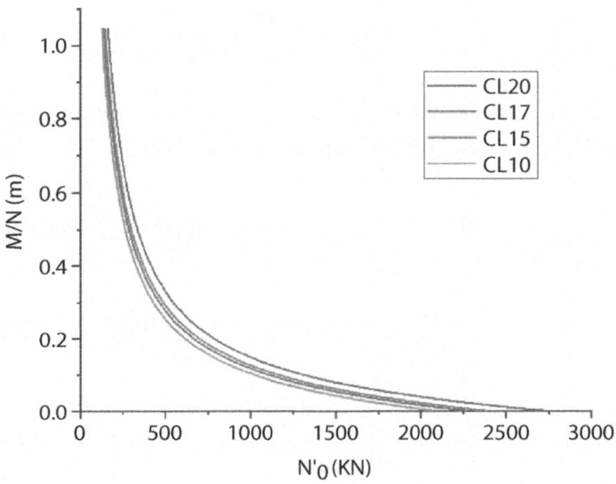


**Figure 12.18** The relationship between the ratio of bending to tension and the combined load for various crack lengths.

under different load conditions. At lower combined loads, the bending to tension ratio decreases more rapidly compared to higher equivalent loads. Towards the higher combined load values, the bending to tension ratio seems to stabilize or change more gradually.

This could suggest that the material or structure reaches a certain point where further increases in the ratio of bending to tension have less impact on the final combined load amount. This underscores the importance of taking into account both tension and bending when assessing the load capacity of a pipe with a through-walled crack. However, Equations (12.7) and (12.8) did not consider any crack in the pipe, and Equations (12.9) and (12.10) take into account the crack depth equal to the pipe thickness. While the current study assumes the crack depth is less than the pipe wall thickness. Plastic limit loads must be accurately determined if engineering assessment of cracked pipes is to be successful. Therefore, in this thesis, the analytical solution for calculating the plastic load limit for pipes subjected to both axial tension and bending at the same time is employed. Based on stress field equilibrium and finite element limit analyses, Kim *et al.* [35] modified Equations (12.10) and (12.11) to study the crack depth effect as follows:

$$N'_L = 2\pi\sigma_y R_m t\left(1 - \frac{\theta a}{4t} - \frac{2}{\pi}\sin^{-1}\left(\frac{a}{t}\frac{f(\theta)}{2\theta}\right)\right) \tag{12.15}$$

$$M'_L = 4\sigma_y R_m^2 t\left(\cos\left(\frac{\pi\theta a}{8t}\right) - \frac{a}{t}\frac{f(\theta)}{2\theta}\right) \tag{12.16}$$

where $N'_L$ and $M'_L$ are plastic limit load for axial tension and bending moment under Tresca condition, respectively, and

$$f(\theta) = 0.7854\theta^2 - 0.0982\theta^4 + 0.0041\theta^6 - 0.000085\theta^8 \tag{12.17}$$

Besides, for the pipe under combined tension and bending with a crack, we have

$$\frac{N'_L}{N_L}\frac{N}{2\pi\sigma_y R_m t} = 1 - \frac{\theta a}{4t} - \frac{2}{\pi}\sin^{-1}\left(\frac{M'_L}{M_L}\frac{M}{4\sigma_y R_m^2 t} + \frac{a}{t}\frac{f(\theta)}{2\theta}\right) \tag{12.18}$$

where $N_L$ and $M_L$ are the plastic limit load for axial tension and bending moment under Mises condition, respectively, and they can be calculated as

$$N_L = 2\pi\sigma_y R_m t\left(1 + A_1\left(\frac{a}{t}\right) + A_2\left(\frac{a}{t}\right)^2\right) \tag{12.19}$$

$$M_L = 4\sigma_y R_i^2 t\left(1 + B_1\left(\frac{a}{t}\right) + B_2\left(\frac{a}{t}\right)^2\right) \tag{12.20}$$

where $R_i$ is the inner radius of the pipe and

$$A_1 = 0.066 - 0.038\left(\frac{\theta}{\pi}\right) - 0.960\left(\frac{\theta}{\pi}\right)^2$$

$$A_2 = -0.060 - 1.525\left(\frac{\theta}{\pi}\right) + 1.427\left(\frac{\theta}{\pi}\right)^2$$

$$B_1 = 0.074 - 0.169\left(\frac{\theta}{\pi}\right)$$

$$B_2 = -0.086 - 1.013\left(\frac{\theta}{\pi}\right)$$

The theoretical equations were solved, and pertinent data was extracted using PYTHON. These data have to be compared with numerical results to validate the finite element model. As previously stated, by using a MATLAB code, the precise amount of yield tension and yield bending is determined from the finite element model. These calculated values for pipe under pure tension are displayed in Table 12.3 and contrasted with the outcomes of Equations (12.15) and (12.19). Besides, Table 12.4 demonstrates the comparison between the numerical outcomes compared with the results of Equations (2.16) and (12.20) when the pipe is subjected to pure bending. Besides, only for a better understanding of the effect of a crack on the pipe behavior, the results of Equations (12.7) and (12.10) and Equations (12.8) and (12.11) are added into Table 12.3 and Table 12.4, respectively. It is evident that the results are generally consistent with one another. The calculated results from ABAQUS more closely match the theory that takes into

**Table 12.3** The tension load calculated by FE and two theoretical approaches.

| Model | $N_y$ (KN) Equation (5-10) | $N_y$ (KN) Equation (5-15) | $N_y$ (KN) (FE) | $N_y$ (KN) Equation (5-19) | $N_y$ (KN) Equation (5-6) |
|---|---|---|---|---|---|
| CD50/CL10 | 2726.098 | 3136.685 | 3145.919 | 3148.308 | 3402.344 |
| CD50/CL15 | 2381.641 | 3006.764 | 3056.668 | 3078.837 | 3402.344 |
| CD50/CL17 | 2266.521 | 2955.700 | 3003.022 | 3050.492 | 3402.344 |
| CD55/CL15 | 2381.641 | 2967.145 | 2998.189 | 3041.644 | 3402.344 |
| CD50/CL20 | 2075.765 | 2880.292 | 2994.730 | 3007.379 | 3402.344 |
| CD60/CL15 | 2381.641 | 2927.508 | 2931.006 | 3000.316 | 3402.344 |
| CD75/CL15 | 2381.641 | 2808.475 | 2832.353 | 2851.513 | 3402.344 |

**Table 12.4** The bending moment calculated by FE and two theoretical approaches.

| Model | $M_y$ (KN m) Equation (5-11) | $M_y$ (KN m) Equation (5-20) | $M_y$ (KN m) (FE) | $M_y$ (KN m) Equation (5-16) | $M_y$ (KN m) Equation (5-7) |
|---|---|---|---|---|---|
| CD50/CL10 | 171.443 | 181.304 | 185.256 | 192.841 | 205.770 |
| CD50/CL15 | 153.376 | 178.186 | 182.421 | 186.374 | 205.770 |
| CD50/CL17 | 146.104 | 176.938 | 180.813 | 183.822 | 205.770 |
| CD55/CL15 | 153.376 | 176.328 | 179.495 | 184.337 | 205.770 |
| CD50/CL20 | 135.225 | 175.067 | 178.253 | 180.052 | 205.770 |
| CD60/CL15 | 153.376 | 174.250 | 176.502 | 182.284 | 205.770 |
| CD75/CL15 | 153.376 | 166.699 | 168.069 | 176.017 | 205.770 |

account the plastic limit load for axial tension and bending moment under von Mises stress.

Thus, the theoretical results are more conservative; however, the results show higher values for the tension and bending when the moment and load are obtained by considering the pipe under Tresca stress. It should be mentioned that the applied theory is more conservative when compared to the Mises condition, but the FE model results are more conservative when we take the Tresca condition into account. Figure 12.19 illustrates a relationship between the bending and tension of all the models. The numerical outcomes and the theories agree fairly well. Furthermore, this figure clearly illustrates the impact of both crack length and depth. It is evident that the fracture depth has a greater impact on the pipe's fracture response than the crack length. Cracks in pipes under tension have a bigger effect than cracks in pipes under bending, as numerous demonstrations have demonstrated.

According to the proposed theories, the pipe's linear elastic mechanics can be analyzed, and the numerical model can be validated from the theories under elastic conditions. However, the fracture response of a steel pipe in the plastic region is complex in the presence of a circumferential crack. The behavior of cracked pipes under combined loading can be assessed based on the following theories. Due to the circumferential crack in the pipe, the theoretical equations cannot accurately predict the behavior of the pipe in the plastic region, since that is the purpose of this study. Typically, the proposed theory can accurately predict the relationship between tension and bending when there is a linear relationship between them, but when there is a nonlinear relationship, the numerical model is preferred. Due to this, calculating the fracture response of the pipe in the plastic region is not possible through theories, so the numerical model validation is just possible in the elastic resolution of the pipe material. When a pipe is subjected to a combined loading, the following equations aid in calculating

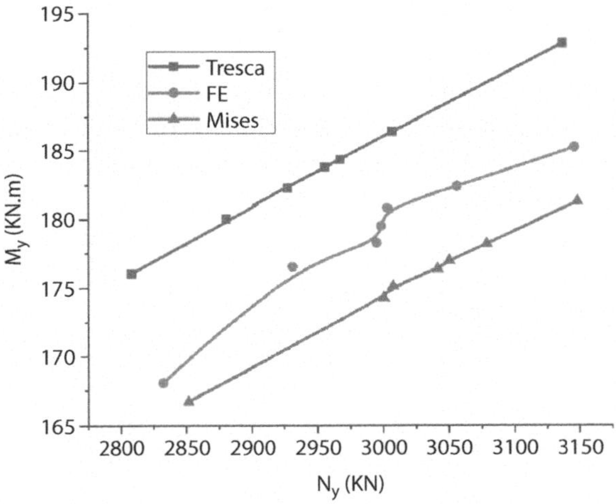


**Figure 12.19** The comparison between the theoretical and numerical methods.

the yield tension and bending. The plastic limit bending moment can be expressed the following way when the axial tension is constant:

$$M_L^N = 4\sigma_y R_m^2 t \frac{M_L}{M_L'} \left\{ \sin\left[ \frac{\pi}{2} \left( 1 - \frac{a}{t} \frac{\theta}{4} - \frac{N}{2\pi\sigma_y R_m t} \frac{N_L'}{N_L} \right) \right] - \frac{a}{t} \frac{f(\theta)}{2\theta} \right\} \tag{12.21}$$

Let the ratio of axial tension to bending moment be represented as follows when the axial tension increases linearly with the bending moment:

$$\xi = \frac{\sigma_a}{\sigma_b} \tag{12.22}$$

where $\sigma_a$ and $\sigma_b$ are axial stresses induced by axial tension and bending moment, respectively. In addition, the uniform axial stress and maximum bending stress can be defined respectively as below:

$$\sigma_a = \frac{N}{\pi\left(R_o^2 - R_i^2\right)} \tag{12.23}$$

$$\sigma_b = \frac{4MR_o}{\pi\left(R_o^4 - R_i^4\right)} \tag{12.24}$$

where $R_o$ is the outer radius of the pipe. Equations (12.23) and (12.24) can be substituted in Equation (12.22) and the plastic limit bending moment can be determined, and subsequently, by substituting Equation (12.22) into Equation (12.18), we have:

$$M_L^N(\xi) = 4\sigma_y R_m^2 t \frac{M_L}{M_L'} \left\{ \sin\left[ \frac{\pi}{2} \left( 1 - \frac{a}{t} \frac{\theta}{4} - \frac{4M_L^N R_o \xi}{2\pi\sigma_y R_m t\left(R_o^2 + R_i^2\right)} \frac{N_L'}{N_L} \right) \right] - \frac{a}{t} \frac{f(\theta)}{2\theta} \right\} \tag{12.25}$$

Equation (12.25) cannot be solved analytically, and it has to be resolved by using iterative methods like Newton-Raphson method [36]. In addition, the corresponding plastic limit tension is defined as

$$N_L^M(\xi) = \frac{4M_L^N R_o \xi}{R_o^2 + R_i^2} \tag{12.26}$$

Equation (12.18) has been solved with PYTHON codes; an indication of the results can be found in Figure 12.20. The related amounts for tension and bending in the numerical model are extracted for comparison with the help of MATLAB codes using the same method as previously mentioned, while it should be noted that determining the exact value of the tension and bending from the finite element model when the pipe is under combined loading is a challenging task, and the results could be not too accurate. However, the amounts which had the best agreement with the theoretical equations are chosen and are drawn in Figure 12.20. In fact, there is no certain reason for the small observed inconsistency of the FE results with theory results. The input values match those of the main numerical model, so all results match each other well. This demonstrates the consistency and validity of the applied models. When the combined loading is under the illustrated curve, then the pipe can continue its service. Lastly, the proposed theories extend the scope of fracture behavior analysis to the pipe with the most possible comprehensive approach.

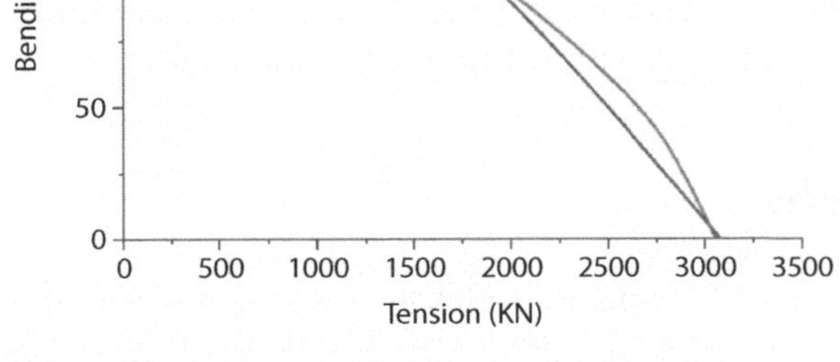


**Figure 12.20** The comparison between the theory and FE results.

## 12.6 Conclusions

The pipe model is placed under simultaneous tension and bending. The variations of curvature with strain under different loading types have been carried out. Since the used pipe is cracked, there is a point between yield strength and ultimate strength, resulting in circumferential crack propagation. These points were derived to obtain the loss efficiency time of the pipe. The parameter values for different types of tension and bending were extracted. The parametric studies have been established to provide a better understanding of the crack depth and crack length influences on the pipe model. Once the crack depth is less than half of the pipe thickness, buckling is more significant than fracture. Nevertheless, as the crack depth rises, more than half of the pipe thickness, the strain, and curvature reduction rate are meaningfully increased, but after that, the decreasing rate is leisurely. For a specific amount of crack length which here is ($2c/\pi D = 0.15$), the strain has a significantly larger influence on the pipe failure capacity than curvature. Additionally, when the crack length becomes lower than this amount, both the tension and bending capacity of the tube decrease sharply. Conversely, everything changes for smaller crack lengths, and the pipe has a much greater collapse capacity. The results showed that both crack depth and crack length have more effect on the ultimate bend strength of the steel pipe rather than the ultimate tensile strength. Finally, it was seen that the ultimate tensile strength is more impressible from the crack depth than the crack length. Besides, related theoretical equations are presented to validate the numerical model. First, the ratio of Nu/Ny, Mu/My, and κu/κy are chosen for comparison. The results showed the theory obtained larger amounts of this ratio, since they have not considered any imperfection in the tube, and by increasing the crack size, the final ratio becomes lower. Four equations for calculating the Ny and My were derived. One considered the pipe without a crack; the other one assumed the crack depth same as the wall thickness of the pipe. The third and fourth equations consider both crack depth and crack length, but one under the Tresca condition and the other one under the Mises condition. The results were in good agreement with the finite element results.

## Bibliography


1. Kashani, M. and Young, R., Installation load consideration in ultra-deepwater pipeline sizing. *J. Transp. Eng.*, 131, 8, 632–639, 2005, doi: 10.1061/(ASCE)0733-947X(2005)131:8(632.

2. Mousselli, A., *O shore Pipeline Design, Analysis, and Methods*, Pennwell Books, Tulsa, Oklahoma, 1981.
3. British Energy, *Assessment of the integrity of structures containing defects*, revision 4 Edition, British Energy Generation Ltd., London, UK, 2001.
4. British Standards Institution, *BS 7910, Guide on methods for assessing the acceptability of flaws in metallic structures*, British Standards Institution, London, UK, 2000.
5. Ostby, E., Fracture control - O shore pipelines: New strain-based fracture mechanics equations including the effects of biaxial loading, mismatch and misalignment. *Proceedings of the International Conference on O shore Mechanics and Arctic Engineering - OMAE 3 (Omae)*, pp. 649–658, 2005, doi:10.1115/OMAE2005-67518.
6. Tang, H., Fairchild, D., Panico, M., Crapps, J., Cheng, W., Strain capacity prediction of strain-based pipelines, in: *Proceedings of the 2014 10th International Pipeline Conference*, Calgary, Alberta, Canada, pp. 1–12, 2014.
7. Hutchinson, J.W., Fundamentals of the phenomenological theory of nonlinear fracture mechanics. *J. Appl. Mech. Trans. ASME*, 50, 4, 1042–1051, 1983, doi: 10.1115/1.3167187.
8. Hutchinson, J.W., Singular behaviour at the end of a tensile crack in a hardening material. *J. Mech. Phys. Solids*, 16, 13–31, 1968.
9. Paris, P., Tada, H., Zahoor, A., Ernst, H., The Theory of Instability of the Tearing Mode of Elastic-Plastic Crack Growth. *Elastic-Plast. Fract.*, 5, 5–32, 2009, doi: 10.1520/stp35825s.
10. O'Dowd, N.P. and Shih, C.F., Family of crack-tip fields characterized by a triaxiality parameter-I. Structure of fields. *J. Mech. Phys. Solids*, 39, 8, 989–1015, 1991, doi: 10.1016/0022-5096(91) 90049-T.
11. Kumar, V. and German, M., Elastic-Plastic Fracture Analysis of Through-Wall and Surface Flaws in, General Electric Co., Schenectady, NY (USA). Corporate Research and Development Center, 1988.
12. Ahamed, J., Papaspyropoulos, V., Brust, F.W., Wilkowski, G.M., A predictive J-estimation method for circumferentially surface cracked pipes of power-low hardening materials in pure bending, in: *American Society of Mechanical Engineers, Pressure Vessels and Piping Division (Publication) PVP*, Honolulu, Hawaii, vol. 167, pp. 71–85, 1989.
13. Bai, Y., Igland, R., Moan, T., Collapse of thick tubes under combined tension and bending. *J. Constr. Steel Res.*, 32, 3, 233–257, 1995, doi: 10.1016/0143-974X (95)93897-D.
14. Rahman, S. and Brust, F.W., Approximate methods for predicting J-integral of a circumferentially surface-cracked pipe subject to bending. *Int. J. Fract.*, 85, 2, 111–130, 1997, doi: 10.1023/A: 1007322018722.
15. Anderson, T., *Fracture Mechanics - fundamentals and applications*, 2nd Edition, CRC Press, London, 1995.
16. Foxen, J. and Rahman, S., Elastic-plastic analysis of small cracks in tubes under internal pressure and bending. *Nucl. Eng. Des.*, 197, 1, 75–87, 2000, doi: 10.1016/S0029-5493(99)00258-7.

17. Shimanuki, H. and Inoue, T., Assessment of brittle fracture in girth weld joint of pipelines subjected to internal pressure and bending load, in: *International Conference on the application and evaluation of highgrade linepipes in hostile environments*, Yokohama, Japan, 2002.
18. Jayadevan, K.R., Ostby, E., Thaulow, C., Strain based fracture mechanics analysis of pipelines. *Proceedings of the International Conference on Advances in Structural Integrity*, Bangalore, India, pp. 14–17, 2004.
19. Jayadevan, K.R., Ostby, E., Thaulow, C., Fracture response of pipelines subjected to large plastic deformation under tension. *Int. J. Press. Vessels Pip.*, 81, 9, 771–783, 2004, doi: 10.1016/j. ijpvp.2004.04.005.
20. Ostby, E., Jayadevan, K.R., Thaulow, C., Fracture response of pipelines subject to large plastic deformation under bending. *Int. J. Press. Vessels Pip.*, 82, 3, 201–215, 2005, doi: 10.1016/j. ijpvp.2004.08.012.
21. Rozumek, D. and Marciniak, Z., Fatigue properties of notched specimens made of fep04 steel. *Mater. Sci.*, 47, 4, 462–469, 2012, http://dx.doi.org/10.1007/s11003-012-9417-x.
22. Oh, C.S., Kim, N.H., Kim, Y.J., Baek, J.H., Kim, Y.P., Kim, W.S., A finite element ductile failure simulation method using stress-modified fracture strain model. *Eng. Fract. Mech.*, 78, 1, 124–137, 2011, doi: 10.1016/j.engfracmech.2010.10.004. URL http://dx.doi.org/10.1016/j.engfracmech.2010.10.004.
23. Gao, X., Faleskog, J., Shih, C.F., Dodds, R.H., Ductile tearing in part through cracks: experiments and cell-model predictions. *Eng. Fract. Mech.*, 59, 6, 761–777, 1998, doi: 10.1016/S0013-7944(97) 00174-4.
24. Bai, Y. and Wierzbicki, T., Application of extended Mohr-Coulomb criterion to ductile fracture. *Int. J. Fract.*, 161, 1, 1–20, 2010, doi: 10.1007/s10704-009-9422-8.
25. Wylllie, D., *Foundations on rock*, second ed Edition, E & FN Spon, London, 1999.
26. Pack, K., Luo, M., Wierzbicki, T., Sandia fracture challenge: Blind prediction and full calibration to enhance fracture predictability. *Int. J. Fract.*, 186, 1–2, 155–175, 2014, doi: 10.1007/s10704-013-9923-3.
27. Pack, K. and Marcadet, S.J., Numerical failure analysis of three-point bending on martensitic hat assembly using advanced plasticity and fracture models for complex loading. *Int. J. Solids Struct.*, 85-86, 144–159, 2016, doi: 10.1016/j.ijsolstr.2016.02.014. URL http://dx.doi.org/10.1016/j.ijsolstr.2016.02.014.
28. Rice, J.R. and Tracey, D.M., On the ductile enlargement of voids in triaxial stress fields. *J. Mech. Phys. Solids*, 17, 3, 201–217, 1969, doi: 10.1016/0022-5096(69)90033-7.
29. Tvergaard, V. and Needleman, A., Analysis of the cup-cone fracture in a round tensile bar. *Acta Metall.*, 32, 1, 157–169, 1984, http://dx.doi.org/10.1016/0001-6160(84)90213-X.

30. Bai, Y. and Wierzbicki, T., A new model of metal plasticity and fracture with pressure and Lode dependence. *Int. J. Plast.*, 24, 6, 1071–1096, 2008, doi: 10.1016/j.ijplas.2007.09.004.
31. Hibbitt, H.D., Karlsson, B.I., Sorensen, P., ABAQUS Theory Manual, 2006.
32. Anderson, T.L., *Fracture Mechanics - Fundamentals and Applications*, CRC Press, London, 1995.
33. Bai, Y., Igland, R.T., Moan, T., Tube Collapse Under Combined External Pressure, Tension and Bending. *Mar. Struct.*, 32, 3, 389–410, 1997.
34. Kim, Y.J., Shim, D.J., Huh, N.S., Kim, Y.J., Plastic limit pressures for cracked pipes using finite element limit analyses. *Int. J. Press. Vessels Pip.*, 79, 321–330, 2002.
35. Kim, Y.J., Shim, D.J., Huh, N.S., Kim, Y.J., Finite element based plastic limit loads for cylinders with part-through surface cracks under combined loading. *Int. J. Press. Vessels Pip.*, 80, 527–40, 2003.
36. Li, C.-Q., Fu, G., Yang, W., Yang, S., Derivation of elastic fracture toughness for ductile metal pipes with circumferential external cracks under combined tension and bending. *Eng. Fract. Mech.*, 178, 39–49, 2017.

# 13

# Mechanical Behavior of Pipes with Crack under Combined Tension and Internal Pressure

***Abstract***
This chapter investigates the fracture behavior of X80 steel offshore pipelines subjected to combined tension and torsion, emphasizing scenarios common during deepwater installation and operation. The study begins with a review of fracture mechanics models and highlights the limitations of traditional criteria like J-integral and CTOD under large plastic strains. A finite element model using ABAQUS/Explicit is developed to simulate ductile crack growth under various load ratios, validated by tension-torsion interaction equations. The pipe, modeled with a surface elliptical crack, is subjected to five different loading cases to derive strain-twist and tension-torsion relationships. Results show a significant influence of crack depth and length on strain capacity and twist angle, with clear threshold values beyond which structural integrity is compromised. Deeper and longer cracks lead to earlier failure, with crack depth being more critical in the plastic regime. Parametric studies reveal critical values for $a/t = 0.5$ and $2c/\pi D = 0.15$, informing design safety limits. The chapter concludes that accurate assessment of fracture response under multi-axial loading is essential and recommends the modified Mohr-Coulomb (MMC) criterion for improved ductile failure prediction in complex pipeline applications.

***Keywords***: Steel pipelines, fracture mechanics, tension-torsion, crack depth, crack length, finite element analysis, MMC criterion

## 13.1 Introduction

The main transport carrier in the oil and gas industries is applied by offshore pipelines. The increasing energy demand has made companies explore oil and gas in the ultra-deep region, which causes severe loading on the pipe structure. Steel pipes play a pivotal role in the offshore industry due to their lower initial investment needs and easier installation.

Yong Bai. Plasticity and Ductile Fracture in Structural Engineering, (373–398) © 2026 Scrivener Publishing LLC

In addition, their compatibility with severe environmental conditions, particularly in deep waters, for conveying the crude oil and natural gas from subsea to vessel or coast, is another advantage point of them. The pipes are under external pressure and combined external and internal pressure under the installation and application process, respectively. In moderately shallow waters, only the internal pressure is important for the bursting failure of the pipe laid [1]. Nowadays, the depth of exploration exceeded more than 3000 m [2]; by these increasing water depths, pipes suffer external hydrostatic pressure, and forasmuch as during installation, the pipe is empty and not subject to any internal pressure, and the collapse can happen [3]. Since offshore pipelines are being used in long-distance transmits, the pipes can face defects, especially in material and welding. Besides, during lifetime service, pipes could be damaged or cracked due to corrosion. Hereupon, studying the cracks in pipelines is an important subject. The types and amounts of the cracks are based on the manufacturing process [4]. By considering the plastic strain, when the internal pressure and external pressure are accompanied by tension, the pipelines are vulnerable to fracture, which may cause the collapse of the pipes and lead to a huge economic loss. During the process of normal reeling operation, the plastic strain in the pipes can be up to 4 percent [5]. ASME B31G [6] is the most useful standard to obtain the bursting failure pressure of corroded or intact pipes. The combined loading like internal pressure and longitudinal axial compression and safety of internally pressurized corroded and intact pipelines were assessed by the methods of DNV-RP-F101 [7]. In order to assess the fracture resistance by calculating the rate of strain, BS 7910 [8] and R6 [9] standards are used. However, these standards are established for elastic response, and they are limited to small plastic strains. Therefore, a precise fracture evaluation method that is compatible with the deformation capacity of offshore pipelines under external and internal pressure is required. The frequently used fracture parameters are CTOD (crack-tip opening displacement) and J-integral (energy release rate). Hutchinson's approach [10] describes the severity of crack growth at initiate and stable states and also assesses the unstable state under load histories. Nonetheless, the J is not applicable anymore if the crack starts to expand [11]. Hence, the J-integral R-curve approach for unstable cracks was presented by Paris *et al.* [12]. In addition, the CTOD and J parameters are widely known for their validation in the crack-tip fields characterized in certain cases [13]. Based on the classical beam theory and deformation theory of plasticity, Rahman and Burst [14] suggested a new method to calculate the J-integral for cracked pipes. The plane defect of pipe is assumed with limited length and depth persistence; thus, they are applicable for both elastic and plastic conditions. This method approach is load base, and this is its benefit since loads have more effect on the fracture parameters than strains [15]. The fracture response is increased when the internal pressure in the pipes induces hoop stresses; this was an issue that no one had considered until Foxen and Rahman [16]. They also proved the hoop stresses increase J and CTOD for high-hardening materials. However, reaching precious answers for fracture response was not possible, so three dimensions finite element simulations of surface cracked pipes were performed. Hence, Shimanuki and Inoue [17] applied an equivalent CTOD concept to the model simulation of brittle fracture in girth weld joints of pipelines. Their model assumed that the surface cracks have elliptical form, and the pipes are under bending and axial forces, in the presence of internal pressure. They assessed the X80 steel pipe and understood that in the elastic region, the fracture persistence is fixed; but in the plastic region, the CTOD surpasses the critical level. Nevertheless, their work has some weaknesses. For example, loading on definite CTOD or J is uncertain, and the effects of crack depth, crack length, radius-to-thickness ratio, and material hardening are not considered. According to Jayadevan *et al.* [18], at an equivalent axial strain, the driving force under tension is higher than bending, so designing based on axial strain is conservative. Østby [19] named this method as SBD and determined that the permissible strain of this method is not dependent on cracks geometry or material properties. Tang *et al.* [20] established a numerical model to validate the tensile strain capacity of steel pipes. Sandvik *et al.* [21] found that the pipe ductile capacity decreases significantly in the presence of internal pressure with tension load coming up to an unpressurized pipe. The effect of internal pressure on CTOD was assessed by Dake *et al.* [22]. Jia *et al.* [24] modified the limit load solution of Kastner *et al.* [25] based on the von Mises yield criterion to be adapted for a pipeline with a circumferential crack under combined internal pressure and axial plastic strain. The influence of internal pressure on the crack behavior of pipe with establishing a 3D elastic-plastic model was investigated by Zhang *et al.* [23]. Based on the nominal strain, Zhao *et al.* [26] constructed a corresponding failure assessment curve including the effect of internal pressure. The used ductile models in recent studies have many deficiencies in obtaining the fracture behavior of steel pipelines. Consequently, Bai and Wierzbicki [27], using ductile metals, presented a modified version of the Mohr-Coulomb criterion which had been used before as an uncoupled method for rock and soil [28]. The MMC method determines the variation in fracture strain at the small change of stress triaxiality by way of in-plane stress function. Pack *et al.* [29, 30] analyzed the operativity of the MMC criterion to attain the start of failure and crack extension in smaller components and metal sheet troubles. The unique feature of this criterion in describing the state of stresses in the 3D model, where strain to fracture is assumed as a function of stress triaxiality, is providing the estimate of the start of failure for pressure sensitivity rates and different loading conditions. The MMC criterion for API X80 steel is evaluated by Han *et al.* [31]. They employed a finite element model based on the experiment of five various notched specimens covering a wide range of stress triaxiality and lode dependence. The examined fracture criterion showed appropriate compatibility to determine the fracture response of the API X80 pipeline steel. Derived from this numerical-experimental calibration method, Saneian *et al.* [32, 33] established a 3D elastic-plastic model by applying the MMC method, to determine the fracture response of cracked X80 steel pipe under simultaneous tension and torsion and tension and bending, separately. The resistance curve of the pipe under different ratios of strain and twist and strain and curvature was drawn by them. However, internal and external pressure was neglected in their research. Hence, the outcome results are not comprehensive enough, and this topic still needs supplementary studies. The present paper used the modified Mohr-Coulomb (MMC) criterion to analyze the fracture behavior of pipelines under external pressure and combined internal and external pressure, by establishing a 3D FE model in ABAQUS software. In order to prevent the occurrence of buckling before fracture, tension load is added to the model. Nevertheless, the misalignment and weld mismatch of the pipe has not been considered in this study. Finally, to analyze the fracture behavior of the pipe with different crack depths and lengths, which have the largest influence on the pipe mechanical behavior, parametric studies have been conducted.

## 13.2 Finite Element Procedure

Similar to the two previous chapters, the ABAQUS software is chosen to develop a finite element model to simulate the fracture response of a steel pipe under combined loading. The pipe is modeled as a solid element in the way that is best suited for this method [38]. This section assesses a surface cracked pipe under simultaneous tension and internal pressure. Figure 13.1 illustrates the schematic of the side section of the pipe under tension and internal pressure. The pipe is located along the x-axis and is perpendicular to the y-axis. The strain is applied on both ends of the pipe model, and the pressure is applied to the entire inner surface of the tube.

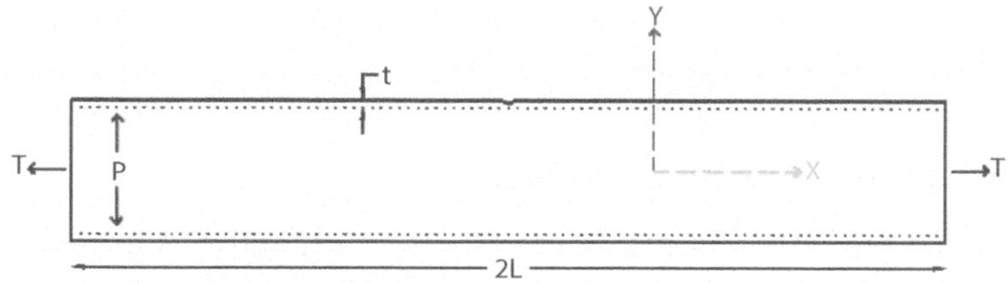


**Figure 13.1** The section of the pipe under tension and internal pressure.

### 13.2.1 The Pipe Model

The schematic of the pipe section is illustrated in Figure 13.2; the outer diameter and thickness of the pipe are shown by D and t, respectively; and also a and 2c parameters indicate the crack depth and crack length, respectively. The maximum crack driving force takes place in the center of the crack tip so the shape of the crack has no significant effect on this section. Hence, an elliptical shape is considered for the crack. However, the external surface of the crack is assumed with uniform depth along the peripheral crack length, and the crack end radius is taken into account as the same as the crack depth. In addition, with regard to the three axes, the crack is positioned on the pipe symmetrically. For the simulation, the outer diameter and thickness of the pipe are chosen as 200 and 10 mm, respectively. In this way, D/t is going to be 20. The crack depth to thickness ratio (a/t) and the crack length to perimeter ratio ($2c/\pi D$) are considered 0.5

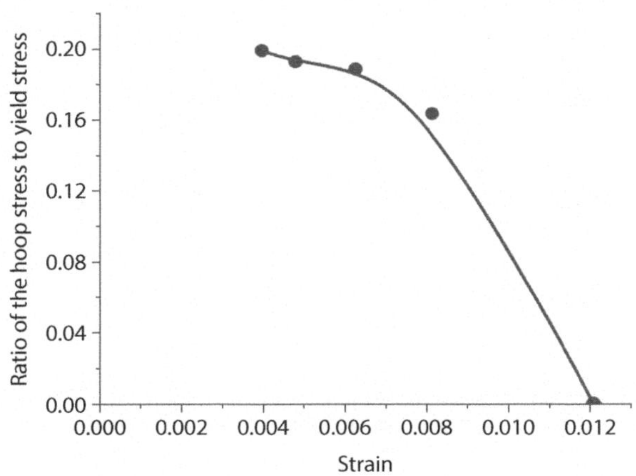


**Figure 13.2** The strain-ratio of the hoop stress to yield stress relationship.

and 0.15, respectively. Besides, the total length of the pipe is assumed as six times the pipe diameter (2L=6D).

### 13.2.2 Finite Element Modeling

The studied pipe has a partial flaw in the surface, and it is loaded under tension and internal pressure simultaneously. The ABAQUS has the ability to include the failure criterion through a user-defined subroutine. Based on the codes of a user-defined subroutine, when the elements reach a critical amount, calculated by Equation (10-10), they will be removed and the stresses become zero. Then, the ductile crack growth is characterized by an incremental element removal method. A quarter of the pipe has been modeled, because of the symmetry deformations and loads. The boundary conditions are prescribed on pipe edges at the top and bottom of the model and are exerted at X=0 and Z=0, symmetrically, as shown in Figure 13.3. The reference point is chosen at the center of the right end of the pipe for considering the coupling relationship, as displayed in Figure 13.4. It is assumed that the right end plane remains plane under loading but the cross-section is free to deform. A typical FE mesh is used around the half circumference and along the longitudinal direction of the pipes demonstrated in Figure 13.5. The crack tip has meshed denser in comparison with the other sections of the model. A mesh study was carried out through trial analysis and was found to be adequate. The displacement ($\Delta L$) is applied to

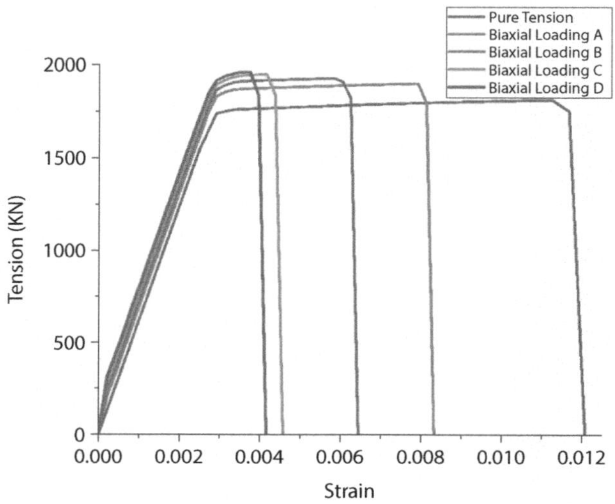


**Figure 13.3** The tension–strain relationship.

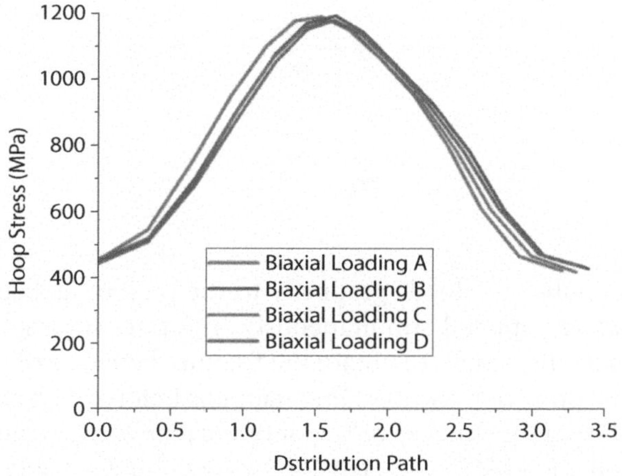


**Figure 13.4** The hoop stress distribution path among the pipe thickness in the cracked area at the time before a fracture occurs.

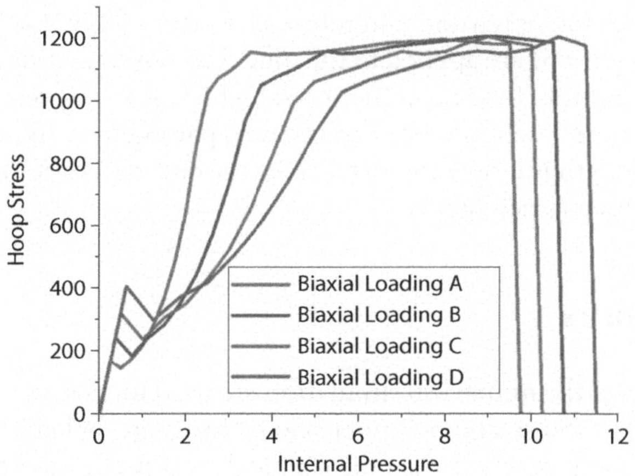


**Figure 13.5** The internal pressure–hoop stress relationship.

the reference point and the internal pressure (P), resulting in the ratio of the hoop stress to the yield stress ($\gamma$) is applied on the inner surface of the pipe, respectively. Besides, the hoop stress ($\sigma_h$) of the pipe can be calculated as

$$\varepsilon = \frac{\Delta L}{L} \tag{13.1}$$

$$\gamma = \frac{\sigma_h}{\sigma_y} \tag{13.2}$$

$$\sigma_h = \frac{PD}{2t} \tag{13.3}$$

This study applies the loads gradually to the pipe model, and all combined loadings are applied simultaneously. The pure internal pressure is not applied in the literature, to obtain the fracture behavior of the pipeline, before. The reason is that the buckling happens before the fracture, so the result for analyzing the fracture behavior of the pipe is not accurate. Hence, the model is only loaded in pure tension by specifying a displacement ($\varepsilon/\gamma = +\infty$) at the uncracked end of the pipe. Then, by considering a constant displacement, to study the effect of biaxial loading on crack, the applied amount of internal pressure differs from 10 to 25 MPa, resulting in the ratio of the hoop stress to the yield stress ($\gamma$). These four different types of concurrent loading of tension and internal pressure (A: $\gamma/\varepsilon = 2$, B: $\gamma/\varepsilon = 3$, C: $\gamma/\varepsilon = 4$, D: $\gamma/\varepsilon = 5$) are applied to the model for investigating the fracture response of multiple loadings. The model which is employed, with rational limits, can precisely predict the fracture parameters. In addition, the chosen pipe length is long enough to make the end effects of the calculated fracture parameters negligible.

## 13.3 Results

The outcomes of the numerical simulation are used to compare the fracture behavior of the X80 steel pipe under biaxial loadings. When the pipeline's rupture pressure is reached, the pipe wall will rapidly rip and become damaged. The relationship between strain and the ratio of the hoop stress to yield stress for all types of loadings is derived in Table 13.1 and illustrated in Figure 13.2. The illustrated figure can assist engineers in determining the collapse limit of a cracked pipe under combined tension and internal pressure, since the pipe in service is never under pure internal pressure. Hence, if the strain according to the hoop stress is below the curve, the pipe can still be used without any issues. As can be seen, strain primarily determines how the pipe model fractures, with an increase in the internal pressure value that causes hoop stress having no discernible impact on crack growth. Figure 13.3 displays the relationship between tension and

**Table 13.1** The parameters values for different types of loading.

| Loading type | Displacement (mm) | Internal pressure (MPa) | Strain | $\sigma_h/\sigma_y$ |
|---|---|---|---|---|
| Pure tension combine loading A | 7.25<br>4.875 | 0<br>9.75 | 0.012<br>0.0081 | 0<br>0.163 |
| Combine loading B | 3.75 | 11.25 | 0.0062 | 0.188 |
| Combine loading C | 2.875 | 11.5 | 0.0048 | 0.193 |
| Combine loading D | 2.375 | 11.875 | 0.004 | 0.199 |

strain. In the elastic behavior stage of the pipe, by increasing the strain, tension is increased intensely, but at specific amounts, the pipe behavior is changed to the plastic stage, so raising the strain has no noteworthy effect on the tension, and it remains almost constant until the pipe model entirely lost its load-carrying capacity, resulting in circumferential crack propagation. The plastic behavior of the pipe is significantly influenced by the hoop stress. It is clear that the pipe stays in the plastic zone less under combined loadings with higher applied internal pressure. Furthermore, the resolution of the plastic region exceeds that of the elastic region in the absence of any internal pressure applied to the pipe. The distribution path of the hoop stress at this time is demonstrated in Figure 13.4. Both the distribution path and the peak value of the hoop stress are similar for all combined loadings. Therefore, to find out the hoop stress of the pipe, the middle element of the cracked section of the pipe has been chosen, then the internal pressure hoop stress relationship at this element and the time before the fracture happens is shown in Figure 13.5. The hoop stress increases quickly until the element is perfect; however, when the chosen element begins to deform, the hoop stress first slightly decreases before gradually beginning to rise again. The curve's misbehavior is more pronounced and its slope is reduced in proportion to the applied internal pressure. The crack failure process begins when the pipe loses its element in the crack section and no additional strain or internal pressure is applied to it. These are the positions where the pipe can no longer be used. By utilizing the applied fracture mechanic model, engineers can accurately evaluate the progression and effects of cracking in pipelines. Table 13.1 displays these points' precise values.

## 13.4 Parametric Studies

After analyzing the results in the previous section, an extensive parametric study applying the finite element method is carried out to assess the influencing mechanisms on the ultimate hoop stress of the pipe model under simultaneous tension and internal pressure, when the crack geometries are different in each modeling process. It should be noted that the crack depth and crack length are the parameters that have the most effect on the fracture behavior of pipelines. Therefore, their effects are studied in this section.

### 13.4.1 Crack Depth

A three-dimensional finite element model is adopted to study the effects of crack depth (a), on the fracture response of the pipes. Three various crack depth ratios were taken to obtain the behavior of the pipe model for varying crack depth. In order to prevent the pipe model from being subjected to buckling, all of the crack depths were chosen to be more than half of the pipe thickness (a/t = 0.55, and a/t = 0.75). In addition, the comparisons corresponding to five different types of loadings include pure tension, and four types of combined tension and internal pressure (A: $\gamma/\varepsilon = 2$, B: $\gamma/\varepsilon = 3$, C: $\gamma/\varepsilon = 4$, and D: $\gamma/\varepsilon = 5$) are conducted, similar to the main model. Lastly, the relationship between strain and the ratio of hoop stress to yield stress with dissimilar crack depth is displayed in Figure 13.6. The outcome shows

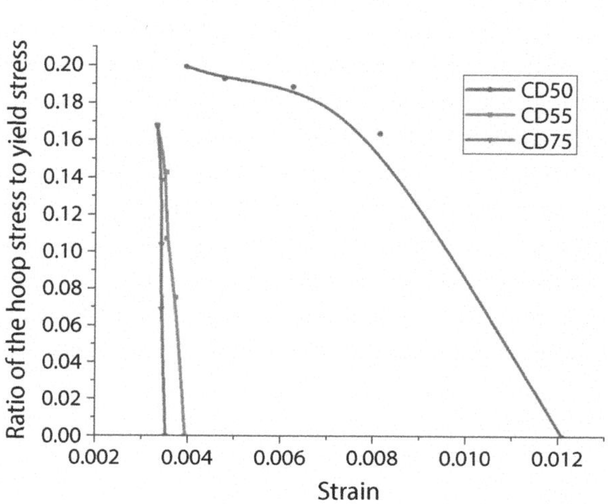


**Figure 13.6** The comparison between the different crack depths.

that when the crack depth increased, the strain goes down sharply. Hence, the crack geometry has a noteworthy influence on the strain capacity of the pipe. As we can see in Figure 13.6, as soon as the deepness of the crack is going to be more than half of the pipe thickness, the strain capacity of the pipe decreases hard, then the decreasing rate becomes leisurely. Besides, by increasing the crack depth, the strain reduction rate is meaningfully reduced. It shows that the strain capacity of the pipe is greatly influenced by the depth of the crack. The pipe has a critical crack depth that is greater than half of the tested pipe model's thickness. It should be noted that when the depth of the crack exceeds 5 mm, the failure strain almost stays constant for all combined loadings, and only the internal pressure varies for each loading. It is clear that for CD75, the curve is almost parallel to the y-axis, indicating that when the crack depth is large, internal pressure does not significantly contribute to the pipe's fracture failure.

### 13.4.2 Crack Length

In this part of the study, the crack length effect on the pipe behavior is investigated by using a 3D FE method. In order to study the behavior of the pipe model for dissimilar crack lengths, two different crack length ratios were adopted. It has been tried to choose the best crack lengths, so one of them is smaller than the main crack length ($2c/\pi D = 0.13$) and one is larger ($2c/\pi D = 0.17$). Same as the previous section, the model is considered to be under different types of loadings (pure tension and three types of combined tension and internal pressure). Finally, the relationship between strain and the ratio of the hoop stress to the yield stress is demonstrated in Figure 13.7. This comparison has a big difference from the previous section; here, one of the crack lengths has been chosen to be smaller than the main model. As it is illustrated in Figure 13.7, when the pipe length becomes small (CL13), the fracture behavior of the pipeline is not as smooth as larger crack lengths. The reason is that as much as the crack size becomes slight, the effect of buckling is more significant, so the pipe has not had a pure fracture behavior. As a matter of fact, the crack length has a noteworthy effect on the fracture behavior of the X80 steel pipeline. When the crack length is large (CL17), the crack growth rate becomes high; therefore, increasing the strain amount and internal pressure does not affect the pipe behavior too much. It could be concluded that once the dimensions of the crack are high enough, the internal pressure does not have a substantial influence on the pipe model, and actually, here, tension plays the main role. As can be seen, the CL17 curve has the same condition as the CD75 curve in the previous section.

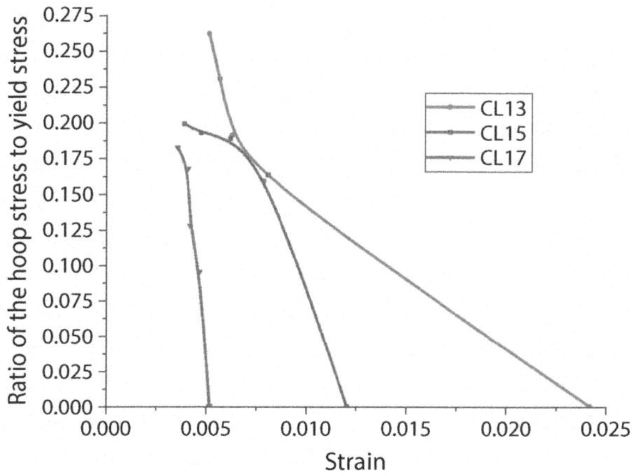


**Figure 13.7** The comparison between the different crack lengths.

## 13.5 Theoretical Solutions

The ASME B31G [6] and DNV RP-F101 [7] codes are commonly used to calculate the residual strength of the internal pressure resistance of pipelines with initial defects.

From the ASME B13G, we have

$$P_{burst} = \frac{2tf_u}{(D-t)} \frac{(1-0.85(a/t))}{\left(1-0.85\left(\frac{a/t}{Z}\right)\right)} \tag{13.4}$$

$$z \le 50 \Rightarrow Z = \sqrt{1+0.6275z-0.003375z^2}$$
$$z > 50 \Rightarrow Z = 0.032z+3.3$$

$$z = \frac{c^2}{Dt}$$

where $f_u$ is the ultimate strength of the pipeline material. To determine whether a pipeline has failed, the ASME B31G specification only provides a calculation of the failure stress. The method used in this study is consistent with the DNV RP-F101 model, which indicates that when the intermediate stress in a pipe wall reaches its failure stress, the pipe has failed. With respect to the internal pressure, the corresponding burst pressure value is obtained from Equation (13.4). In addition, the DNV RP-F101 model is presented as below:

$$P_{burst} = \frac{2tf_u}{(D-t)} \frac{\left(1-0.85(a/t)\right)}{\left(1-0.85\left(\frac{a/t}{Q}\right)\right)} \tag{13.5}$$

$$Q = \sqrt{1+0.31\left(\frac{c}{\sqrt{Dt}}\right)^2}$$

By looking at the aforementioned models, two parameters related to the defect size, which are the crack depth and crack length, affect the residual strength of the pipeline in addition to its basic parameters, which include the outer diameter $D$, wall thickness $t$, and material ultimate strength $f_u$. The crack width has no significant effect on the results of our model, and the effect of the crack size on the residual strength of the pipe is determined by using normalized values of crack depth and crack length which are expressed as $a_{norm} = a/t$ and $c_{norm} = c^2/Dt$, respectively. An analysis of a 21-inch X80 steel pipe is provided in this section. The steel pipe has an outer diameter of $D = 534\ mm$ and a wall thickness of $t = 27\ mm$. The material's ultimate strength is $f_u = 650MPa$ on average.

The burst pressure surface of the steel pipe under $c_{norm}$ and $a_{norm}$ is determined by ASME B31G as illustrated in Figure 13.8 and DNV RP-F101 as demonstrated in Figure 13.9. In this study, the range of $c_{norm}$ is chosen to be between 0 and 30 for the examination. It is obvious that there is a similarity between the two normative models' surfaces. The model predicts a decrease in burst pressure as $c_{norm}$ and $a_{norm}$ increase. The burst pressure is significantly different when $c_{norm}$ is small, and when it is equal to zero, the burst pressure values predicted by the two models are the same, indicating that there is no pipeline defect. Besides, when $a_{norm}$ is equal to one, it means the pipeline defect extends through the pipe wall; at this time, the blast pressure in ASME B31G is not 0 but in DNVRP-F101 is.

The variation of the burst pressure with the crack length while the crack depths are equal to 0.3, 0.5, and 0.7 is depicted in Figure 13.10. It could be observed when the crack length is less than 5, the bursting pressure changes swiftly, and this change becomes too slow once the crack length

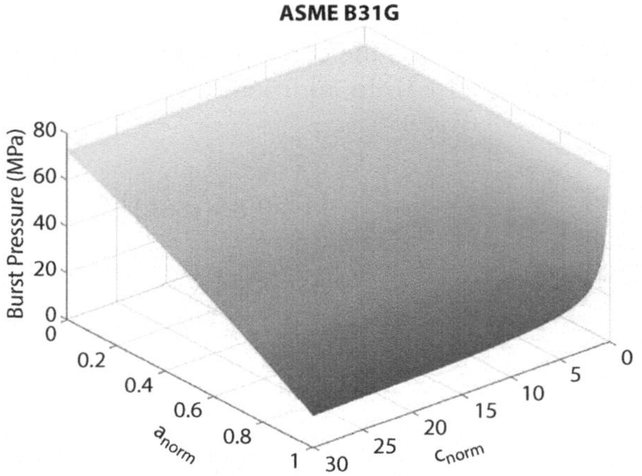


**Figure 13.8** Burst pressure surface for the ASME B31G model.

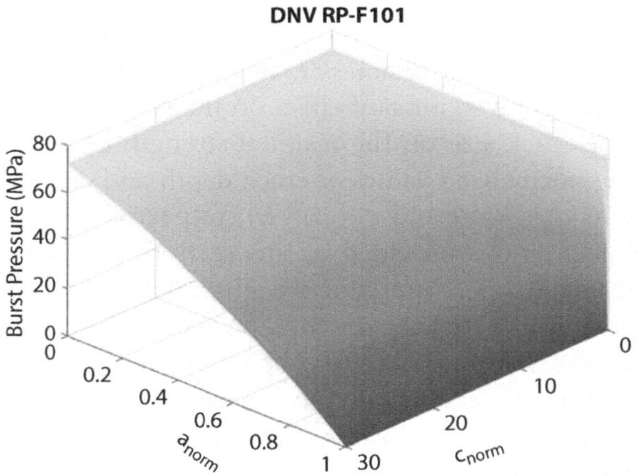


**Figure 13.9** Burst pressure surface for the DNV RP-F101 model.

is larger than this amount. As $c_{norm}$ increases, ASME B31G decreases less than the DBV RP-F101. In fact, the difference between two specifications' predicted values increases with growing crack depth and crack length.

Figure 13.11 shows the relationship between the crack depth and burst pressure, while the crack lengths are 3, 10, and 30. It is evident that the burst pressure predicted by both specifications exhibits a downward trend as $a_{norm}$ increases, with the DNV RP-F101 specification showing a faster decrease. Furthermore, the ASME B31G specification does not have a burst pressure

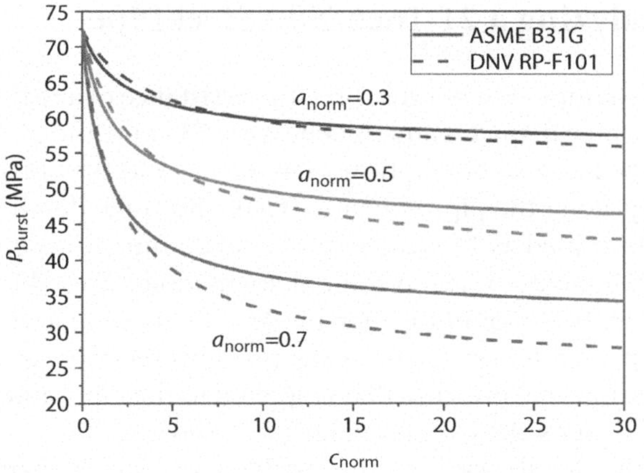


**Figure 13.10** Burst pressure–crack length relationship.

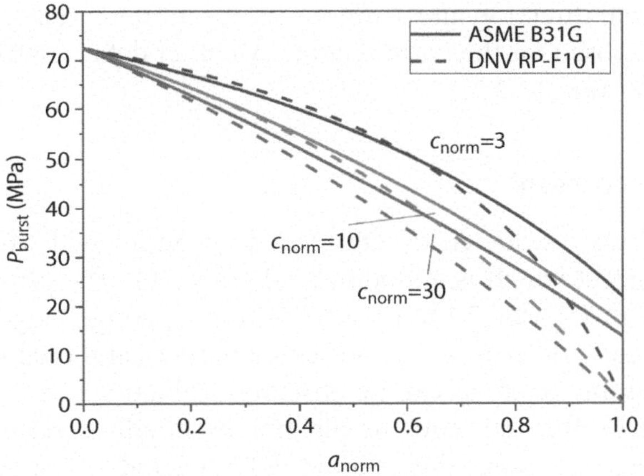


**Figure 13.11** Burst pressure–crack depth relationship.

of zero when the crack depth is equal to one, whereas the DNV RP-F101 specification does. As can be seen from the above analysis, the predicted burst pressures of ASME B31G and DNV RP-F101 are relatively close when both $c_{norm}$ and $a_{norm}$ are small; however, when $c_{norm}$ and $a_{norm}$ become larger, DNV RP-F101's predicted value is lower than ASMEB31G's, particularly when $c_{norm} > 20$ or $a_{norm} > 0.7$. This represents a significant difference between these two specifications, which is about 20%.

## 13.6 Analyzing a 21-Inch X80 Steel Pipe

This section examines the internal pressure resistance of a real 21-inch X80 steel pipe. The pipe diameter and thickness are 534 mm and 27 mm, respectively. The pipe has a semi-elliptical axial defect, and the crack is running longitudinally along the pipe's outer surface. The crack depth, length, and width are determined as 14 mm, 80 mm, and 20 mm. The detailed pipe is demonstrated in Figure 13.12. A finite element model is established to analyze the fracture behavior of the pipe model. The input material properties and fracture parameters are same as the previous sections. In this section, the maximum internal pressure that is applied to the pipe's internal surface is 80 MPa. The pipe's chosen overall length is 800 mm.

The meshing details used for this analysis are denser than those used in Figure 13.5, but the details are still the same as what was mentioned in Section 13.2.2. Here, the maximum mesh size is set to 10 mm, and Figure 13.13 displays a schematic of the meshed model. A smaller mesh size of 2 mm in the axial direction and 1 mm in the circumferential and thickness directions is chosen for the crack section. All other details of the model are the same as Section 13.2.2.

### 13.6.1 Discussion

A burst test was conducted by Oh *et al.* [39] on an API X65 steel pipe with initial defect which is a longitudinal crack, and the pipe after blasting is shown in Figure 13.14. They found out as gouge length increases; the burst pressure decreases. The presented numerical model in this study also shows a same result as can be observed in Figure 13.15. The Figures 13.14 and 13.15 illustrate how the pipeline eventually creates a region of centralized damage along the original defect's length. The center of the crack expands outwardly in a sharp manner, while other parts only slightly expand outward. The failure mode seen in the real test matches the one

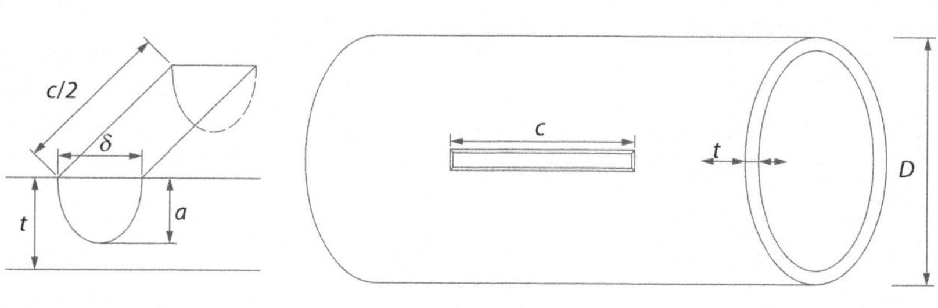


**Figure 13.12** The examined 21-in pipe with initial defect.

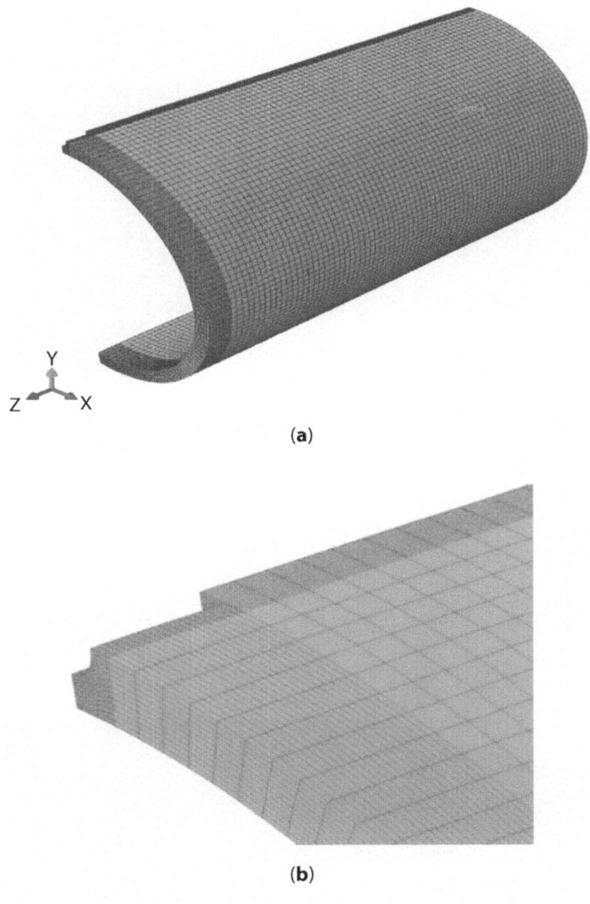


**Figure 13.13** The detailed mesh.

predicted by the model. There are three stages to the damage that internal pressure causes to a pipeline: in the first, damage starts to occur at the center of the gap when the pipe diameter expands due to internal pressure; in the second stage, after the first unit fails at the center of the gap, the units continue to fail along the axial and thickness directions of the pipeline until all failures occur, i.e., the pipe wall penetrates, and in the third stage, the units fail rapidly along the axial direction of the pipeline and the pipe wall is clearly separated.

According to ASME B31G and DNV RP F101 standards, Table 13.2 compares the pressure in the first stage when the first unit fails and the pressure in the second stage when wall thickness penetration occurs. The third stage

**Figure 13.14** Gouge defect after the test.

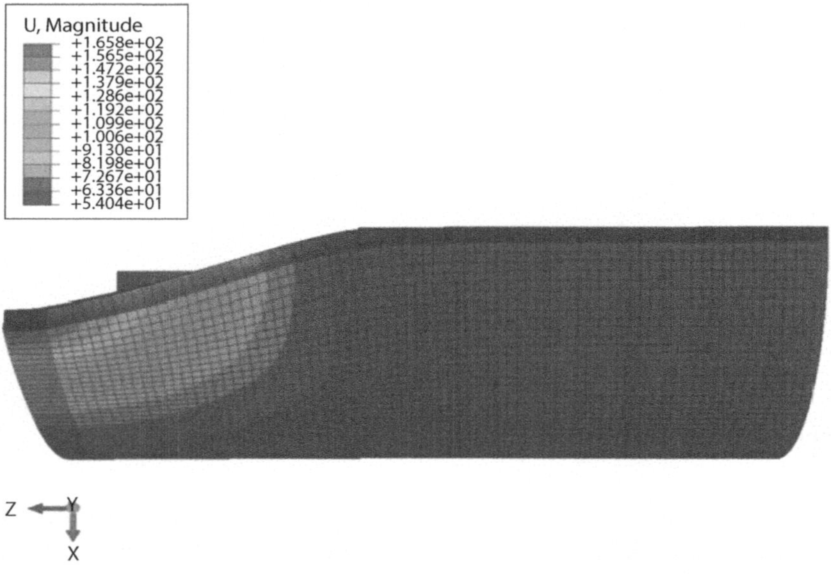


**Figure 13.15** Numerical result of the defect after the test.

**Table 13.2** The comparison between numerical and theories results.

| Theoretical models (*MPa*) | | Numerical results (*MPa*) | |
|---|---|---|---|
| **ASME B31G** | **DNV RP F101** | **The first unit fails** | **Wall thickness penetration** |
| 63.46 | 65.98 | 68 | 68.4 |

pressure is correlated with the axial expansion of the crack, so the pressure of this stage is not taken into account since this thesis only addresses the pipeline's failure pressure under internal pressure. The obtained pressures are very close to each other. The finite element models' predicted values differ slightly from standards' calculations. The model's calculation results show a slight increase over the specification's calculation results. The defect shape in this study may not be the most critical one, and the standards do not define the defect's shape. Secondly, the theories do not take the steel pipe's ductile fracture process into account. In fact, the error between the extracted values is less than 8%, which proves the model's calculation result is accurate. Besides, the pipeline's burst pressure is determined by the pressure at the second stage, which is when the pipe wall penetrates, and its equal to 68.4 *MPa*. The FE result of the employed model at this pressure is displayed in Figure 13.16. As can be seen, the unit failure range expands along the pipe wall thickness direction and length direction.

The change curves for the pipeline's axial change and the defect's opening displacement with internal pressure are displayed in Figures 13.17 and 13.18, respectively. The point at which the pipeline fails is indicated by the "X" mark. The parameters represented by $\Delta L$ and $\Delta\delta$ in the figures are taken from the model, which represent a quarter of the pipe and have not been transformed to represent the pipe's entire deformation. Since the pipe shrinks along the positive Z-axis, Figure 13.17 shows that, at low

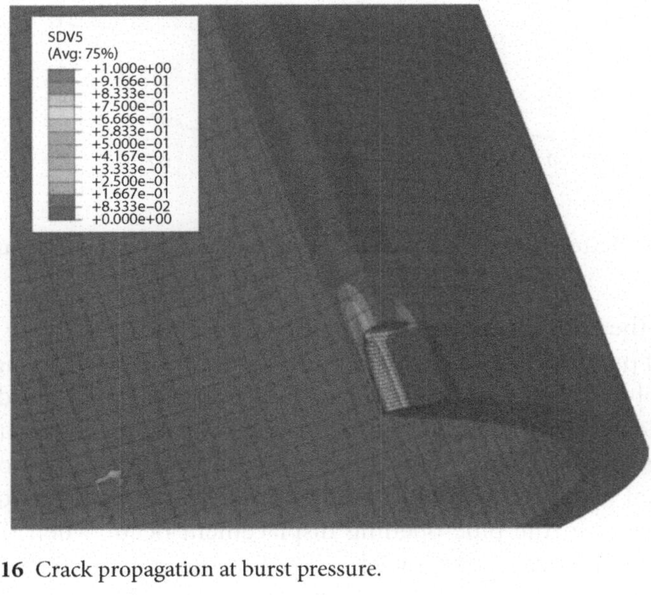


**Figure 13.16** Crack propagation at burst pressure.

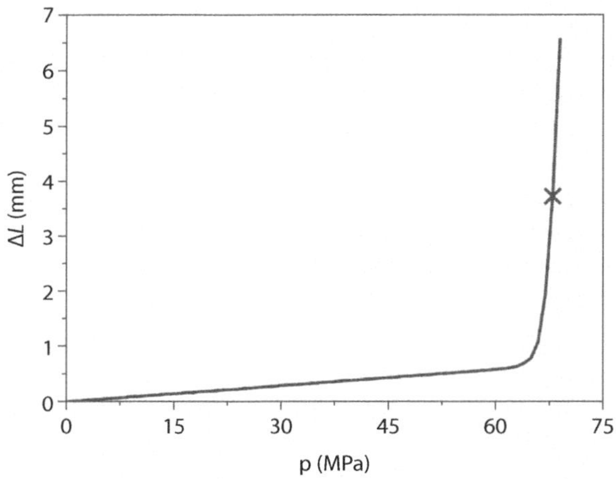


**Figure 13.17** The relationship between internal pressure and axial displacement.

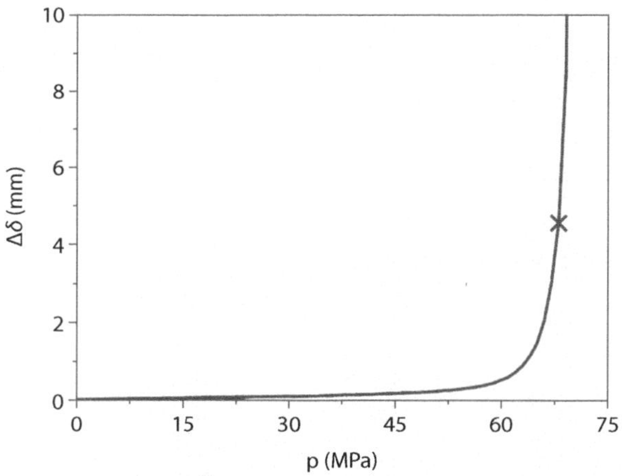


**Figure 13.18** The relationship between internal pressure and opening displacement.

pressures, the axial shrinkage of the pipe changes linearly with the increase in internal pressure. However, at high pressures, the axial shrinkage of the pipe sharply increases, indicating that the defect has experienced significant plastic deformation and that the pipeline failed quickly. Figure 13.18 also shows that very little change in pipe opening displacement occurs at low internal pressure. A significant amount of plastic deformation and a sharp increase in the pipe opening displacement occur when the internal pressure reaches approximately 65 MPa. The pipe then experiences penetration failure (opening displacement of approximately 5 mm), at which point the opening displacement approaches vertical. This is because X80 pipeline steel has a poor hardening capacity. It should be concluded that once the steel reaches its breaking point, significant deformation will result from even a small load. Besides, the historical curves of the stress triaxiality and Lode angle parameters in the crack development area were extracted in order to analyze the stress state in the crack development area. The stress state modification curves in the length and thickness directions are depicted in Figures 13.19 and 13.20, respectively, along with the normalized equivalent plastic strain or damage variable (Dv). The center of the pipe's outer defect is where the crack originated, and the stress state at various separations from the center is taken out for analysis. The figures show that the parameters of the Lode angle and stress triaxiality at the crack initiation position essentially do not change with damage.

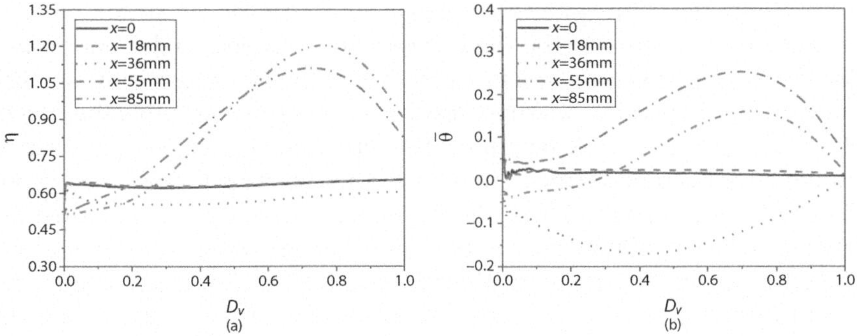


**Figure 13.19** The relationship between damage variable and (a) stress triaxiality, (b) Lode angle parameter; in the longitudinal direction.

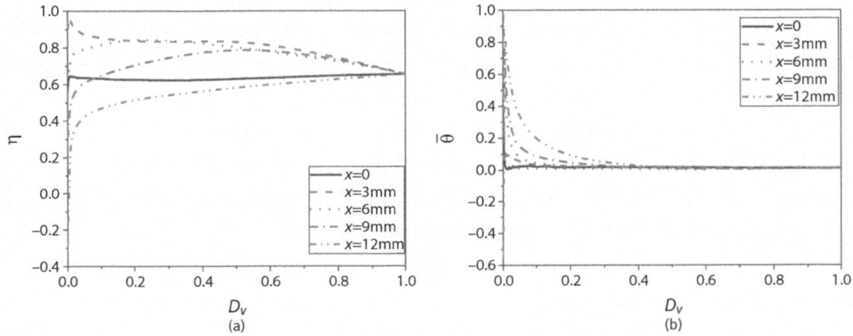


**Figure 13.20** The relationship between damage variable and (a) stress triaxiality, (b) Lode angle parameter; in the circumferential direction.

A plane strain tensile state is indicated by the stress triaxiality of roughly 0.6 and the Lode angle parameter of roughly 0. The greater the change in stress state, regardless of length or thickness direction, the farther away from the location where the crack initiated. The material in the defective area sustains localized damage as the load increases, and the crack widens in the defect's direction. The internal pressure can still rise at this point because a new fracture surface will be forming.

## 13.7 Conclusions

First to analyze the hoop stress of X80 steel pipe the internal pressure is added with tension to the model. For comparison, the variations of internal pressure with tension for similar cracked pipes under different loadings, pure tension, and four types of biaxial loadings have been carried out. Accordingly, the failure points of the pipe model for various strains and the ratio of the hoop stress to the yield stress values were derived. It was found that the tension affects the fracture behavior of the pipe much more significantly in comparison to internal pressure. In order to determine the effects of crack depth and crack length on the pipe model behavior, parametric studies have been performed. As soon as the crack depth increases, more than half of the pipe thickness, the strain, and internal pressure reduction rate are meaningfully increased, but after that the reduction rate is slow. The outcome was when the depth of the crack becomes large; the internal pressure effect of the fracture response is more significant. It also happened the same for the larger crack lengths. The crack depth affects the value of the ratio of the hoop stress to the yield stress more than the crack length. Besides, a real 21-in pipe is modeled by ABAQUS with an axial defect, to consider the fracture response of a real pipe size with a longitudinal crack. Then, to validate the numerical model for the internal pressure, the empirical formulas in ASME B31G and DNV RP-F101 are examined. The ASME B31G and DNV RP-F101 standards yield similar findings when the crack length and depth are minor. The bursting pressure determined by the DNV standard reduces much more quickly than that of the ASME standard as the depth of the crack grows. The bursting pressure obtained by ASME is not equal to the bursting pressure obtained by DNV when dnorm = 1. The damage model can obtain the process of penetration failure of the pipe wall under internal pressure and more precisely compute the internal pressure resistance of the X80 steel pipe with an initial axial crack. The center of the pipe wall's exterior crack is where the fracture in the pipe begins. This point is in a plane strain tensile state since the stress triaxiality at this point prior to failure is roughly 0.6 and the Lode angle parameter is almost 0.

## Bibliography


1. Yan, S., Shen, X., Jin, Z., Ye, H., On elastic-plastic collapse of subsea pipelines under external hydrostatic pressure and denting force. *Appl. Res. Ocean J.*, 58, 305–321, 2016, http://dx.doi.org/10.1016/j.apor.2016.04.007.
2. Hillenbrand, H.G., Graef, M.K., Grob-Weege, J., Knauf, G., Marewski, U., Development of line pipe for deepwater applications, in: *Proceedings of the 12th international offshore and polar engineering conference*, Kitakyushu, Japan, pp. 287–294, 2002.
3. Gong, S., Sun, B., Bao, S., Bai, Y., Buckle propagation of offshore pipelines under external pressure. *Mar. Struct. J.*, 29, 115–130, 2012, http://dx.doi.org/10.1016/j.marstruc.2012.10.006.
4. Mousselli, A., *Offshore Pipeline Design, Analysis, and Methods*, Pennwell Books, Tulsa, Oklahoma, 1981.
5. Jia, P., Jing, H., Xu, L., Han, Y., Zhao, L., A modified reference strain method for engineering critical assessment of reeled pipelines. *Mech. Sci.*, 105, 2016.
6. American National Standards Institute (ANSI)/American Society of mechanical Engineers (ASME) ASME B31G, ASME B31G, 1984.
7. DNV Recommended Practice, RP-F101. Corroded Pipelines, Det NorskeVeritas, 1999, April.
8. British Energy, Assessment of the integrity of structures containing defects, revision 4, London, UK, 2001.
9. British Standards Institution, BS 7910, Guide on methods for assessing the acceptability of flaws in metallic structures, London, UK, 2000.
10. Hutchinson, J.W., Fundamentals of the phenomenological theory of nonlinear fracture mechanics. *J. Appl. Mech. Trans. ASME*, 50, 1042–1051, 1983, https://doi.org/10.1115/1.3167187.
11. Hutchinson, J.W., Singular behaviour at the end of a tensile crack in a hardening material. *J. Mech. Phys. Solids*, 16, 13–31, 1968.
12. Paris, P., Tada, H., Zahoor, A., Ernst, H., The Theory of Instability of the Tearing Mode of Elastic-Plastic Crack Growth, in: *Elastic-Plastic Fract.*, 5–36, ASTM Special Technical Publications, 1979. https://doi.org/10.1520/stp35825s.
13. O'Dowd, N.P. and Shih, C.F., Family of crack-tip fields characterized by a triaxiality parameter-I. Structure of fields. *J. Mech. Phys. Solids*, 39, 989–1015, 1991, https://doi.org/10.1016/0022-5096(91)90049-T.
14. Rahman, S. and Brust, F.W., Approximate methods for predicting J-integral of a circumferentially surface-cracked pipe subject to bending. *Int. J. Fract.*, 85, 111–130, 1997, https://doi.org/10.1023/A:1007322018722.

15. Anderson, T., *Fracture Mechanics - fundamentals and applications*, 2nd ed, CRC Press, London, 1995.
16. Foxen, J. and Rahman, S., Elastic-plastic analysis of small cracks in tubes under internal pressure and bending. *Nucl. Eng. Des.*, 197, 75–87, 2000, https://doi.org/10.1016/S0029-5493(99)00258-7.
17. Shimanuki, H. and Inoue, T., Assessment of brittle fracture in girth weld joint of pipelines subjected to internal pressure and bending load, in: *Int. Conf. Appl. Eval. Highgrade Linepipes Hostile Environ.*, Yokohama, Japan, 2002.
18. Jayadevan, K.R., Ostby, E., Thaulow, C., Strain-based fracture mechanics analysis of pipelines. *Proc. Int. Conf. Adv. Struct. Integrity*, Bangalore, India, pp. 14–17, 2004.
19. Østby, E., Fracture control - Offshore pipelines: New strain-based fracture mechanics equations including the effects of biaxial loading, mismatch and misalignment. *Proc. Int. Conf. Offshore Mech. Arct. Eng. - OMAE*, vol. 3, pp. 649–658, 2005, https://doi.org/10.1115/OMAE2005-67518.
20. Tang, H., Fairchild, D., Panico, M., Crapps, J., Cheng, W., Strain Capacity Prediction of Strain-Based Pipelines. Huang, in: *Proc. 2014 10th Int. Pipeline Conf.*, Calgary, Alberta, Canada, pp. 1–12, 2014.
21. Sandvik, A., Østby, E., Thaulow, C., A probabilistic fracture mechanics model including 3D ductile tearing of bi-axially loaded pipes with surface cracks. *Eng. Fract. Mech.*, 75, 76–96, 2008, https://doi.org/10.1016/j.engfracmech.2007.02.015.
22. Dake, Y., Sridhar, I., Zhongmin, X., Kumar, S.B., Fracture capacity of girth welded pipelines with 3D surface cracks subjected to biaxial loading conditions. *Pressure Vessels Pip.*, 92, 115e126, 2012, https://doi:10.1016/j.ijpvp.2011.10.019.
23. Zhang, Y.M., Ariffin, M.Z., Xiao, Z.M., Zhang, W.G., Huang, Z.H., Nonlinear elastic-plastic stress investigation for two interacting 3-D cracks in offshore pipelines. *Fatigue Fract. Eng. Mater. Struct.*, 1–11, 2014, https://doi.org/10.1111/ffe.12259.
24. Jia, P., Jing, H., Xu, L., Han, Y., Zhao, L., A modified fracture assessment method for pipelines under combined inner pressure and large-scale axial plastic strain. *Theor. Appl. Fract. Mech.*, 2016, http://dx.doi.org/10.1016/j.tafmec.2016.10.007.
25. Kastner, W.E., Röhrich, E., Schmitt, W., Steinbuch, R., Critical crack sizes in ductile piping. *Press. Vessels Pip.*, 9, 197–219, 1981, https://doi.org/10.1016/0308-0161(81)90002-8.
26. Zhao, X., Xu, L., Jing, H., Han, Y., Zhao, L., Cao, J., Lv, Y., Song, Z., A modification of reference strain approach for thin-walled submarine pipelines under large-scale plastic strain and internal pressure. *Thin Walled Struct.*, 140, 182–194, 2019, https://doi.org/10.1016/j.tws.2019.03.038.

27. Bai, Y. and Wierzbicki, T., Application of extended Mohr-Coulomb criterion to ductile fracture. *Fracture*, 161, 1–20, 2010, https://doi.org/10.1007/s10704-009-9422-8.
28. Wylllie, D., *Foundations on rock,* Second ed, E & FN Spon, London, 1999.
29. Pack, K., Luo, M., Wierzbicki, T., Sandia fracture challenge: Blind prediction and full calibration to enhance fracture predictability. *Fracture*, 186, 155–175, 2014, https://doi.org/10.1007/s10704-013-9923-3.
30. Pack, K. and Marcadet, S.J., Numerical failure analysis of three-point bending on martensitic hat assembly using advanced plasticity and fracture models for complex loading. *Solids Struct.*, 85–86, 144–159, 2016, https://doi.org/10.1016/j.ijsolstr.2016.02.014.
31. Han, P., Cheng, P., Yuan, S., Bai, Y., Characterization of ductile fracture criterion for API X80 pipeline steel based on a phenomenological approach. *Thin Walled Struct.*, 164, 107254, 2021, https://doi.org/10.1016/j.tws.2020.107254.
32. Saneian, M., Han, P., Jin, S., Bai, Y., Fracture response of steel pipelines under combined tension and torsion. *Thin Walled Struct.*, 154, 106870, 2020, https://doi.org/10.1016/j.tws.2020.106870.
33. Saneian, M., Han, P., Jin, S., Bai, Y., Fracture response of steel pipelines under combined tension and bending. *Thin Walled Struct.*, 155, 106987, 2020, https://doi.org/10.1016/j.tws.2020.106987.
34. Rice, J.R. and Tracey, D.M., On the ductile enlargement of voids in triaxial stress fields*. *J. Mech. Phys. Solids*, 17, 201–217, 1969, https://doi.org/10.1016/0022-5096(69)90033-7.
35. Bai, Y. and Wierzbicki, T., A new model of metal plasticity and fracture with pressure and Lode dependence. *Int. J. Plast.*, 24, 1071–1096, 2008, https://doi.org/10.1016/j.ijplas.2007.09.004.
36. American Petroleum Institute, *API Spec 5L, Specification for Line Pipe*, 44th ed, API Publishing Services, 1220 L Street, N.W. Washington D.C, 2005.
37. Swift, H.W., Plastic Instability Under. *J. Mech. Phys. Solids*, 1, 1–18, 1952.
38. Raju JCNJ, I.S., Stress-intensity factors for internal and external surface cracks in cylindrical vessels. *J. Pressure Vessel Technol.*, 104, 4, 1982.
39. Oh, C.-S., Kim, N.-H., Kim, Y.-J., Baek, J.-H., Kim, Y.-P., Kim, W.-S., A finite element ductile failure simulation method using stress-modified fracture strain model. *Eng. Fract. Mech.*, 78, 1, 124–137, 2011.

# 14

# The Hoop Stress Failure Analysis of Defective X80 Steel Pipes


***Abstract***

This chapter presents a numerical investigation of the fracture behavior of X80 steel pipelines containing surface cracks subjected to combined axial tension and internal pressure. Offshore pipelines, especially in ultra-deepwater applications, experience complex loading during installation and operation, making accurate fracture prediction critical. A 3D finite element model using ABAQUS and a user-defined VUMAT based on the modified Mohr-Coulomb (MMC) criterion is developed to simulate ductile crack growth. The pipe is modeled with elliptical surface cracks and subjected to five loading conditions, including pure tension and four combined tension-internal pressure scenarios. Results indicate that axial strain is the dominant factor in fracture, while internal pressure primarily affects the hoop stress distribution and modifies the plastic behavior. Parametric studies on crack depth and length reveal that increasing either significantly reduces strain capacity, with crack depth showing a stronger influence on failure behavior. At high crack sizes, internal pressure has a diminishing effect on pipe integrity. This study provides valuable insights into the interplay between loading conditions and crack geometry, supporting the design and safety evaluation of high-strength offshore pipelines under complex biaxial loading conditions.



***Keywords*:** Steel pipelines, internal pressure, crack geometry, ductile fracture, finite element analysis, MMC criterion, biaxial loading


## 14.1 Introduction

Pipelines are the leading transport carriers in the oil and gas industries. Steel pipes are still the most common in petrochemicals for the transportation of industrial materials because of their lower costs and easier installation. In addition, their compatibility with severe environmental conditions, particularly in deep waters, for conveying crude oil and natural gas from subsea to vessel or coast is another advantage of theirs. The pipes are under


Yong Bai. Plasticity and Ductile Fracture in Structural Engineering, (399–426) © 2026 Scrivener Publishing LLC

external pressure and combined external and internal pressure under the installation and application process, respectively. In moderately shallow waters, only the internal pressure is essential for the bursting failure of the pipe laid [1]. Pipelines are being used in long-distance transmissions, and they may face defects, especially in the material and welding. The manufacturing process has a noticeable effect on the type and number of cracks [2]. Besides, during lifetime service, pipes could be damaged or cracked due to corrosion; therefore, studying the cracks in pipelines is an important subject. Considering the plastic strain, the pipelines are vulnerable to fracture during the service, which may cause the pipes to collapse and lead to substantial economic loss. In order to assess the fracture resistance by calculating the rate of strain, API 579 [3], BS 7910 [4], and R6 [5] standards are used. However, these standards are established for elastic response, and they are limited to small plastic strains, while during the process of normal reeling operation, the plastic strain in the pipes can be up to 4% [6]. Therefore, a precise fracture evaluation method that is compatible with the deformation capacity of pipelines under internal pressure is required.

Within the field of fracture mechanics, some methodologies are based on the micromechanical characterization of the material and have demonstrated considerable efficacy in addressing the ductile fracture of structural steel. These models characterize the ductile failure process by considering the processes of void nucleation, growth, and coalescence in materials. A micromechanical model that explains the impact of the expansion of spherical voids on the properties of materials is proposed by Gurson [7]. Numerous studies have concentrated on extending the Gurson model to account for ductile fracture under diverse conditions. Chu and Needleman [8] introduced models of void nucleation that are governed by either local stress or plastic strain, employing a two-parameter criterion for void nucleation. Tvergaard [9] enhanced the application of the Gurson model by introducing two parameters that adjust for the influence of material strain hardening and void interaction. For ductile metals containing non-spherical voids, Gologanu *et al.* [10, 11] extended the Gurson model to derive an approximate model. Tvergaard and Needleman [12] integrated ultimate material failure due to void coalescence into the constitutive model by making the yield function dependent on the void volume fraction. The Gurson model, as further modified by Tvergaard and Needleman, is referred to as the GTN model and is recognized as one of the most frequently cited frameworks for modeling void nucleation, growth, and coalescence. Following the introduction of the GTN damage model, a substantial body of research has been devoted to the fracture analysis of various materials utilizing this model. Benseddiq and Imad [13] conducted numerical simulations using the GTN model and carried out experimental tests on both CT specimens and notched round specimens. Liao *et al.* [14] conducted a study on plastic deformation and damage behavior in tensile bars by employing the GTN model. Sirinakorn *et al.* [15] carried out a study on the local deformation and damage characteristics of dual-phase steel, employing the GTN model to characterize damage progression within the microstructure. Wang *et al.* [16] identified parameters of the GTN model for high-strength steel, employing response surface methodology. Through multiple numerical simulations guided by a central composite design, they pinpointed four crucial parameters in the GTN model.

The Rice–Tracey (RT) criterion [17] addresses the expansion of a spherical void within an infinite medium. It was assumed that the material demonstrated perfect plasticity and complied with the von Mises yield criterion along with its corresponding flow rule. In order to consider the damage growth in shear-dominated stress states at low-stress triaxiality, and introduce the Lode dependence, the original RT criterion is extended by Gruben *et al.* [18] by following along the same lines as Nahshon and Hutchinson [19]. Due to the deficiencies in the methods for determining the fracture behavior of steel pipelines, Bai [20] presented a modified version of the Mohr-Coulomb criterion that is well suited for materials that fail within the elastic range or under plasticity of small strains, such as rock and soil [21]. The modified Mohr-Coulomb method determines the variation in fracture strain at the small change of stress triaxiality by way of an in-plane stress function. Pack *et al.* [22] analyzed the operativity of the MMC criterion to attain the start of failure and crack extension in smaller components and metal sheet troubles. The unique feature of this criterion in describing the state of stresses in the 3D model, where strain to fracture is assumed as a function of stress triaxiality, is providing the estimate of the start of failure for pressure sensitivity rates and different loading conditions [23]. Kofiani *et al.* [24] extended the applicability of the MMC model to high-stress triaxiality by characterizing anisotropic plasticity and fracture in API X70 offshore pipes through a comprehensive experimental-computational approach, using optimized FE models of six fracture specimens, and successfully predicting fracture initiation and propagation under various loading conditions, including shear. Paredes *et al.* [25] presents a step-by-step phenomenological methodology for analyzing stress state evolution in growing cracks in X70 pipeline steels, involving calibration of crack initiation through tensile, shear, and compression tests, and similarly, the plastic behavior of X65 pipeline steels was assessed through MMC-based fracture modeling by Paredes *et al* [26]. The MMC criterion for API X80 steel is evaluated by Han *et al.* [27]. They employed a finite element model based on the experiment of five various notched specimens covering a wide range of stress triaxiality and lode dependence. The examined fracture criterion showed the appropriate compatibility to determine the response to fracture of API X80 pipeline steel. Using this numerical experimental calibration method, Saneian *et al.* [28, 29] established a 3D elastic-plastic model applying the MMC method to determine the fracture response of a cracked X80 steel pipe under simultaneous tension-torsion and tension-bending separately. They drew the resistance curve of the pipe under different ratios of strain and twist and strain and curvature. However, internal pressure was neglected in their research.

For the evaluation of burst pressure in defective pipelines, current standards such as ASME B31G [30] and DNV RP-F101 [31] provide semi-theoretical and semi-empirical formulas. ASME B31G also states that such formulas are generally used for preliminary assessments, and detailed evaluations should follow using numerical simulation methods. Pipeline failure under internal pressure is typically accompanied by wall separation and tearing. To accurately simulate this process, fracture factors of the material must be considered in the finite element model. This study examines the resistance to internal pressure exhibited by X80 steel pipelines possessing initial axial defects by employing a damage model. A numerical model is established through ABAQUS software, and the fracture response of the model is compared with the theoretical equations. Besides, the burst performance of the pipe under varying defect dimensions, tensile loads, and bending circumstances is investigated.

## 14.2 Damage Mechanics Models

Damage mechanics theory is an approach that can obtain the ductile fracture, which considers the effects of strain variables and local stress by allowing the evolution of damage. Crack initiation and propagation are well described in both coupled and uncoupled damage models. The difference between these two types of models is illustrated in Figure 14.1. The coupled method describes the failure process as a sequence of nucleation, growth, and coalescence of the void, effectively coupling the evolution of damage with the plastic behavior of the material. The uncoupled fracture model is made up of a fracture criterion and a rule for the evolution of the damage. The fracture criterion specifies the fracture strain under a given stress state, whereas the damage evolution rule delineates the accumulation of damage associated with the incremental plastic strain. To assess the effectiveness of damage mechanics models in predicting pipeline fracture

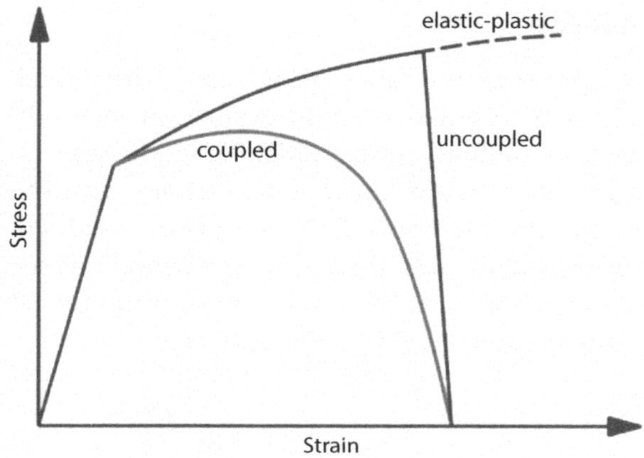


**Figure 14.1** The relationship between coupled–uncoupled damage mechanics models.

behavior, the GTN model is investigated, along with two recently developed uncoupled fracture criteria, the MMC and ERT models in Chapter 10, and data is used for further study in this Chapter.

## 14.3 The Pipe Model

The burst pressure analysis of the X80 steel pipe under internal pressure is performed based on a damage model, using a real 21-inch X80 steel pipe with an initial crack. The pipe has an outer diameter of D0 = 534 mm and a wall thickness of t = 27 mm. The initial defect is a semi-elliptical axial flaw located on the outer surface of the pipe wall. The specific shape of the crack and the geometry of the pipe are shown in Figure 14.2. The crack dimensions are defined by the crack width (δ = 20 mm), crack length (l = 80 mm), and crack depth (d = 14 mm). The mechanical properties are based on the result obtained from the experiments in Chapter 10.

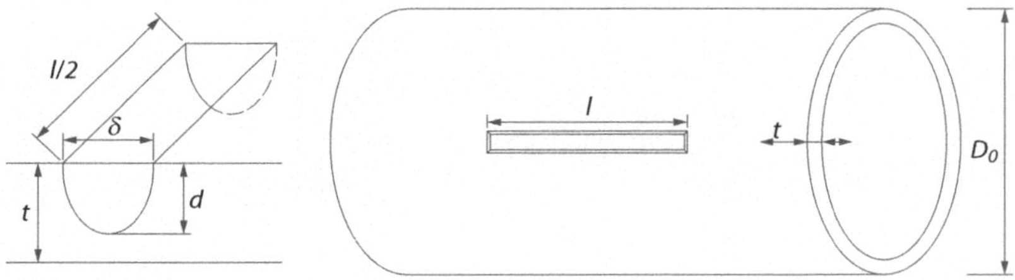


**Figure 14.2** Schematic diagram of the pipe with an initial axial crack.

### 14.3.1 Standards

For the internal pressure resistance of pipelines with initial defects, the most commonly used residual strength evaluation methods include the standard calculation methods from ASME B31G [30] and DNV RP-F101 [31]. The formulas provided in these standards are semi-theoretical and semi-empirical models. The ASME B31G standard includes both the original B31G criterion and the modified B31G criterion. This section focuses on introducing the modified ASME B31G criterion and the DNV RP-F101 criterion only. The modified ASME B31G model is

$$P_{\text{burst}} = \frac{2tf_u}{(D_0 - t)} \frac{\left(1 - 0.85\dfrac{d}{t}\right)}{\left(1 - 0.85\dfrac{d}{tM}\right)} \tag{14.1}$$

$$z = \frac{l^2}{D_0 t} \tag{14.2}$$

$$M = \begin{cases} \sqrt{1 + 0.6275z - 0.003375z^2}, & z \leq 50 \\ 0.032z + 3.3, & z > 50 \end{cases} \tag{14.3}$$

It should be noted that in the ASME B31G standard, only the failure stress is provided for determining whether the pipeline has failed. To maintain consistency with the DNV RP-F101 model, this paper adopts the criterion that pipeline failure is considered to occur when the mid-wall stress reaches the failure stress, and the corresponding internal pressure at that point is taken as the burst pressure, leading to Equation 14.1. DNV RP-F101 model is

$$P_{\text{burst}} = \frac{2tf_u}{(D_0 - t)} \left( \frac{\left(1 - \dfrac{d}{t}\right)}{\left(1 - \dfrac{d/t}{Q}\right)} \right) \tag{14.4}$$

$$Q = \sqrt{1 + 0.31\left(\frac{l}{\sqrt{D_0 t}}\right)^2} \tag{14.5}$$

Upon examining the ASME B31G and DNV RP-F101 standard models, it is observed that in addition to basic pipeline parameters such as outer diameter ($D_0$), wall thickness ($t$), and ultimate tensile strength ($f_u$), two defect-related parameters, defect length ($l$), and defect depth ($d$) have significant influence on the residual strength. These parameters are expressed in dimensionless form as $l_{\text{norm}} = \frac{l}{\sqrt{D_0 t}}$ and $d_{\text{norm}} = \frac{d}{t}$. It is important to note that the defect width does not affect the calculated results in either of the standard models. Figure 14.3 presents the burst pressure surfaces calculated using the ASME B31G and DNV RP-F101 models with varying $l_{norm}$ and $d_{norm}$ values, where the range of $l_{norm}$ is limited to 0–30. From the figure, it is clear that both models show similar trends in their surface predictions. As $l_{norm}$ and $d_{norm}$ increase, the predicted burst pressure decreases, with more rapid changes occurring when $l_{norm}$ is small. When $l_{norm} = 0$, representing a pipe without any defects, the burst pressure predicted by both models is the same. However, when $d_{norm} = 1$, indicating a through-wall defect, the ASME B31G model still predicts a non-zero burst pressure, while the DNV RP-F101 model gives a burst pressure of zero. Figure 14.3 shows the variation in burst pressure with respect to $l_{norm}$ at fixed defect depths of $d_{norm} = 0.3$, 0.5, and 0.7. The results show that when $l_{norm}$ is less than 5, the burst pressure decreases rapidly. As $l_{norm}$ becomes greater than

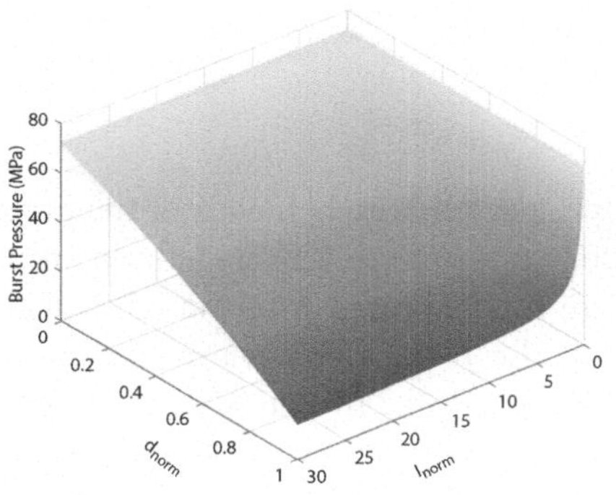


(**a**) ASME B31G

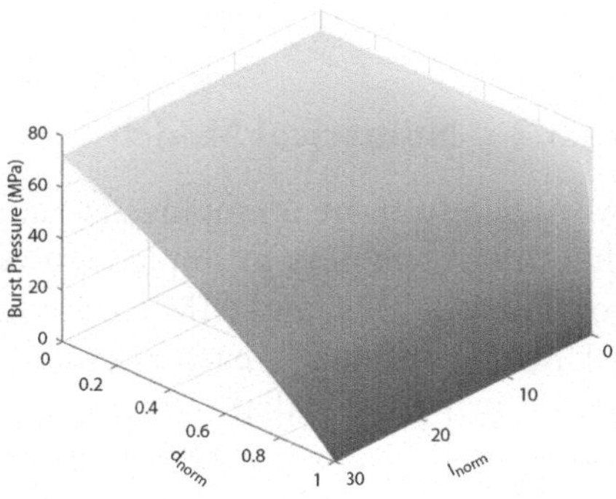


(**b**) DNV RP-F101

**Figure 14.3** Burst pressure surface of the standard models.

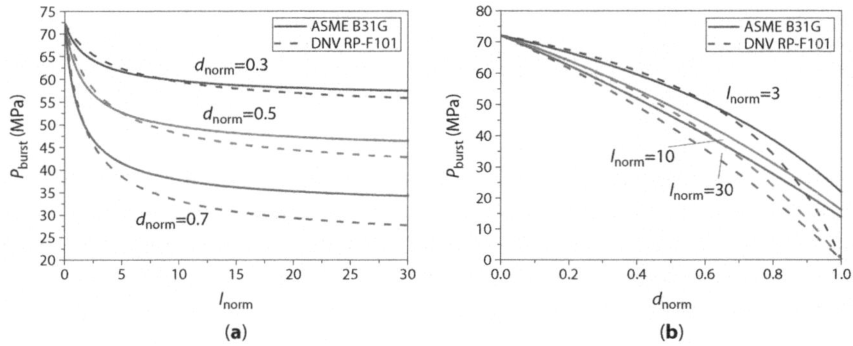


**Figure 14.4** (a) defect length and (b) defect depth effects on bursting pressure.

5, the decrease in burst pressure slows significantly. Additionally, as $l_{norm}$ increases, the decline in burst pressure predicted by the ASME B31G model is less steep than that predicted by DNV RP-F101.

The difference between the two models also becomes more pronounced with increasing defect depth, especially at larger defect lengths. Figure 14.4 shows the variation in burst pressure with $d_{norm}$ at fixed values of $l_{norm}$ = 3, 10, and 30. As $d_{norm}$ increases, both models predict decreasing burst pressure values. However, the DNV RP-F101 model shows a faster rate of decline, and when $d_{norm}$ = 1, it predicts a burst pressure of zero, while the ASME B31G model still gives a non-zero result. Based on this analysis, it can be concluded that when both $l_{norm}$ and $d_{norm}$ are small, the predicted burst pressures from the two models are relatively close. When either $l_{norm}$ or $d_{norm}$ is large, the DNV RP-F101 model gives lower predictions than the ASME B31G model. In particular, when $d_{norm}$ > 0.7 and $l_{norm}$ > 20, the difference between the two models becomes significant, reaching approximately 20%.

## 14.3.2 Numerical Model

ABAQUS software is employed to analyze the fracture behavior of the pipe due to its capability to implement failure criteria *via* a user-defined subroutine. The ERT and MMC criteria are incorporated into the ABAQUS VUMAT subroutine, each coupled with the $J_2$ plasticity framework and solved using the radial return mapping algorithm. In contrast, the GTN model is applied using ABAQUS's built-in "porous metal plasticity" material model. It should be noted that ductile crack propagation is simulated through an element deletion approach.

Due to symmetry, only one-quarter of the pipe is modeled in the FE analysis. An axisymmetric boundary condition along the Z-axis (ZSYMM) is applied at the defect location on the left end surface of the model and along the X-axis (XSYMM) is also applied on the side surfaces of the model. To facilitate control of the boundary conditions, a reference point (RP1) is placed at the center of the cross-section on the defect-free right end of the model. All degrees of freedom at that end are kinematically coupled to RP1. All degrees of freedom at the reference point are constrained except for the Z-axis direction. When studying the effects of axial tension or bending, the tensile load can be directly applied to RP1. Internal pressure is applied as a uniformly distributed radial load on the inner surface of the pipe. In this section, the maximum internal pressure is set to 80 MPa. The total length of the model is set to 800 mm.

The model uses solid elements of type C3D8R, which are reduced integration elements. With sufficiently refined meshing, potential hourglass issues associated with this element type can be avoided. The mesh configuration is shown in Figure 14.5. A finer mesh is used around the crack region, while a coarser mesh is applied near the right end of the model. The maximum element size is approximately 10 mm. The appropriate mesh size for the GTN and ERT models is 0.4 mm, while the MMC model requires a finer mesh of 0.1 mm. However, applying such small elements in the crack region, relative to the size of the pipe, would significantly slow down the computation. Since pipeline bursting is an instantaneous failure process without slow crack propagation, two types of mesh configurations are applied at the defect location, Fine mesh: 0.4 mm uniform mesh. Non-uniform mesh: 2 mm in the axial direction, 1 mm in the circumferential

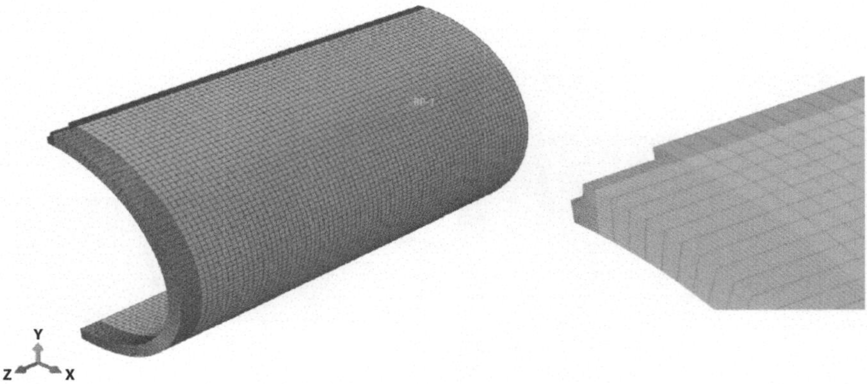


**Figure 14.5** The mesh details of the pipe model.

direction, and 1 mm in the thickness direction. Among these, only the ERT model is simulated with the fine 0.4-mm mesh, while the ERT, MMC, and GTN models are analyzed using the coarser mesh. Figure 14.5 shows the coarse mesh layout for the 21-inch steel pipe model.

A quasi-static analysis is performed using ABAQUS/Explicit, where internal pressure and corresponding axial tensile force are applied simultaneously during the loading step. Throughout the failure process and a short period afterwards, the ratio of kinetic energy to internal energy is maintained below 10% to ensure the accuracy of the results. After this point, the kinetic energy of the model increases sharply. This is because the burst of the pipe under internal pressure occurs swiftly. Once the burst pressure is reached, the pipe wall experiences sudden tearing and failure, which is not a static process.

### 14.3.3 Results

Figure 14.6 compares the deformed finite element model after pipe failure with the experimental results of a burst test on an X65 steel pipe with an initial axial defect [40]. As can be seen, a core fracture zone is formed along the length of the initial defect. The center of the defect expands outward sharply, while the surrounding regions experience relatively less outward deformation. The failure mode predicted by the model closely matches the observations from the physical experiment. The failure of a pipeline under internal pressure can generally be described in three progressive stages. Initially, the pipe diameter begins to expand under the influence of internal pressure until localized damage initiates at the center of the notch. Following this, once the first finite element fails at the notch center, damage propagates along both the axial and thickness directions of

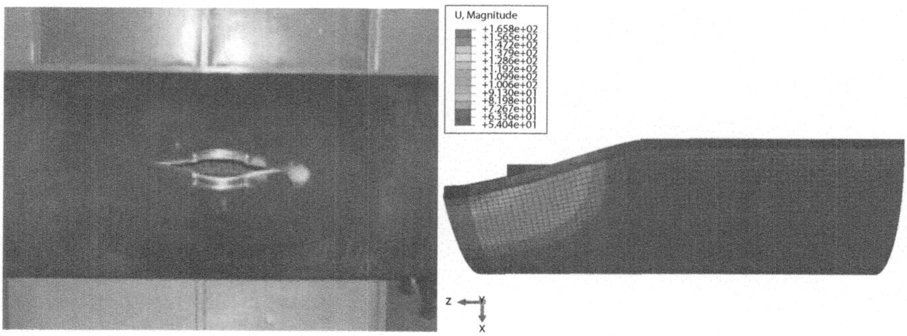


**Figure 14.6** Failure mode of pipeline under internal pressure.

the pipe wall. This progression continues until all elements through the wall thickness at the notch center fail, resulting in complete penetration of the pipe wall. Subsequently, rapid failure occurs along the axial direction, leading to pronounced wall separation. Table 14.1 compares the internal pressures corresponding to the onset of the first element failure and to the full wall penetration with predictions from the ASME B31G and DNV RP F101 standards. Since the internal pressure during the third stage does not significantly increase and is primarily dependent on the axial crack extension, this study focuses only on the failure pressure associated with the first two stages and does not report the third-stage pressure.

Table 14.1 reveals several important observations. First, the model using a fine mesh size of 0.4 mm, suitable for simulating crack propagation in the ERT model, yields results for both the initial element failure and full wall penetration pressures that are very close to those obtained from coarser mesh models. Second, the pressures predicted by all three models at the onset of first element failure and the point of through-thickness penetration are highly consistent with one another. Finally, the predicted pressures are slightly higher than those calculated using the ASME B31G and DNV RP F101 standards, with a maximum deviation within 8%, indicating that the model predictions are accurate and reliable. The reason for the first two observed phenomena may be that under internal pressure, once the pipe wall elements enter a plastic state, even a slight increase in internal pressure

**Table 14.1** Comparison of burst pressures from FEA models and design codes (Unit: MPa).

| **Design code predictions** | | | |
|---|---|---|---|
| **ASME B31G** | | **DNV RP-F101** | |
| 63.46 | | 64.87 | |
| **FE: first element failure** | | | |
| **GTN** | **MMC** | **ERT** | **ERT*** |
| 68.1 | 68.0 | 68.2 | 68.4 |
| **FE: through-thickness penetration** | | | |
| **GTN** | **MMC** | **ERT** | **ERT*** |
| 68.3 | 68.4 | 68.6 | 68.6 |

Note: *indicates the result calculated using a fine mesh (0.4 mm) at the defect location.

can cause a significant increment in plastic strain. Therefore, despite different mesh sizes or fracture strains used in different models, the large plastic strain increment makes the variation between them negligible. As a result, mesh size and model differences have little impact on the pressure result when the first element fails. Furthermore, once the first element fails, the crack rapidly propagates under internal pressure, resulting in a sudden burst rather than a stable crack propagation. Consequently, the pressure at which through-thickness failure occurs is also essentially the same across models. The results of the three models are relatively consistent. The model-predicted pressures are slightly higher than those calculated by design codes, possibly due to (1) the codes not defining the exact shape of the defect whereas the defect shape used in this study may not be the most critical one and (2) the codes not accounting for the ductile fracture process of the steel pipe. Since the results from the three models are largely consistent, the following analysis uses only the results from the MMC model as an example. Additionally, the pressure at which through-thickness failure occurs (i.e., the second stage) is considered the pipeline's burst pressure.

Figure 14.7 shows the element failure conditions under different internal pressures, corresponding to internal pressures of 68 MPa, 68.2 MPa, 68.4 MPa, 68.6 MPa, 68.8 MPa, and 69 MPa, respectively. As shown in Figure 14.7(a), no element failure has occurred at this stage, but the element damage at the notch gradually decreases outward from the center. When the internal pressure reaches 68.2 MPa, as shown in Figure 14.7(b), element failure begins to occur, although some elements still remain intact along the wall thickness direction. In Figure 14.7(c), the range of element failure continues to expand along both the wall thickness and longitudinal directions. By Figure 14.7(d), all the elements at the defect center along the wall thickness direction have failed, indicating structural failure of the pipe. However, as the pressure continues to increase, as seen in Figures 14.7(e) and 14.7(f), the crack propagation speed along the pipe's longitudinal direction increases significantly, and rapid tearing occurs along the length of the pipe wall. The pressure at which the pipe becomes fully penetrated in the thickness direction is taken as the burst pressure, which is determined to be 68.4 MPa for this pipe.

Figures 14.8 and 14.9 show the curves of axial deformation of the pipeline and the opening displacement at the defect location versus internal pressure, respectively. The points marked with an "X" indicate the internal pressure at which burst failure occurs. In the figures, both $\Delta L$ and $\Delta\delta$ represent the results from the quarter model and have not been scaled to represent the full pipeline deformation. Since the pipeline contracts along the

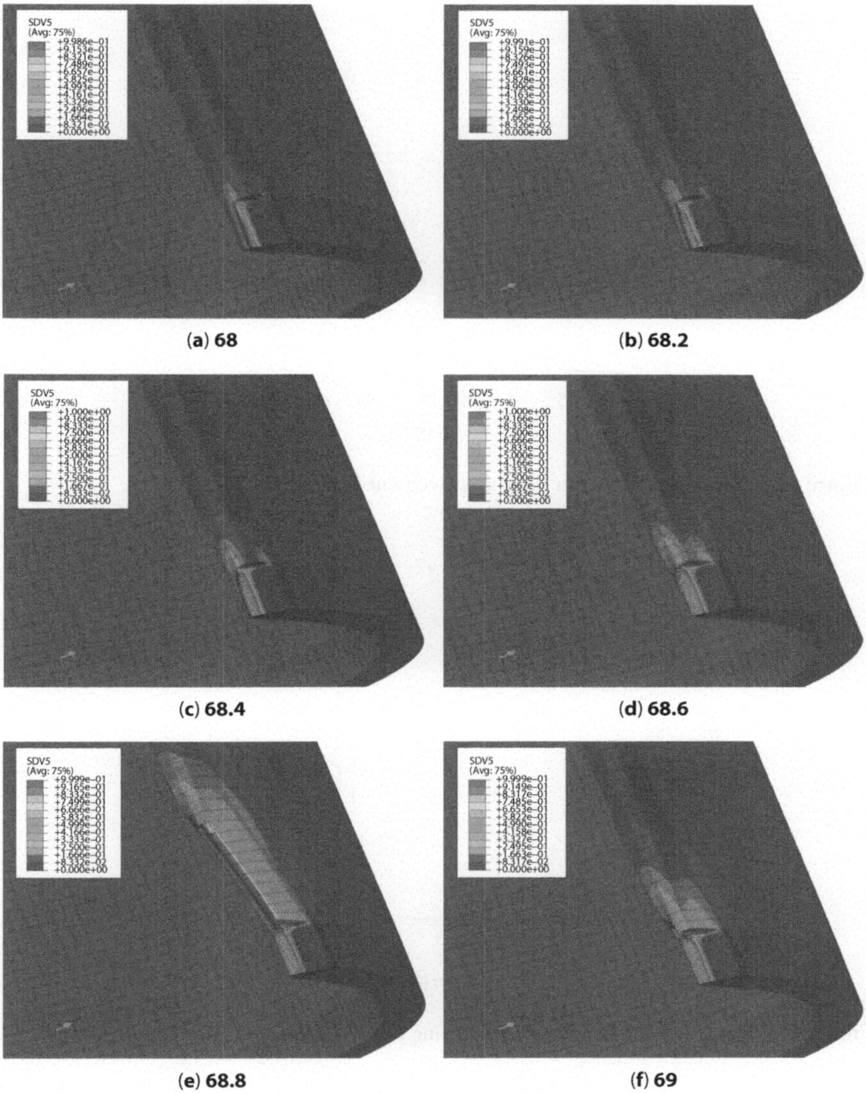


**Figure 14.7** Crack propagation behavior with internal pressure variation.

positive direction of the Z-axis, Figure 14.8 shows that at lower pressures, the axial contraction increases linearly with internal pressure. However, beyond a certain pressure, the axial contraction rapidly increases, indicating that significant plastic deformation has developed at the defect

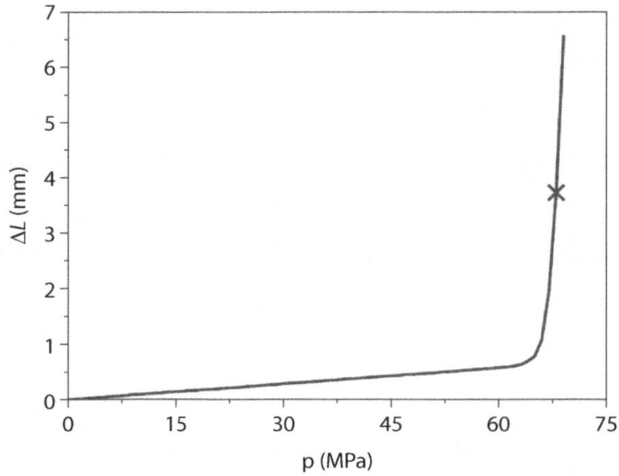


**Figure 14.8** Variation of pipeline axial displacement with internal pressure.

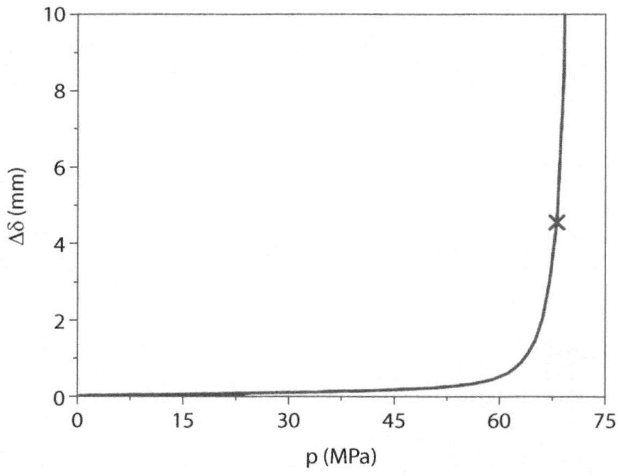


**Figure 14.9** Variation of pipeline crack opening displacement with internal pressure.

location, eventually leading to sudden failure of the pipeline. Similarly, from Figure 14.9, it can be seen that when the internal pressure is low, the change in opening displacement is minimal. Once the pressure reaches approximately 65 MPa, substantial plastic deformation begins to occur, and the opening displacement increases sharply until the pipeline undergoes through-wall failure (with an opening displacement of about 5 mm).

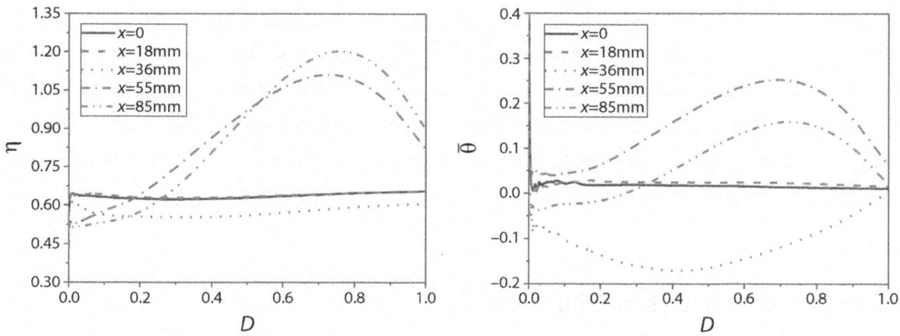


**Figure 14.10** Variation of the stress state at different points along the length direction.

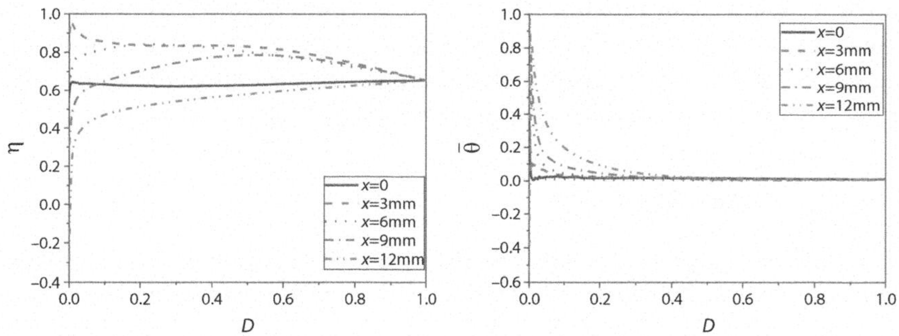


**Figure 14.11** Variation of the stress state at different points along the thickness direction.

After this point, the opening displacement becomes nearly vertical. This is because X80 pipeline steel has relatively weak hardening characteristics, and once failure occurs, even a small additional load can result in significant deformation.

To analyze the stress state in the crack propagation region, the historical curves of stress triaxiality and Lode angle parameter in this region were extracted. Figure 14.10 and Figure 14.11, respectively, show how the stress state along the length and thickness directions changes with the normalized equivalent plastic strain, i.e., the damage variable 5D. The crack initiates from the outer surface defect center of the pipeline. Using this as the reference point, the stress state at various distances from the center was analyzed. It can be observed from the figures that at the crack initiation point, both stress triaxiality and the Lode angle parameter remain almost unchanged with increasing damage. The stress triaxiality is about 0.6, and the Lode angle parameter is approximately 0, indicating a plane strain tensile state. In both the length and thickness directions, the farther the distance from the crack initiation point, the more significant the variation in the stress state. As the load increases, local failure occurs in the material around the defect, and the crack further propagates along the defect direction. At this stage, new fracture surfaces are formed, which explains why the internal pressure can still increase despite ongoing damage.

## 14.4 Parametric Studies

### 14.4.1 Burst Resistance of X80 Steel Pipe with Different Initial Defects

In the assessment of pipeline defects, the remaining strength of the pipeline is a very important factor. This section analyzes the internal pressure resistance of X80 steel pipes with different axial initial defects, focusing primarily on two main defect parameters, crack length and crack depth.

#### *14.4.1.1 Crack Length*

This section analyzes 21-inch X80 steel pipes with initial crack lengths of 30 mm, 80 mm, 200 mm, 400 mm, and 600 mm, corresponding to normalized lengths ($l_{norm}$) of 0.062, 0.44, 2.77, 11.10, and 24.97, respectively. The initial defect depth is 14 mm, with a normalized depth ($d_{norm}$) of 0.52, and the crack width is 20 mm. Table 14.2 compares the simulation results

**Table 14.2** Comparison of burst pressure under different crack lengths (Unit: MPa).

| Crack length | FEM (a) | ASME B31G (b) | DNV RP-F101 (c) | Error (%) | | |
|---|---|---|---|---|---|---|
| | | | | (a) and (b) | (a) and (c) | (b) and (c) |
| $l = 30$ mm | 72.00 | 68.21 | 68.53 | 5.3 | 4.8 | 0.5 |
| $l = 50$ mm | 68.00 | 63.46 | 64.87 | 6.7 | 4.6 | 2.2 |
| $l = 200$ mm | 59.00 | 53.78 | 52.86 | 8.8 | 10.4 | 1.7 |
| $l = 400$ mm | 50.00 | 46.12 | 44.21 | 7.8 | 11.6 | 4.1 |
| $l = 600$ mm | 45.75 | 43.77 | 40.42 | 4.3 | 11.7 | 7.6 |

with the ASME standard results for different crack lengths. As seen in the table, the simulation results agree relatively well with the ASME standard. However, as the defect length increases, the error between the simulation results and the ASME and DNV standards also increases.

As shown in Figure 14.12, the burst pressure of the pipeline decreases with the increase in defect length, while the crack mouth opening displacement at failure remains approximately 5 mm. This indicates that the defect length does not affect the displacement at the point of failure. Figure 14.13 compares the burst pressure predicted by the model with the results from standard formulas for different defect lengths. The model accurately captures the nonlinear decreasing trend of burst pressure with increasing normalized defect length. When $l_{norm} < 5$, the burst pressure decreases rapidly with increasing $l_{norm}$. When $l_{norm} > 15$, the burst pressure decreases more gradually.

The largest error between the model and the standards occurs around $l_{norm} = 8$, with a maximum discrepancy of less than 10%. The model predictions are slightly higher than those of the standards, which could be due to the lack of a clearly defined defect shape and the absence of ductile fracture considerations in the standards. When $l_{norm} = 25$, the model result closely matches the ASME standard, but is still significantly higher than the DNV standard.

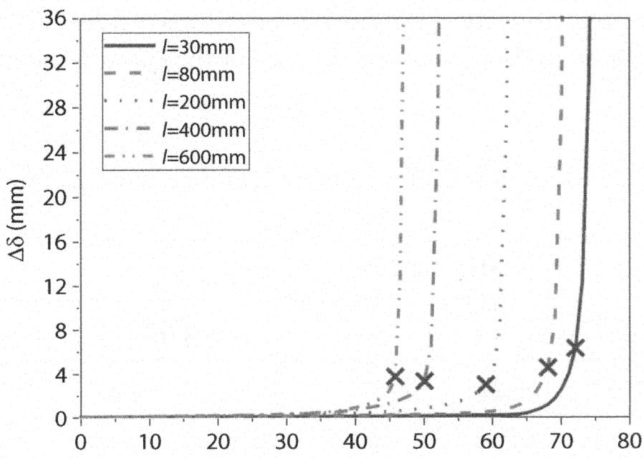


**Figure 14.12** Variation of displacement with internal pressure effect of crack length.

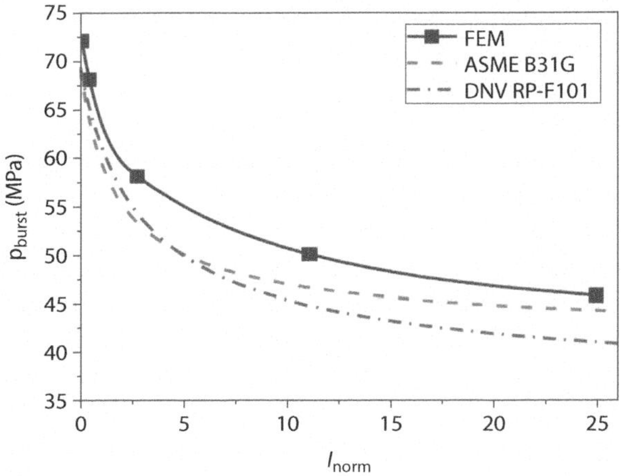


**Figure 14.13** Variation of crack opening displacement with internal pressure effect of crack length.

### *14.4.1.2 Crack Depth*

This section analyzes 21-inch X80 steel pipes with initial defect depths of 5 mm, 10 mm, 14 mm, 18 mm, and 22 mm, corresponding to normalized depths ($d_{norm}$) of 0.19, 0.37, 0.52, 0.67, and 0.81, respectively. The initial defect length is 200 mm, corresponding to a normalized length ($l_{norm}$) of 2.77; the defect width is 20 mm. Table 14.3 presents a comparison between the model results and the standard predictions under different defect depths. It can be seen that when the defect depth is small, the model results

**Table 14.3** Comparison of burst pressure under different crack depths (Unit: MPa).

| Crack length | FEM (a) | ASME B31G (b) | DNV RP-F101 (c) | Error (%) | | |
|---|---|---|---|---|---|---|
| | | | | (a) and (b) | (a) and (c) | (b) and (c) |
| $d = 5$ mm | 70.00 | 65.27 | 68.21 | 7.8 | 6.8 | 0.5 |
| $d = 10$ mm | 65.00 | 59.84 | 63.46 | 9.8 | 7.9 | 2.2 |
| $d = 14$ mm | 59.00 | 53.78 | 52.86 | 8.8 | 10.4 | 1.7 |
| $d = 18$ mm | 52.50 | 45.14 | 45.73 | 12.9 | 14.0 | 1.3 |
| $d = 22$ mm | 42.75 | 36.71 | 31.85 | 14.1 | 25.0 | 13.2 |

are close to both standards; when the defect depth is large, the model results align more closely with the ASME standard. Figure 14.14 shows the variation of the opening displacement at the defect location under different defect depths as internal pressure increases. The burst pressure of the pipe decreases as the defect depth increases, but the defect depth has no effect on the opening displacement at the time of pipe failure.

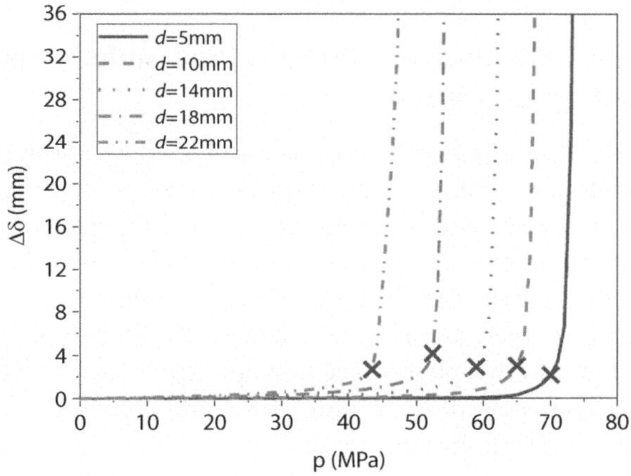


**Figure 14.14** Variation of displacement with internal pressure effect of crack depth.

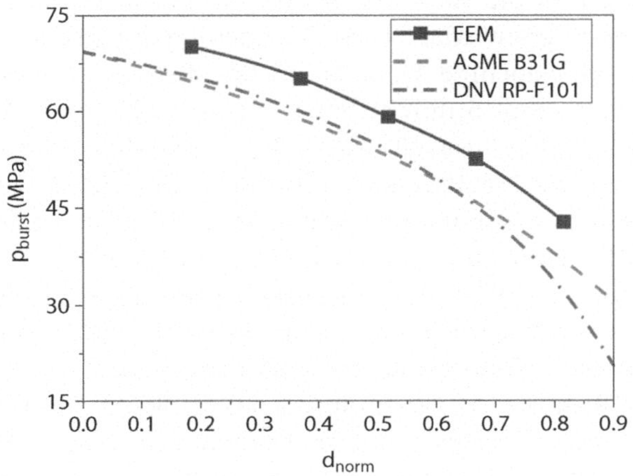


**Figure 14.15** Variation of crack opening displacement with internal pressure effect of crack depth.

Figure 14.15 presents a comparison between the model-predicted burst pressures and standard empirical formulas for various defect depths. The maximum deviation between the model predictions and the empirical formulas is within 10%. The trend predicted by the ASME formula closely matches the model results, where the burst pressure decreases more rapidly with increasing dnorm. When the defect depth exceeds dnorm > 0.7, the DNV prediction shows a faster rate of decline compared to the simulation results.

### 14.4.2 Burst Resistance of X80 Steel Pipe under Combined Loading Conditions

During service, oil and gas pipelines are subjected to combined external loads such as bending, axial force, internal pressure, and external pressure. Internal pressure typically does not occur in isolation. This section investigates the burst pressure of X80 pipeline under the influence of axial force and bending curvature. The initial defect length is set to 200 mm, the defect depth to 14 mm, and the defect width to 20 mm. The loading steps are divided into two phases: the first applies an axial force or bending curvature at reference point RP-1, and the second applies uniform internal pressure until pipeline failure occurs.

#### *14.4.2.1 Axial Loads*

Figure 14.16 shows the variation of opening displacement at the defect location under different axial forces. The opening displacement here refers to the quarter-pipe opening. Upon failure, the displacement increases rapidly and reaches about 5 mm at burst. Figure 14.17 illustrates the burst pressures under different axial loads. The results show that tensile force slightly increases the burst pressure of the pipe; for example, a tensile force of 5000 kN results in a burst pressure of about 63 MPa, while 10,000 kN yields 63.3 MPa, indicating an increase of approximately 5%. In contrast, compressive force significantly reduces burst pressure: at an axial compressive force of 10,000 kN, the burst pressure drops to 38 MPa, a reduction of about 37% compared to the pressure without axial load.

Figure 14.18 shows the variation of axial displacement of the pipeline under different axial forces as internal pressure increases. Negative values of $\Delta L$ indicate axial elongation of the pipeline. As seen in the figure, the initial segment of the curves exhibits linear behavior, which is due to the pipeline remaining within the elastic range under the applied axial force. During the application of internal pressure, the axial displacement

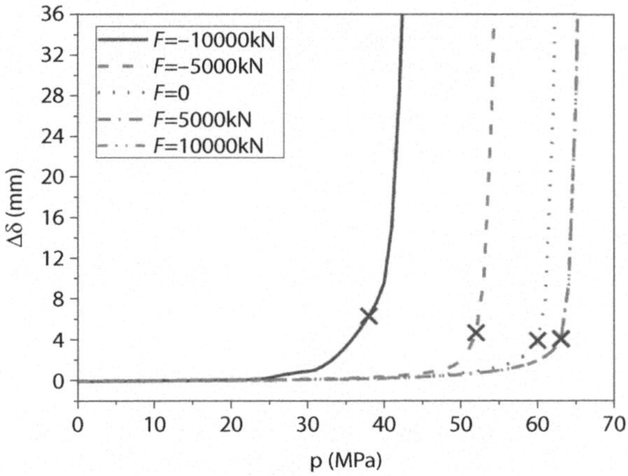


**Figure 14.16** Variation of opening displacement with internal pressure–axial force.

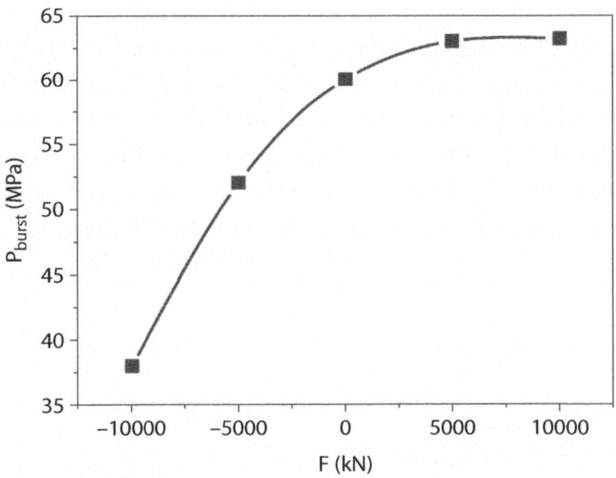


**Figure 14.17** Variation of opening displacement with internal pressure–axial force.

contracts inward at a similar rate. When a burst occurs, the pipeline undergoes a sudden contraction.

### *14.4.2.2 Bending*

In this section, a 21-inch X80 steel pipe is analyzed under initial bending angles of 0.01 rad, 0.005 rad, 0, 0.005 rad, and 0.01 rad. The corresponding bending curvatures are 0.0125 $m^{-1}$, 0.00625 $m^{-1}$, 0, 0.00625 $m^{-1}$, and 0.0125

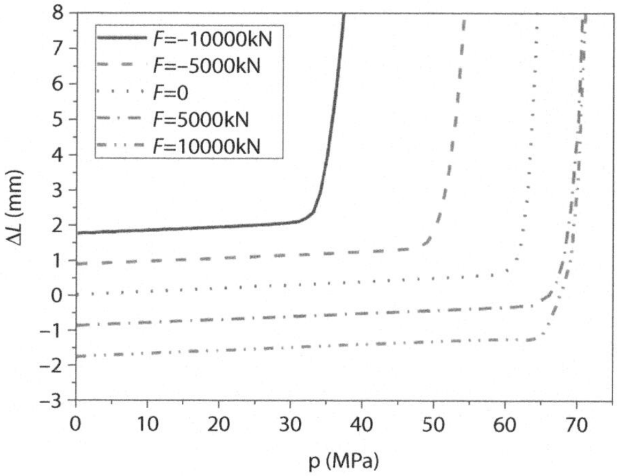


**Figure 14.18** Variation of axial force with internal pressure under different axial displacements.

$m^{-1}$, respectively. The negative sign (−) indicates that the bending direction causes compressive stress along the axial direction at the defect location. Figures 14.19 and 14.20 show the variation of the opening displacement at the defect under different bending curvatures and the corresponding burst pressure under those conditions. From the figures, it can be observed that the influence of initial bending curvature on burst pressure is similar to

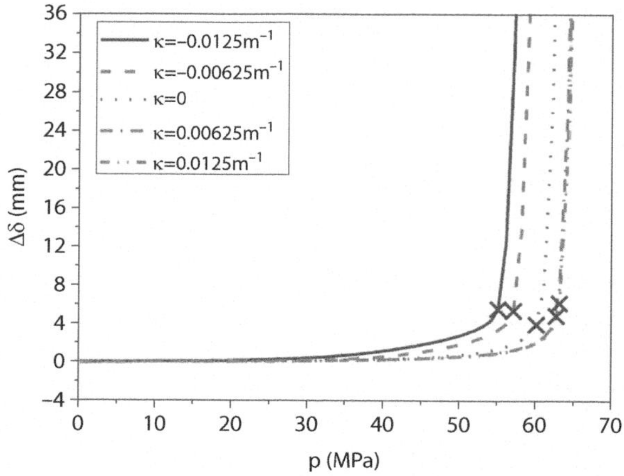


**Figure 14.19** Variation of opening displacement with internal pressure–bending curvature.

that of axial force. When bending causes tensile stress at the defect location, the pipe's burst pressure increases. However, the increase in burst pressure does not significantly rise with greater curvature. At a curvature of 0.0125 $m^{-1}$, the burst pressure is approximately 63 MPa. Conversely, when bending causes compressive stress at the defect location, the burst pressure of the pipe significantly decreases, reaching about 55 MPa at a curvature of 0.0125 $m^{-1}$.

Figure 14.21 shows the variation of the pipe's bending moment under different bending curvatures as internal pressure increases. The minus sign

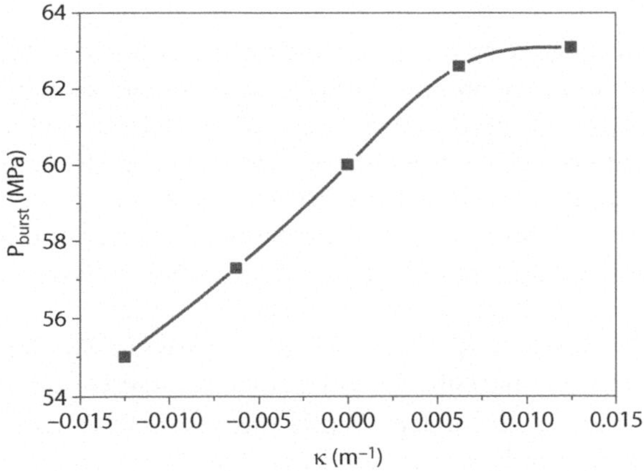


**Figure 14.20** Variation of burst pressure with bending curvature.

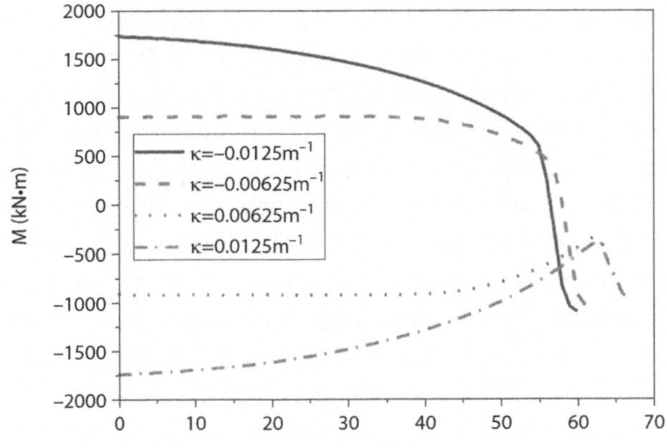


**Figure 14.21** Variation of bending moment with internal pressure under different bending curvatures.

(−) indicates elongation along the axial direction. As seen in the figure, the initial portion of each curve exhibits linear behavior, indicating that the pipe remains in the elastic state under the initial curvature. When the curvature is 0.00625 $m^{-1}$, the bending moment remains nearly constant under increasing internal pressure before eventually decreasing until failure. When the curvature is 0.0125 $m^{-1}$, the bending moment gradually decreases under internal pressure until failure occurs.

## 14.5 Conclusions

This paper first compares the empirical formulas from ASME B31G and DNV RP-F101 and then applies three damage models to analyze the internal pressure resistance of X80 steel pipes with initial axial defects. Since the predictions from the three models are close, only the MMC model is used in subsequent analyses to investigate the effects of defect length and depth on burst pressure. Additionally, the impact of axial force and bending curvature on burst pressure is studied. The main conclusions are as follows:

1. ASME B31G and DNV RP-F101 provide similar burst pressure predictions for small defect lengths and depths. However, as the defect depth increases, DNV predicts a faster drop in burst pressure than ASME. When dnorm = 1, DNV predicts zero burst pressure, while ASME does not.
2. Damage models can accurately simulate the internal pressure resistance of X80 steel pipes with initial axial defects and the through-wall failure process. Crack initiation occurs at the center of the outer wall defect. At this location, the stress triaxiality is approximately 0.6 and the Lode angle parameter is nearly 0, indicating a plane strain tensile state.
3. When the defect depth is between 5 and 22 mm and the defect length is between 30 and 600 mm, the stress triaxiality and Lode angle parameter at the crack initiation location remain nearly constant. Thus, the variation in burst pressure depends mainly on the increase in plastic strain.
4. Burst pressure is highly sensitive to defect length when the length is small and less sensitive at greater lengths. Larger defect depth leads to significantly lower burst pressure. The effects of axial force and bending curvature are similar: when tensile stress is applied at the defect, burst pressure increases slightly; under compressive stress, it decreases significantly.

## Bibliography


1. Yan, S.T., Shen, X.L., Jin, Z.J., Ye, H., On elastic-plastic collapse of subsea pipelines under external hydrostatic pressure and denting force. *Appl. Ocean Res.*, 58, 305–321, 2016.
2. Mousselli, A., *Offshore pipeline design, analysis, and methods*, Penn Well Books, Tulsa, OK, 1981.
3. A. Engineers, Api 579-1/asme ffs-1 fitness-for-service, American Society of Mechanical Engineers, 2016.
4. BSI, BS 7910: Guide to methods for assessing the acceptability of flaws in metallic structures, British Standards Institution, 2005.
5. EDF, R6: Assessment of the Integrity of Structures Containing Defects, Revision 4, British Energy Generation Limited, 2001.
6. Jia, P., Jing, H., Xu, L., Han, Y., *et al.*, A modified reference strain method for engineering critical assessment of reeled pipelines. *Int. J. Mech. Sci.*, 105, 23–31, 2016.
7. Gurson, A.L., Continuum theory of ductile rupture by void nucleation and growth: Part i—yield criteria and flow rules for porous ductile media. *J. Eng. Mater. Technol.*, 99, 1, 2–15, 1977.
8. Chu, C. and Needleman, A., Void nucleation effects in biaxially stretched sheets. *J. Eng. Mater. Technol.*, 102, 3, 249–256, 1980.
9. Tvergaard, V., On localization in ductile materials containing spherical voids. *Int. J. Fract.*, 18, 237–252, 1982.
10. Gologanu, M., Leblond, J.-B., Devaux, J., Approximate models for ductile metals containing non-spherical voids—case of axisymmetric prolate ellipsoidal cavities. *J. Mech. Phys. Solids*, 41, 11, 1723–1754, 1993.
11. Gologanu, M., Leblond, J.-B., Devaux, J., Approximate models for ductile metals containing nonspherical voids—case of axisymmetric oblate ellipsoidal cavities. *J. Eng. Mater. Technol.*, 116, 3, 290–297, 1994.
12. Tvergaard, V. and Needleman, A., Analysis of the cup-cone fracture in a round tensile bar. *Acta Metall.*, 32, 1, 157–169, 1984.
13. Benseddiq, N. and Imad, A., A ductile fracture analysis using a local damage model. *Int. J. Press. Vessels Pip.*, 85, 4, 219–227, 2008.
14. Liao, R., Sun, Y., Liu, J., Zhang, W., Applicability of damage models for failure analysis of threaded bolts. *Eng. Fract. Mech.*, 78, 3, 514–524, 2011.
15. Sirinakorn, T., Wongwises, S., Uthaisangsuk, V., A study of local deformation and damage of dual phase steel. *Mater. Des.*, 64, 729–742, 2014.
16. Wang, L.Y. and Li, L., Parameter identification of gtn model using response surface methodology for high-strength steel br1500hs. *J. Mater. Eng. Perform.*, 26, 3831–3838, 2017.
17. Rice, J.R. and Tracey, D.M., On the ductile enlargement of voids in triaxial stress fields. *J. Mech. Phys. Solids*, 17, 3, 201–217, 1969.

18. Gruben, G., Hopperstad, O., Børvik, T., Evaluation of uncoupled duc- tile fracture criteria for the dual-phase steel docol 600dl. *Int. J. Mech. Sci.*, 62, 1, 133–146, 2012.
19. Nahshon, K. and Hutchinson, J., Modification of the gurson model for shear failure. *Eur. J. Mech. A. Solids*, 27, 1, 1–17, 2008.
20. Bai, Y., Effect of loading history on necking and fracture, Ph.D. thesis, Massachusetts Institute of Technology, Cambridge, USA, 2007.
21. Wyllie, D.C., *Foundations on rock: engineering practice*, CRC Press, Boca Raton, Florida, USA, 2003.
22. Pack, K., Luo, M., Wierzbicki, T., Sandia fracture challenge: blind prediction and full calibration to enhance fracture predictability. *Int. J. Fract.*, 186, 155–175, 2014.
23. Pack, K. and Marcadet, S.J., Numerical failure analysis of three-point bending on martensitic hat assembly using advanced plasticity and fracture models for complex loading. *Int. J. Solids Struct.*, 85, 144–159, 2016.
24. Kofiani, K., Nonn, A., Wierzbicki, T., New calibration method for high and low triaxiality and validation on sent specimens of api x70. *Int. J. Press. Vessels Pip.*, 111, 187–201, 2013.
25. Paredes, M., Wierzbicki, T., Zelenak, P., Prediction of crack initiation and propagation in x70 pipeline steels. *Eng. Fract. Mech.*, 168, 92–111, 2016.
26. Paredes, M., Lian, J., Wierzbicki, T., Cristea, M.E., Münstermann, S., Darcis, P., Modeling of plasticity and fracture behavior of x65 steels: seam weld and seamless pipes. *Int. J. Fract.*, 213, 17–36, 2018.
27. Han, P., Cheng, P., Yuan, S., Bai, Y., Characterization of ductile frac- ture criterion for api x80 pipeline steel based on a phenomenological approach. *Thin Walled Struct.*, 164, 107254, 2021.
28. Saneian, M., Han, P., Jin, S., Bai, Y., Fracture response of steel pipelines under combined tension and torsion. *Thin Walled Struct.*, 154, 106870, 2020.
29. Saneian, M., Han, P., Jin, S., Bai, Y., Fracture response of steel pipelines under combined tension and bending. *Thin Walled Struct.*, 155, 106987, 2020.
30. ASME, B31G: manual for determining the remaining strength of corroded pipelines: a supplement to ANSI/ASME B31 code for pressure piping, American Society of Mechanical Engineers, 1991.
31. DNV, RP F101: recommended practice RP F101 corroded pipelines, Det Norske Veritas, 1999.
32. Bai, Y. and Wierzbicki, T., A new model of metal plasticity and fracture with pressure and lode dependence. *Int. J. Plast.*, 24, 6, 1071–1096, 2008.
33. Bai, Y. and Wierzbicki, T., Application of extended mohr–coulomb criterion to ductile fracture. *Int. J. Fract.*, 161, 1, 1–20, 2010.
34. ASTM, E8/E8M-16a: standard test methods for tension testing of metallic materials, ASTM International, West Conshohocken, PA, USA, 2016.
35. Swift, H., Plastic instability under plane stress. *J. Mech. Phys. Solids*, 1, 1, 1–18, 1952.

36. Tvergaard, V., Influence of voids on shear band instabilities under plane strain conditions. *Int. J. Fract.*, 17, 389–407, 1981.
37. Cho, W.Y., Kim, K.S., Lee, J.H., Yoon, Y.C., Inelastic analysis of the sent specimen of x80 pipe for strain capacity evaluation, in: *International Conference on Offshore Mechanics and Arctic Engineering*, vol. 49965, American Society of Mechanical Engineers, p. V005T04A034, 2016.
38. Qiang, B. and Wang, X., Ductile crack growth behaviors at different locations of a weld joint for an x80 pipeline steel: A numerical investigation using gtn models. *Eng. Fract. Mech.*, 213, 264–279, 2019.
39. GB/T21143-2014, Metallic materials-unified method of test for determination of quasistatic fracture toughness, National Standards of People's Republic of China, 2014.
40. Oh, C.S., Kim, N.H., Kim, Y.J., Baek, J.H., Kim, Y.P., Kim, W.S., A finite element ductile failure simulation method using stress-modified fracture strain model. *Eng. Fract. Mech.*, 78, 1, 124–137, 2011.

# 15

# Fracture Study of X80 Steel Based on Phase Field Method


***Abstract***

This chapter explores the fracture behavior of API X80 pipeline steel using a phase-field method (PFM), an advanced computational technique that enables simulation of complex crack propagation without predefined paths. Traditional fracture mechanics approaches, such as CTOD, J-integral, and cohesive zone models, are discussed for context, along with their limitations in modeling ductile fracture and arbitrary crack paths. Experimental tests—including uniaxial tensile and notched specimen fracture tests—are performed to extract mechanical properties and validate the numerical models. A 2D finite element model is developed in ABAQUS, incorporating both elastic and elastic–plastic material properties. The PFM is implemented through a staggered algorithm using UMAT and UEL subroutines. Simulation results reveal that the elastic-only model captures sudden fracture well, while the plastic model produces a plateau in the force–displacement curve, consistent with ductile behavior. However, the current PFM code is better suited for brittle fracture and does not simulate ductile crack growth adequately. The study highlights the accuracy of PFM in predicting crack initiation and urges further development for ductile fracture applications, especially in high-performance steels like X80.



***Keywords*:** Phase-field method, X80 steel, fracture simulation, crack propagation, finite element modeling, elastic–plastic behavior, experimental validation


## 15.1 Introduction

X80 steel, a high-strength low-alloy (HSLA) grade, has been extensively studied due to its favorable combination of strength, toughness, and weldability, making it a material of choice in various structural applications. Understanding its fracture behavior is critical for preventing catastrophic failure, especially in environments prone to corrosion or mechanical


Yong Bai. Plasticity and Ductile Fracture in Structural Engineering, (427–440) © 2026 Scrivener Publishing LLC

overload. Traditional fracture analysis methods, such as linear elastic fracture mechanics (LEFM), have been used to characterize crack initiation and growth in metals. However, LEFM assumes small-scale yielding and linear elastic behavior, which are not always valid for high-strength steels like X80, especially under ductile fracture conditions [1]. To overcome these limitations, cohesive zone models (CZMs) and elastic–plastic fracture mechanics (EPFM) have been employed. For example, studies have used CZM to describe the initiation and progression of microcracks in X80 steel under tensile loading, providing improved predictions of crack resistance curves (R-curves) [2]. The response of fracture mechanics depends on CTOD (crack-tip opening displacement) and the J-integral (energy release rate). This mono-parametric approach describes the severity of crack growth at initiation and stable states and assesses the unstable state of ductile materials as follows [3]. Nevertheless, if the crack starts to expand, the J-integral is not applicable anymore [4]. Therefore, Paris *et al.* [5] presented the J-integral R-curve approach for unstable cracks. Moreover, the CTOD and J parameters are well known for their validation in the crack-tip fields characterized in certain cases [6]. Nyhus *et al.* [7] and Brocks *et al.* [8] proposed normalization methods to make the resistance curve independent of loading mode and structural geometry. This method does not apply to real operations, because it has not modeled tearing explicitly, and another problem of this approach was its disability to capture the plasticity that occurs in a tearing crack locality [9]. Bai [10] presented a modified version of the Mohr-Coulomb criterion which had been used before as an uncoupled method for rock and soil. Modified Mohr-Coulomb (MMC) criterion obtained the alteration in fracture strain at the small variety of stress triaxiality and predicted the fracture strain by way of a stress triaxiality function in-plane stress. Saneian *et al.* [11, 12] established a 3D elastic-plastic model by applying the MMC method, to determine the fracture response of cracked X80 steel pipe under simultaneous tension and torsion and tension and bending. Nevertheless, discrete models such as CZM typically require predefined crack paths or complex remeshing, limiting their ability to capture arbitrary crack propagation. To address this, diffuse fracture models such as the phase-field method (PFM) have been introduced. PFM treats cracks as a continuous damage process governed by energy minimization principles and can simulate complex crack phenomena without explicit crack tracking [13]. Recent research has demonstrated the effectiveness of PFM in simulating fracture behavior in steels. Wu *et al.* [14] applied a phase-field model to study ductile fracture in high-strength steels and confirmed that the model could reproduce experimental crack patterns with high accuracy. Guo *et al.* [15] implemented a PFM framework for simulating hydrogen-induced cracking in X80 steel, showing good agreement with experimental observations and highlighting the method's ability to capture environmentally assisted cracking. Moreover, experimental investigations have revealed that the fracture behavior of X80 steel is strongly influenced by microstructural features, such as grain size, bainitic phases, and inclusions, which affect both the crack initiation toughness and propagation resistance [16]. The integration of microstructural characteristics into computational models remains an active area of research, aiming to enhance the predictive capabilities of phase-field-based simulations. Fang *et al.* [17] developed a variational framework for phase-field fracture modeling integrated with multi-surface plasticity and investigated how different yield criteria and hardening laws influence the material response. Subsequently, they implemented phase-field modeling for elastoplastic fracture in Abaqus using a staggered algorithm *via* UEL and UMAT subroutines [18]. Despite significant progress, the literature still lacks studies that comprehensively validate phase-field models against experimental fracture data specific to X80 steel under controlled laboratory conditions. The current study addresses this gap by implementing and validating a phase-field model for X80 fracture using experimental results, offering a more robust tool for predicting material failure.

## 15.2 Experiments

Material tests and fracture tests on X80 steel are performed in this section. Corresponding to API Spec 5L [1], X80 pipeline steel is utilized in this model. All specimens were extracted from a 21-inch API X80 UOE pipeline using electrical discharge machining. All tests are conducted at room temperature on an Instron universal test machine shown in Figure 15.1, with a maximum load of 250 kN. The measurement precision of the force value is 0.5%. Besides, an extensometer has been used to measure the displacements.

### 15.2.1 Uniaxial Tensile Test

The large-diameter UOE pipes used to create the work's specimens were subjected to rolling and cold forming, two processes that unavoidably changed the metal structure's anisotropy and plasticity. Consequently, tests are conducted on flat tensile specimens in three distinct directions. The ASTM [2] standard tensile specimens are cut from various orientations. The specimens have a thickness of 5 mm. Figure 15.2 provides a

**Figure 15.1** Instron universal test machine.

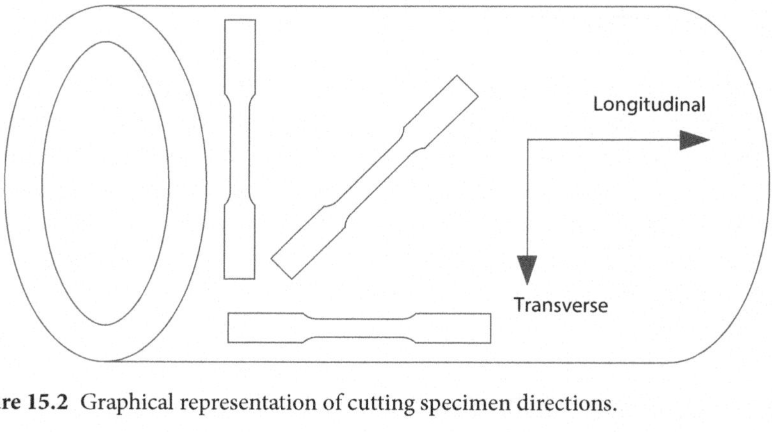


**Figure 15.2** Graphical representation of cutting specimen directions.

schematic representation of the specimens. In the present model, material behavior is isotropic. The longitudinal direction of the pipeline is shown by "L 0°," parallel to the transverse direction is displayed by "T 90°," and between longitudinal and transverse direction is demonstrated by "LT 45°." The mechanical properties of the X80 offshore pipeline are chosen for the simulation as shown in Table 15.1. According to the specification ASTM E8/E8M [3], API X80 steel has no obvious yield plateau, so the stress corresponding to the residual plastic strain amount of 0.2% is considered yield stress. To ensure repeatability, three tests are run for every direction. A load cell and an extensometer measure and record the elongation of the gauge section and the load, respectively. The loading rate is set at 0.5 mm/min. The specimens before and after the test are shown in Figure 15.3. As the load on the specimen increases, they continue to elongate. Once the load reaches a certain point, the test section of the specimen undergoes necking, which is followed by a fracture in the necked area. Figure 15.4(a) presents the engineering stress–strain curves, and the true stress–strain curves are presented in Figure 15.4(b). These curves are sketched up to 5% engineering strain for specimens from three directions because necking occurs in the specimen when the engineering stress reaches 5%. Figure 15.4(a) and Figure 15.4(b) show that at the initial yield strength point, the pipe material shows three distinct directions and a specific orientation-dependent

**Table 15.1** Mechanical tensile properties at room temperature of the API X80 steel.

| Young's modulus E (MPa) | Poisson's ratio v | Strain hardening exponent n | Yield strength (MPa) | | | Ultimate strength (MPa) | | |
|---|---|---|---|---|---|---|---|---|
| | | | L 0° | LT 45° | T 90° | L 0° | LT 45° | T 90° |
| 210,000 | 0.3 | 0.095 | 580 | 570 | 653 | 650 | 630 | 668 |

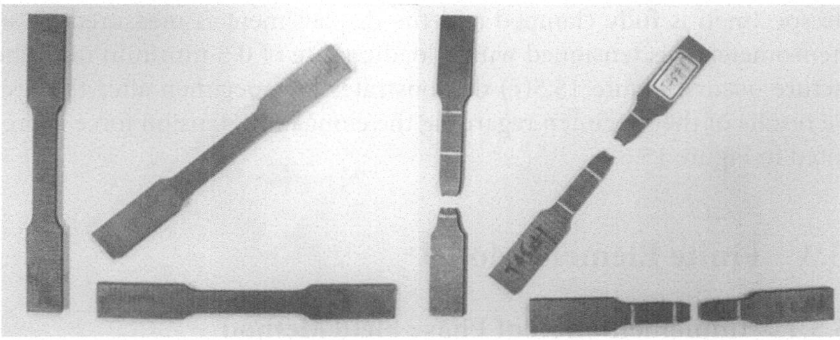


**Figure 15.3** The specimen (a) before and (b) after the test.

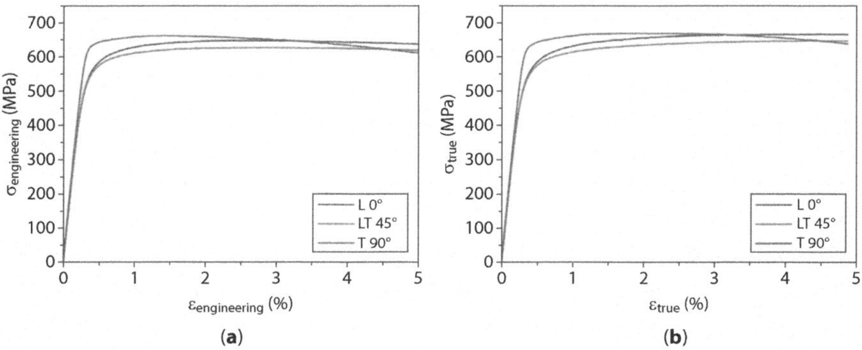


**Figure 15.4** Material test curves. (a) Engineering stress–strain curves and (b) true stress–strain curves.

behavior; however, as deformation rises, the disparity diminishes. When the strain is greater than 3%, the greatest difference between the flow stress in the three orientations is 3.3%. The difference in stress in huge strain is insignificant since the fracture happens after strain localization as particle cracking. Since anisotropy might result in macroscopic delamination cracks, this analysis assumes that the material is isotropic [4, 5].

### 15.2.2 Fracture Test of Notched Specimens

A specimen with a center hole is made to capture the fracture characteristic of the material, as shown in Figure 15.5(a). The detailed physical dimensions of the specimens are shown in Figure 15.5(b). The specimen is only cut out in the longitudinal direction of the pipe. The fracture test for the specimen is repeated three times to get reliable and repeatable results. The specimen is fully clamped and the displacement is measured by an extensometer. It is tensioned with a loading rate of 0.5 mm/min until the fracture occurs. Figure 15.5(c) demonstrates the specimen after the test. The results of the specimen regarding the elongation–tension force is presented in Figure 15.6.

## 15.3 Finite Element Model

### 15.3.1 Implementation of Phase Field Method

The implementation of the phase field method used in this study is based on Fang *et al.* [6]. For the completeness of this paper, the basics are introduced

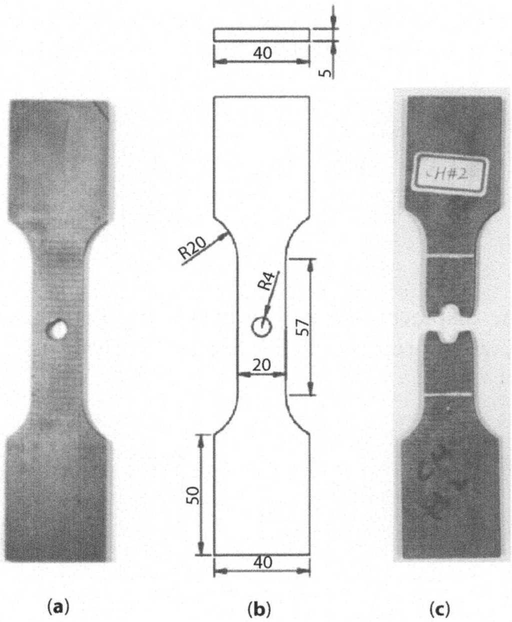


**Figure 15.5** Fracture specimen. (a) Notched sample and (b) the geometry details of the sample (c) after the test.

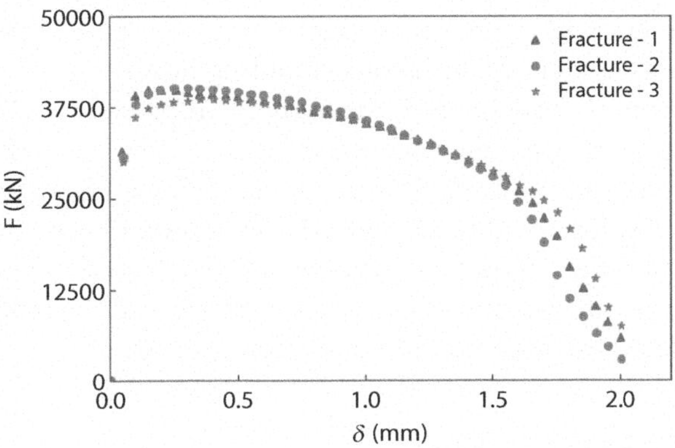


**Figure 15.6** The elongation–tension force curves for fracture specimens.

in this section. The displacement field and phase field are decoupled by the authors, and staggered algorithm is employed to proceed the calculation. In other words, the displacement field can be solved by freezing the phase field at the last loading increment. Likewise, the phase field can be solved by freezing the displacement field and internal plastic parameters at the last loading increment. The corresponding Newton-Raphson iteration may be written as

$$\begin{Bmatrix} \mathbf{u} \\ \mathrm{d} \end{Bmatrix}_{n+1} = \begin{Bmatrix} \mathbf{u} \\ d \end{Bmatrix}_{n} - \begin{bmatrix} \mathbf{K}^{uu} & 0 \\ 0 & \mathbf{K}^{dd} \end{bmatrix}^{-1} \begin{Bmatrix} \mathbf{r}^{u} \\ r^{d} \end{Bmatrix}_{n} \tag{15.1}$$

in which $\mathbf{u}$ is the displacement field and $d$ being the phase field. $\mathbf{K}$ is the stiffness matrix, while r represents the residual vectors. The stiffness matrix is

$$\mathbf{K}_{ij}^{uu} = \frac{\partial r_i^u}{\partial u_j} = \int_\Omega \left[ (1-d)^2 + k \right] (\mathbf{B}_i^u)^{\mathrm{T}} \mathbf{E}^{\mathrm{epc}} \mathbf{B}_j^u \,\mathrm{d}\Omega \tag{15.2}$$

$$\mathbf{K}_{ij}^{dd} = \frac{\partial r_i^d}{\partial d_j} = \int_\Omega \left\{ g_f l_c (\mathbf{B}_i^d)^{\mathrm{T}} \mathbf{B}_j^d + \left[ \frac{g_f}{l_c} + 2\mathcal{H} \right] N_i N_j \right\} \mathrm{d}\Omega \tag{15.3}$$

where $\mathbf{B}_i^u$ and $\mathbf{B}_i^d$ are the matrices with spatial derivatives of the corresponding shape functions. $\mathbf{E}^{\mathrm{epc}}$ is the consistent Jacobian matrix, $g_f$ the crack surface energy, $l_c$ the crack scale parameter, and $N$ stands for the shape functions associated with corresponding nodes.

To implement the calculation, the authors employed two layers to deal with the displacement field and phase field in software ABAQUS. The user-defined subroutine UMAT is called at each Gauss point in the first layer for the displacement field, and UEL is called at each element in the second layer for the phase field. The corresponding two elements in the two layers share information *via* a common block, and the relevant parameters are updated throughout the calculation process.

### 15.3.2 Details of the Model

Finite element model is built for the analysis of the fracture in the notch sample. To reduce the calculation cost, a 2D model is established and the gripping section of the dog-bone is deleted. The final model with the mesh result after mesh sensitivity study is shown in Figure 15.7(a). CPE4 is employed for the first layer of finite elements in capturing the displacement field, while a user-defined element is coded in the subroutine for the

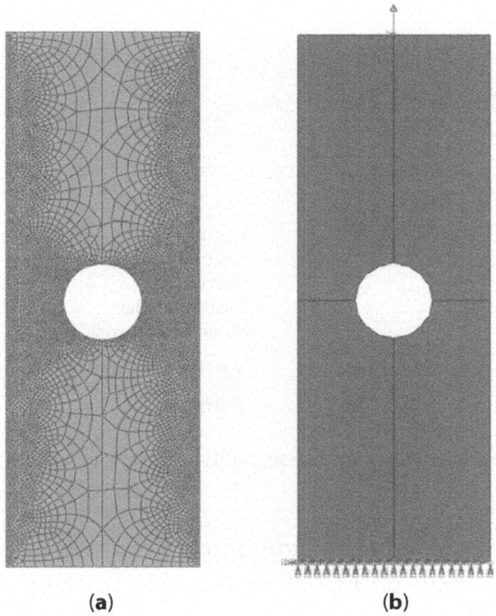

Figure 15.7 FE model. (a) Mesh results; (b) boundary conditions.

second layer of finite elements in capturing the phase field. The total number of nodes is 8886 and the total number of elements is 8454 for the first layer of finite elements. Therefore, the amount of the nodes and elements for the whole finite element model after using phase field method doubles on the given values.

As shown in Figure 15.7(b), the bottom of the model is totally fixed. The nodes on the upper line are coupled to its middle point where a displacement is applied. Static general algorithm is employed in the FE model, in order to simulate the static loading process of the fracture test. As the phase field method developed here is for brittle material where the geometry deformation is accounted as linear, the geometry nonlinearity is checked off in the simulation process.

### 15.3.3 Results and Discussions

The simulation is performed regarding two types of material property: one with only elastic (Case 1) and the other one with nonlinear behavior (Case 2). The elongation–tension force curves from these two cases together with the test results are organized and presented in Figure 15.8. It can be observed that the tension force increases with the elongation in the beginning phase for both cases. The phase field is less than one, which means the

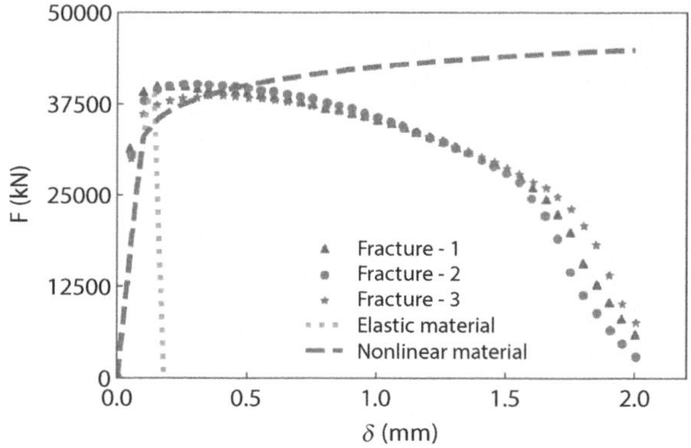


**Figure 15.8** The comparison between the test and simulation results.

material is still intact. When the elongation keeps on increasing, the curve of Case 1 reaches a peak point before it suddenly drops down. This peak point represents the fracture of the sample, as shown by the phase field distribution in Figure 15.9(a). The shape of the phase field distribution is similar to the fracture shape of the sample shown in Figure 15.5(c). After the peak point, the sample is too weak to sustain more loading, and the corresponding drops down immediately. As for Case 2 where the material plasticity is introduced in the model, the corresponding curve has an obvious plateau, and the curve does not drop down like that in Case 1. Meanwhile, the nonlinear geometry is not checked on in the simulation process, and the plateau does not have a decreasing trend like that in the test curve. The phase field distribution at the end of the simulation of Case 2 is also illustrated in Figure 15.9(b).

The phase field of the critical node in both models are marked and presented in Figure 15.9. The phase field variation of these two points along with the sample elongation is then shown in Figure 15.10. It can be observed that the phase field value of Case 1 keeps on increasing until it reaches the highest value of 1, which means this node has already entered the fracture state. When more nodes in this model enter into this state, the fracture area propagates through the sample around the circle notch, and the structure loses its ability to resist more load. While in Case 2, the phase field value of the most critical node stops increasing at a small elongation. This situation happens because the phase field code is developed for brittle material, and the ductile material behavior cannot be simulated accordingly. This also triggers us to further develop the code for ductile materials for more abroad engineering applications.

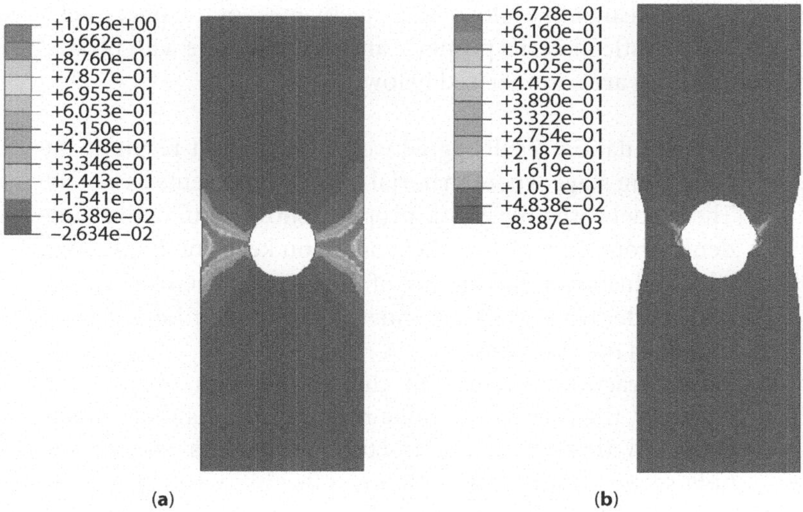


**Figure 15.9** Phase field of two cases. (a) Elastic material; (b) nonlinear material.

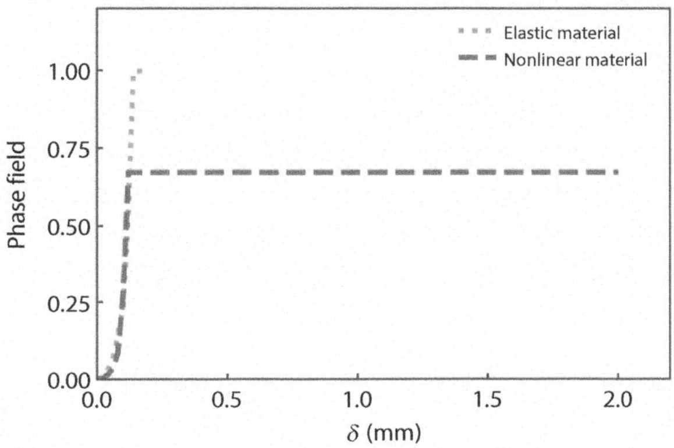


**Figure 15.10** Phase field values of the two models along with the elongation.

## 15.4 Conclusions

This chapter analyzed the fracture of X80 steel based on test and numerical methods. Material test is performed to obtain the nonlinear behavior of X80 steel for the input into the numerical model. Fracture test is then performed on a flaw sample with a notched hole in the center. The fracture test data is used to compare with the simulation results. Two types of FE models are established to evaluate the effectiveness of the phase field code: one with only elastic material property and the other one with plastic property. Key findings are summarized below:

1. The simulation results agree well with the test result at the beginning stage where material mainly represents as elastic. The model with only elastic property shows a curve that suddenly drops down when the elongation keeps on increasing, which illustrates the fracture of the sample whereas the second model with plasticity shows a curve with long plateau that does not drop down.
2. When fracture happens in the model with only elastic property, the phase field value reaches to 1. As more nodes approach this value, the fracture propagates around the hole section, and the sample loses its ability to resist more tension.
3. The second FE model is not able to reach to a fracture state as this method is not developed for ductile material. The phase field value of the most critical node in the model does not reach to 1 either.

The phase field method is only developed for brittle material where strong geometry nonlinearity is not accounted. However, most of the materials used in practical engineering are made of ductile material that can be under large deformation. Therefore, it is necessary to further develop the phase field method for more abroad engineering applications.

## References

1. Anderson, T.L., *Fracture Mechanics: Fundamentals and Applications*, CRC Press, Boca Raton, Florida, USA, 2017.
2. Li, H., Zhang, Z., Sun, X., Cohesive zone modeling of ductile fracture initiation in X80 pipeline steel under tensile loading. *Eng. Fract. Mech.*, 110, 349–363, 2013.
3. Hutchinson, J.W., Fundamentals of the phenomenological theory of nonlinear fracture mechanics. *J. Appl. Mech.*, 50, 4, 1042–1051, 1983.
4. Hutchinson, J.W., Singular behaviour at the end of a tensile crack in a hardening material. *J. Mech. Phys. Solids*, 16, 13–31, 1968.

5. Pari, P.C., Tada, H., Zahoor, A., Ernst, H., The Theory of Instability of the Tearing Mode of Elastic-Plastic Crack Growth, in: *Elastic-Plastic Fracture*, pp. 5–36, ASTM International, West Conshohocken, Pennsylvania, USA, 2009.
6. O'Dowd, N. and Shih, C.F., Family of crack-tip fields characterized by a triaxiality parameter - I. Structure of fields. *J. Mech. Phys. Solids*, 39, 8, 989–1015, 1991.
7. Nyhus, B., Zhang, Z., Thaulow, C., Normalisation of material crack resistance curves by the stress. *Mater. Sci.*, 2013.
8. Brocks, W., Anuschewski, P., Hellmann, D., A concept for scaling j r-curves by plastic constraint factors. *Int. J. Fract.*, 130, 455–469, 2004.
9. Fairchild, D.P., Crapps, J.M., Cheng, W., Tang, H., Shafrova, S., Full-scale pipe strain test quality and safety factor determination for strain-based engineering critical assessment. *International Pipeline Conference*, American Society of Mechanical Engineers Digital Collection, 2016.
10. Bai, Y., Effect of Loading History on Necking and Fracture, PhD thesis, Massachusetts Institute of Technology, Massachusetts, USA, 2008.
11. Saneian, M., Han, P., Jin, S., Bai, Y., Fracture response of steel pipelines under combined tension and torsion. *Thin Walled Struct.*, 154, 106870, 2020.
12. Saneian, M., Han, P., Jin, S., Bai, Y., Fracture response of steel pipelines under combined tension and bending. *Thin Walled Struct.*, 155, 106987, 2020.
13. Miehe, C., Hofacker, M., Welschinger, F., A phase field model for rate-independent crack propagation: Robust algorithmic implementation based on operator splits. *Comput. Methods Appl. Mech. Eng.*, 199, 45–48, 2765–2778, 2010.
14. Wu, J.Y., Nguyen, V.P., Bordas, S., Phase-field modeling of fracture. *Adv. Appl. Mech.*, 53, 1–87, 2020.
15. Guo, Q., Zhang, L., Wang, J., Liu, Q., Modeling hydrogen-induced cracking in X80 pipeline steel using a coupled phase-field and diffusion model. *Int. J. Hydrogen Energy*, 46, 15, 10158–10172, 2021.
16. Zhao, Y., Liu, C., Han, J., Effects of microstructure on fracture toughness of X80 pipeline steel. *Mater. Sci. Eng., A*, 631, 82–89, 2015.
17. Fang, J., Wu, C., Li, J., Liu, Q., Wu, C., Sun, G., *et al.*, Phase field fracture in elastoplastic solids: Variational formulation for multi-surface plasticity and effects of plastic yield surfaces and hardening. *Int. J. Mech. Sci.*, 156, 382–396, 2019.
18. Fang, J., Wu, C., Rabczuk, T., Wu, C., Ma, C., Sun, G., Li, Q., Phase field fracture in elastoplastic solids: abaqus implementation and case studies. *Theor. Appl. Fract. Mech.*, 103, 102252, 2019.

# Index

# Also of Interest

## From the Same Authors

*Autonomous Vehicles Planning and Control,* By Yong Bai & Liang Zhao, ISBN: 9781394355044. Master the future of marine exploration and technology with Autonomous Vehicles Planning and Control, which provides a comprehensive, interdisciplinary guide to the principles, control, and real-world applications of autonomous marine vehicles.

*Structural Reliability in Civil Engineering,* By Wei-Liang Jin, Qian Ye, and Yong Bai, ISBN: 9781119418153. *Structural Reliability in Civil Engineering* gives essential insights into the complexities of uncertainty in engineered structures, along with practical examples and advanced methods, making it an invaluable resource for both theory and real-world application in your civil engineering projects.

*Flexible Pipelines and Power Cables,* by Yong Bai, Shuai Yuan, and Kaien Jiang, ISBN: 9781394287505. This latest volume in the "Advances in Pipes and Pipelines" series focuses on flexible pipelines, and power cables, offering the engineer the most up-to-date and comprehensive coverage of pipeline and cable engineering available today.

*DEEPWATER FLEXIBLE RISERS AND PIPELINES,* By Yong Bai, ISBN: 9781119322726. This second volume in the "Advances in Pipes and Pipelines" series focuses on flexible pipelines, risers, and umbilicals, offering the engineer the most up-to-date and comprehensive coverage of pipeline engineering available today.

*Structural Dynamics*, by Yong Bai and Zhao-Dong Xu, ISBN: 9781119605607. Written by two experts across multiple disciplines, this is the perfect reference on structural dynamics for veteran engineers and introduction to the field for engineering students. *NOW AVAILABLE!*

*Flexible Pipes*, by Qiang Bai, Yong Bai, and Weidong Ruan, ISBN: 9781119041269. Written by one of the most well-respected teams of scientists in the area of pipelines, this revolutionary approach offers the engineer working in the energy industry the theory, analysis, and practical applications for applying new materials and modeling to the design and effective use of flexible pipes. *NOW AVAILABLE!*

## Check out these other titles from Scrivener Publishing

*AI for Autonomous Vehicles: The Future of Driverless Technology*, Edited by Sathiyaraj Rajendran, Munish Sabharwal, Yu-Chen Hu, Rajesh Kumar Dhanaraj, and Balamurugan Balusamy, ISBN: 9781119847465. This title presents the research prospects and security challenges present in the exciting new field of autonomous vehicles by comprehensively introducing advanced concepts of artificial intelligence (AI) in driverless technology.

*AUTONOMOUS VEHICLES VOLUME 1: Using Machine Intelligence*, Edited by Romil Rawat, A. Mary Sowjanya, Syed Imran Patel, Varshali Jaiswal, Imran Khan, and Allam Balaram. ISBN: 9781119871958. Addressing the current challenges, approaches and applications relating to autonomous vehicles, this groundbreaking new volume presents the research and techniques in this growing area, using Internet of Things, Machine Learning, Deep Learning, and Artificial Intelligence.

*AUTONOMOUS VEHICLES VOLUME 2: Smart Vehicles for Communication*, Edited by Romil Rawat, Purvee Bhardwaj, Upinder Kaur, Shrikant Telang, Mukesh Chouhan, and K. Sakthidasan Sankaran, ISBN: 9781394152254. The companion to *Autonomous Vehicles Volume 1: Using Machine Intelligence*, this second volume in the two-volume set covers intelligent techniques utilized for designing, controlling and managing vehicular systems based on advanced algorithms of computing like machine learning, artificial Intelligence, data analytics, and Internet of Things with prediction approaches to avoid accidental damages, security threats, and theft.

*AI AND IOT-BASED INTELLIGENT AUTOMATION IN ROBOTICS*, Edited by Ashutosh Kumar Dubey, Abhishek Kumar, S. Rakesh Kumar, N. Gayathri, Prasenjit Das, ISBN: 9781119711209. The 24 chapters in this book provide a deep overview of robotics and the application of AI and IoT in robotics across several industries such as healthcare, defense, education, etc.

Wiley The manufacturer's authorized representative according to the EU General Product Safety Regulation is Wiley-VCH GmbH, Boschstr. 12, 69469 Weinheim, Germany, e-mail: Product_Safety@wiley.com.

Printed and bound by CPI Group (UK) Ltd, Croydon, CR0 4YY
15/07/2026
02167953-0001